Autofrettage Processes

Technology and Modeling

Autofrettage Processes

Technology and Modeling

Uday S Dixit
Seikh Mustafa Kamal
Rajkumar Shufen

CRC Press
Taylor & Francis Group
Boca Raton London New York

CRC Press is an imprint of the
Taylor & Francis Group, an **informa** business

CRC Press
Taylor & Francis Group
52 Vanderbilt Avenue,
New York, NY 10017

CRC Press is an imprint of Taylor & Francis Group, an Informa business
No claim to original U.S. Government works

Printed on acid-free paper

International Standard Book Number-13: 978-1-138-38854-3 (Hardback)

Visit the Taylor & Francis Web site at
http://www.taylorandfrancis.com

and the CRC Press Web site at
http://www.crcpress.com

To my teachers.

—Uday S Dixit

To my beloved parents, Mr. Kalamsha Seikh and Ms. Arifa Begum.

—Seikh Mustafa Kamal

To my majestic father, Mr. Rajkumar Tikendra, and

loving mother, Ms. Hijam Soroma Devi.

—Rajkumar Shufen

Contents

Preface

Autofrettage is a widely employed process for the strengthening of pressure vessels and similar such components. It has been in industrial practice since the beginning of the 20th century with the invention of the hydraulic autofrettage process. Mechanical autofrettage was invented in the 1960s, whereby a mandrel is pushed through a hollow cylinder to induce non-uniform plastic deformation. Recently, several other types of autofrettage processes have been developed. All types of autofrettage processes achieve the strengthening effect by inducing beneficial compressive residual stresses. Apart from enhancing the static strength, these processes also improve the fatigue strength and resistance to stress–corrosion cracking.

In spite of the wide use of autofrettage in industry, many engineers are unaware of its basic principle. We have come across professionals believing the misconception that strain hardening is the dominant factor in autofrettage. Many engineers fail to realize the importance of residual stresses in general and in autofrettage in particular. One reason for this scenario may be that most of the undergraduate curricula do not include plasticity theory. Without the knowledge of plasticity, it is hard to visualize the role and significance of residual stresses. There are excellent books on plasticity, but they may look frightening to an undergraduate student (or a practicing engineer) due to their detailed coverage. Moreover, those books provide only a brief coverage of autofrettage, focusing mainly on hydraulic autofrettage. Hence, we felt the need to write a dedicated book on autofrettage processes. We feel that autofrettage may become more important in the near future, as sustainable engineering requires high strength-to-weight ratios of its components.

In this book, both technological and theoretical aspects of autofrettage processes have been described. Topics include hydraulic autofrettage, swage autofrettage, explosive autofrettage, thermal autofrettage, and rotational autofrettage. There is some discussion on microstructural aspects also in the context of heat treatment of thermally autofrettaged cylinders. Examples also include the autofrettage of spherical vessels.

We hope that this book will be useful for undergraduate and graduate students of mechanical, chemical, and manufacturing engineering. Autofrettage can also form an excellent case study for learning the fundamentals of plasticity. We hope that this book can be of interest to all the persons interested in plasticity theory. Finally, due to its concise size and focused treatment, it will be very useful for practicing engineers. We welcome any feedback on this book.

Uday S Dixit
Seikh Mustafa Kamal
Rajkumar Shufen

Authors

Dr Uday S Dixit received his bachelor's degree in mechanical engineering from the erstwhile University of Roorkee (now Indian Institute of Technology (IIT) Roorkee) in 1987, his MTech degree in mechanical engineering from IIT Kanpur in 1993, and his PhD in mechanical engineering from IIT Kanpur in 1998. He has worked in two industries—HMT, Pinjore, and INDOMAG Steel Technology, New Delhi—where his main responsibility was designing various machines. Dr Dixit joined the Department of Mechanical Engineering, IIT Guwahati, in 1998, where he is currently a professor. He was also the Officiating Director of the Central Institute of Technology, Kokrajhar (February 2014–May 2015). Dr Dixit has been actively engaged in research in various areas of design and manufacturing over the last twenty-five years. He has authored/co-authored 115 journal papers, 109 conference papers, twenty-seven book chapters, and six books on mechanical engineering. He has also co-edited seven books related to manufacturing. He has guest-edited eleven special issues of journals. Presently, he is an Associate Editor of the *Journal of Institution of Engineers (India), Series C*, and the Regional Editor (Asia) of the *International Journal of Mechatronics and Manufacturing Systems*. He has guided twelve doctoral and forty-four master's students. Dr Dixit has investigated a number of sponsored projects and developed several courses. Presently, he is the Vice-President of the AIMTDR conference.

Dr Seikh Mustafa Kamal obtained a bachelor's degree in mechanical engineering from Jorhat Engineering College, Assam, India in 2009 and a PhD in mechanical engineering from IIT Guwahati in 2016. He has served as Assistant Professor of mechanical engineering at NITS Mirza, Guwahati, for a period of about three years (2009–2012) prior to joining the direct PhD program at IIT Guwahati. Dr Kamal joined the Department of Mechanical Engineering, Tezpur University, Assam, India in 2016, where he is currently an Assistant Professor. He is an active researcher in the areas of autofrettage, plasticity, and stress analysis. During his PhD, he developed a new thermal autofrettage method for strengthening thick-walled pressure vessels under the supervision of Prof Uday S Dixit. His recent research interest is in exploring rotational autofrettage. He has published eight journal papers, seven conference papers, and two book chapters on the area of autofrettage. He has edited conference proceedings (Proceedings of the National Conference on Sustainable Mechanical Engineering: Today and Beyond, 25–26 March, 2017, Tezpur University, India) and guided two master's students.

Mr Rajkumar Shufen received his bachelor's degree in production engineering from the Birla Institute of Technology, Mesra, Ranchi in 2011, and his master's degree in engineering design from the Birla Institute of Technology and Science, Pilani in 2015. He has worked in the US-based firm Fluor Daniel Corporation for two years in Gurugram, India, as an Associate Piping Design Engineer. He has earned significant industrial experience in plant design, piping layouts, pump piping, and pressure vessels. He is currently pursuing a PhD in mechanical engineering from IIT Guwahati. Since beginning his PhD he has published five research journal papers. His research areas include computational solid mechanics, plasticity, pressure vessels, and material science.

1

Introduction to Autofrettage

1.1 What Is Autofrettage?

The thick-walled cylinder or sphere is one of the most important and widely used engineering components. There are several examples where a thick-walled hollow cylindrical or spherical component is subjected to a high magnitude of static/pulsating internal pressure. In such cases, high tensile hoop stresses are setup near the inner surface of the component, whereas the outer surface has relatively smaller hoop stresses. Apart from pressure, a thermal gradient across the wall of the cylinder and sphere can produce significant hoop stresses. Guns and similar types of weapons of strategic importance are typical examples. Usually the designers limit the maximum load that the cylinder or sphere can withstand by the yield strength of the material. If the working load on the inner wall of the component exceeds the yield strength of the material, the material may crack at the inner wall, where the hoop stress developed due to the load is at its highest. In order to increase the load carrying capacity of the thick-walled cylinder or sphere, a process called autofrettage is practiced prior to their use in service conditions.

Autofrettage is a metal-working process, where beneficial compressive residual stresses are induced in the vicinity of the inner wall of a thick-walled cylindrical or spherical vessel. In this process, the non-homogeneous plastic deformation of the inner wall of the vessel is deliberately produced by applying a uniformly distributed load at the inner surface. The load causing the plastic deformation of the inner wall is called the autofrettage load. Due to the application of autofrettage load during operation, the wall of the cylinder splits into two zones. An inner plastic zone extends from the inner surface to a certain intermediate radial position within the wall thickness, beyond which the material up to the outer surface experiences elastic deformation, creating an outer elastic zone. The intermediate radial position is the demarcating line between the two zones and is known as the radius of elastic–plastic interface. The autofrettage load is gradually applied during operation till the desired plastic deformation is achieved at the inner wall. Subsequently, the unloading of the autofrettage load is carried out. During unloading, the applied autofrettage load is gradually reduced to zero. As a consequence, the plastically deformed material at the inner plastic zone tries to remain in the deformed state, and the elastically deformed material at the outer elastic zone tries to retain its original position. This counteraction between the inner plastic zone and the outer elastic zone generates compressive residual stresses at the inner surface of the cylinder and some portion beneath it. Thus, autofrettage is accomplished in the thick-walled cylinder. Now, when the autofrettaged cylinder is put in service to carry a high magnitude of internal pressure or temperature gradient, the compressive residual stresses at the inner side offset the tensile stresses due to the working load. Thus, the load carrying capacity of the cylinder is increased. The probability of crack initiation at the inner wall is also reduced due to the compressive residual stresses, which slow down the growth of cracks (Stacey and Webster, 1988; Perl and Aroné, 1988). This enhances the fatigue life. The process also enhances the stress–corrosion resistance of the cylinder in corrosive environments.

The term "autofrettage" is of French origin, which means "self-hooping." In French, "frettage" means the hooping of a cask or cylindrical container to increase strength against internal pressure. "Auto" means the material itself withstands the stresses without external devices, hence the name "autofrettage." A schematic representation of the basic autofrettage operation for a typical thick-walled cylinder is shown in Figure 1.1. The procedure consists of two basic stages—the loading stage and the unloading stage. The cylinder is loaded with an autofrettage load at the inner surface in the loading stage as shown in Figure 1.1. During loading, the inner portion of the cylinder gets plastically deformed and the outer portion remains in an elastic state. The displacement of the inner surface of the cylinder with respect to the original inner surface (shown by dotted lines) is apparent in an exaggerated way. The unloading stage is shown in Figure 1.1. In this stage, the originally applied

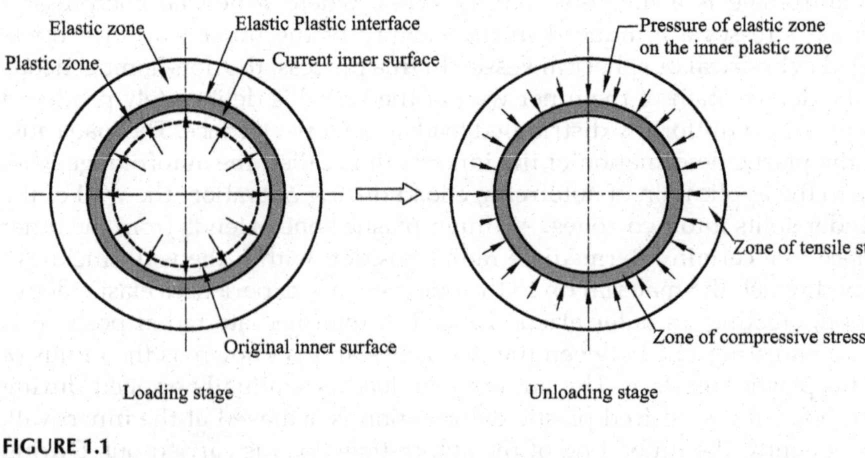

FIGURE 1.1
Operational principle of a typical autofrettage process. With permission from Shufen and Dixit (2018). Copyright ASME.

plastically deforming load is removed from the cylinder. As a result, the elastic recovery of the outer elastic zone takes place, which applies pressure to the inner plastic zone and causes a slight shrinkage of the inner surface of the cylinder in the final configuration of the cylinder. At this stage, high compressive residual stresses are setup in the inner plastic zone, and relatively lesser tensile residual stresses are setup in the outer elastic zone. The tensile residual stresses in the outer elastic zone are self-equilibrating and do not pose any problem as long as their magnitude remains small.

1.2 A Brief History of the Autofrettage Process

In ancient times, gunpowder cannons were used as weapons on the battlefield. These were the earliest form of artillery made in the form of thick-walled tubes. The word "cannon" is derived from the Latin origin "canna," which means tube. The first form of gunpowder cannon barrels were used in China in the 12th century. These were cylindrical bamboo tree trunks filled with gunpowder charges. The bamboo cannons were wrapped tightly with a rope on the outer surface along its length to strengthen them. The bamboo cannons were very low-pressure devices. Afterwards, cast iron tubes were used as an alternative, which were typically heavy and stronger than bamboo tubes. However, due to the poor sealing between the projectile and the cast tube wall along with ineffective gunpowder, these remain comparatively low-pressure devices. Soon after the 1400s, the development of effective gunpowder such as corned powder and the use of heavy spherical projectiles made from stone or iron, increased the combustion pressure inside the cannon during firing. The increase of operating pressure inside the gunpowder cannon resulted in the fracture of the cannon/barrel tubes, sometimes even causing the tubes to split into parts. In the mid-1800s, gun designers realized that if some inward force could be applied to the barrel, the pressure-carrying capacity of the barrel could be enhanced; as a consequence, barrels were "hooped" to pre-compress the material to a certain depth of the barrel's inner wall. Thus, the history of self-hooping or autofrettage dates back to the mid-19th century. Between 1841–1845, professor Treadwell made cannons out of wrought iron material by welding together rings or short hollow cylinders endwise (Wilder, 1865). Each ring or short hollow cylinder was made of bars that were wound upon an arbor spirally, like the winding of a ribbon upon a block. They were provided proper shape by passing through dies. Due to the softness of the wrought iron, Treadwell refined his cannons with a steel lining. This was done by winding wrought iron bars upon a previously formed steel ring. This principle of constructing cannons was used by William Armstrong, who assembled compound cylinders from wrought iron tubes to achieve pre-compression in 1855 (Wilder, 1865). In addition, General Thomas

Rodman of the Ordnance Department of the United States Army developed a pre-compressing method for gun barrels from direct casting using a differential cooling technique (Rodman, 1847). In this technique, the exterior of the cylinder was heated by the application of artificial heat and the inner surface was cooled by cold water. In the early 1900s, L. Jacob, a French artillery officer, suggested a method for pre-stressing monobloc gun barrels by plastic expansion (Jacob, 1907). He created the plastic expansion of the barrel's inner wall by subjecting the barrel to high internal pressurization. The depressurization of the barrel then caused residual stresses, thereby enhancing its pressure-carrying capacity. Jacob termed this method of achieving self-hooping at the bore of the barrel as "autofrettage." Thus, the development of the autofrettage process is credited to him. Soon after this development, L. B. Turner presented a complete mathematical study of the process at King's College, Cambridge, in 1909 (Gibson, 2008). The autofrettage process was adopted by the French in 1923 and soon found its widespread use in the defense industry. Some early analysis of the process was carried out by (Macrae, 1930; Nádai, 1931; Dirmoser, 1931; Hill et al., 1947; MacGregor et al., 1948; Allen and Sopwith, 1951; Steele, 1952). Over the years, several advancements have been achieved in the development of autofrettage processes by researchers. Other new methods of achieving autofrettage in the thick-walled vessels have been developed (Davidson et al., 1962; Mote et al., 1971; Kamal and Dixit, 2015; Zare and Darijani, 2016) and research has been attempted on the modeling and experimentation of such processes.

1.3 Application of the Autofrettage Process

Originally the autofrettage process was developed for military applications and used in the strengthening of cannons and gun barrels. Even in the modern artillery industry, the process is being widely employed in the manufacturing of gun barrels and similar weapon systems. Later on the process found application in several other industries where thick-walled cylindrical or spherical vessels are used for the purpose of withstanding high pressure. The barrel of a gun is a thick-walled metallic cylinder that is designed to discharge projectiles. The amount of work done on the projectile depends on the pressure acting upon its base as it travels along the barrel. The barrel must be able to withstand very high pressure. Thus, the barrel of a gun needs to be autofrettaged. The other examples where autofrettage finds widespread application are as follows:

- High-pressure processing (HPP) equipment in the food industry
- Submarine hulls
- Pressurized water reactor (PWR) in nuclear power plants

- Fuel injection lines in diesel engines
- High-pressure steam/chemical pipelines in the oil and chemical industries.
- Expansion of tubular components down hole in oil and gas wells
- Water jet cutting machines
- High-pressure containers for storing pressurized oil and gas
- Manufacturing of hydrostatic extrusion chambers

FIGURE 1.2
Autofrettage applications: (a) petro-chemical refinery (May, 2015), (b) gun barrel (https://pxhere.com/en/photo/712201), (c) natural gas pipelines and cylindrical pressure vessels (https://www.flickr.com/photos/bilfinger/14074154115), (d) nuclear power plant (https://www.flickr.com/photos/iaea_imagebank/3441138290) and (e) fuel injector (https://www.flickr.com/photos/17968829@N03/22743260562). (All photographs under creative commons license).

Figure 1.2 depicts some interesting examples where autofrettaged pressure vessels are employed. Figure 1.2(a) shows a petro-chemical refinery (May, 2015). Figure 1.2(b) shows an artillery with a gun barrel (https://pxhere.com/en/photo/712201), Figure 1.2(c) shows a natural gas pipeline and cylindrical pressure vessels (https://www.flickr.com/photos/bilfinger/14074154115), Figure 1.2(d) shows a nuclear power plant (https://www.flickr.com/photos/iaea_imagebank/3441138290), and Figure 1.2(e) shows a fuel injector (https://www.flickr.com/photos/17968829@N03/22743260562).

1.4 Classification of Autofrettage Processes

Depending upon the type of applied autofrettage load, autofrettage processes can be classified into the following types:

- Hydraulic Autofrettage
- Swage Autofrettage
- Explosive Autofrettage
- Thermal Autofrettage
- Rotational Autofrettage.

Hydraulic autofrettage utilizes an ultra-high hydraulic pressure at the inner wall of the cylinder to achieve the desired extent of plastic deformation. In swage autofrettage, the required plastic deformation is created by passing a tapered oversized mandrel through the inner wall axially. Autofrettage can also be achieved by detonating an explosive charge inside the vessel, which is called explosive autofrettage (Mote et al., 1971; Ezra et al., 1973; Taylor et al., 2012). However, the process is not commonly practiced due to the involvement of explosives. Thermal autofrettage is achieved by generating a temperature gradient across the wall thickness of the cylinder, which is sufficient to deform the inner wall plastically and cause the subsequent cooling of the cylinder to room temperature (Kamal and Dixit, 2015). The principle of rotational autofrettage is based on the fact that the cylinder experiences plastic deformation at the inner surface when it is rotated at a sufficiently high angular velocity (Zare and Darijani, 2016; Kamal, 2018). The centrifugal force due to rotation acts as the autofrettage load. When the cylinder is brought to rest, beneficial compressive residual stresses are induced at the inner surface of the cylinder. In the following subsections, the various autofrettage processes are briefly described.

1.4.1 Hydraulic Autofrettage

Hydraulic autofrettage is the oldest, most popular autofrettage method. The development of the process is credited to Jacob (1907), a French artillery officer. In hydraulic autofrettage, initially a single thick metallic cylinder of internal radius slightly less than the desired dimension is used. Then the thick-walled cylindrical vessel is subjected to a high hydrostatic pressure, p_i, such that the equivalent stress at the inner wall of the cylinder exceeds the yield strength of the material. Thus, the material at and around the inner radius of the cylinder is subjected to plastic deformation. Pressure is further increased in order to achieve the desired depth of plastic deformation within the cylinder wall. The material near the outer wall of the cylinder deforms elastically. Generally, the cylinder is pressurized by hydraulic oil pumped through a high-pressure hydraulic pump. The ends of the cylinder are closed using rigid plugs or floating pistons to retain the oil inside the cylinder. A solid spacer is usually inserted into the center of the cylinder to reduce the volume of the hydraulic oil that must be pumped into the cylinder to achieve autofrettage (Gibson, 2008).

When the pressurized oil is released from the cylinder, the outer portion of the material, which has undergone elastic deformation, attempts to return to its original dimension whilst the material at and around the inner wall, which has been deformed plastically, attempts to remain in the deformed state. This results in residual compressive stresses at and around the inner wall of the cylinder. Thus, when the cylinder is repressurized, the magnitude of the maximum stress reduces, significantly enhancing its pressure-carrying capacity. As the residual stresses are self-equilibrating, the portion of the cylinder from the outer surface to some intermediate radial position will have tensile residual stresses, but this does not pose any problem as the working pressure produces relatively smaller hoop stresses in the outer portion of the cylinder. A typical thick-walled cylinder undergoing hydraulic autofrettage is shown in Figure 1.3.

In hydraulic autofrettage, the applied pressure is to be very carefully controlled to achieve the desired plastic deformation within the cylinder wall.

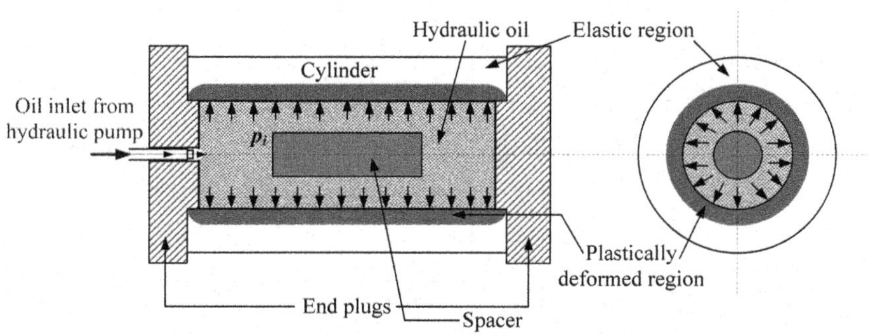

FIGURE 1.3
Hydraulic autofrettage of a thick-walled cylinder (Kamal, 2016).

This makes the process slow. Also, the arrangement of the plugs at the ends of the cylinder should be precisely made to contain the required pressure. The highly-pressurized oil is dangerous if failure should occur during the process. Therefore, much care needs to be taken in applying the high hydraulic pressure. The process is also expensive due to the requirement of a costly hydraulic power pack for pressurization. Moreover, any leakage of the hydraulic oil during the autofrettage process is harmful to the environment. Thus, it is not a greener manufacturing procedure.

Despite certain disadvantages, due to its effectiveness, hydraulic autofrettage has attracted many researchers since its inception. Many researchers have widely investigated the process analytically, numerically, and experimentally. Significant contributions have been made by (Hill et al., 1947; Gao, 1992, 1993, 2003, 2007; Gao et al., 2015; Huang, 2005; Perl, 1998; Perl and Perry, 2006; Parker, 2001a, 2001b, 2005; Parker et al., 1983, 1999; Parker and Kendall, 2003; Jahed and Dubey, 1997; Rees, 1987, 1990; Stacey and Webster 1988).

1.4.2 Swage Autofrettage

To overcome the disadvantages associated with the cost and processing of the hydraulic autofrettage procedure, Davidson et al., (1962) developed a new approach for achieving autofrettage known as swage autofrettage. The method was based on experimental study. The basic principle of the process is to achieve plastic deformation at the inner side of the cylinder by the mechanical interference of an oversized tapered mandrel and the inner wall of the cylinder. The mandrels used in the swage autofrettage process typically consist of two conical sections connected by a short length of constant diameter. The schematic diagram of a typical mandrel and the swage autofrettage process is shown in Figure 1.4. In Figure 1.4(a), θ_F is the forward

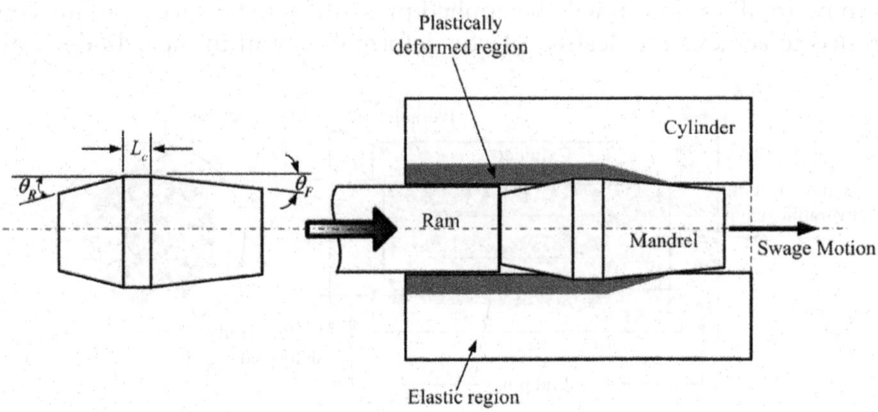

FIGURE 1.4
Schematic diagram of (a) mandrel and (b) swage autofrettage (Kamal, 2016).

cone angle, θ_R is the rear cone angle, and L_c is the length of the straight portion connecting the two cones in the mandrel. The mandrel is prepared in such way that the forward conical section has a smaller slope than the rear section. The forward conical section controls the initial deformation of the cylinder, and the rear conical section helps in the subsequent unloading of the cylinder. The mandrel is usually pushed through the inner wall of the cylinder by means of pressure exerted by the ram of a hydraulic press. As a result, localized plastic deformation of the inner wall (and some portion beneath it) progresses, keeping the outer wall in an elastic state. After the unloading of the mandrel, the compressive residual stresses are generated at and around the inner wall.

Davidson et al., (1962) used three types of setups, namely mechanical push swaging, mechanical pull swaging, and hydraulic push swaging in order to have localized plastic deformation at the inner surface of the cylinder. In mechanical push swaging, a ram was pushed against the mandrel by a hydraulic press. In mechanical pull swaging, an overhead crane was utilized to pull up the mandrel vertically. The hydraulic push swaging used hydraulic pressure applied at the end of the mandrel. The hydraulic fluid for swaging was a mixture of water and glycerin. The front and rear taper angles in the mandrel for all three swaging methods were kept as 1.5° and 3°, respectively, with $L_c = 19$ mm for all three swaging methods. The mandrel was polished to about 0.05–0.4 μm Ra surface roughness.

In the swage autofrettage process, when the mandrel travels through the inner wall of the cylinder, a lot of frictional force is generated between the mandrel and the cylinder. Therefore, the contact between the mandrel and the cylinder needs to be lubricated in order to reduce the sliding friction (O'Hara, 1992). A molybdenum disulfide suspension in oil and copper plating may be used in combination for lubrication during the swaging procedure (Davidson et al., 1962). The method also requires the use of a large and expensive hydraulic press and a mandrel of a material with an excessively high compressive strength. The hydraulic pressure required to push the mandrel through the cylinder is much less in this process than that required by the conventional hydraulic autofrettage. However, the pressure required for driving the ram increases with the increase in the size of the mandrel (Iremonger and Kalsi, 2003).

The swage autofrettage process is less investigated by the researchers as compared to hydraulic autofrettage. The earlier research works on the swage autofrettage were based on the experimental determination of residual stresses in thick-walled cylinders due to swage by using the traditional Sachs boring technique (Davidson et al., 1962). The analytical solution of the swage autofrettage process is difficult due to the transient and localized nature of the process. A simple analysis of the process was reported by Chen (1988) considering a simplified plane–strain problem of the mandrel tube assembly. Additional contributions are made by (O'Hara, 1992; Iremonger and Kalsi, 2003; Bihamta et al., 2007; Perry and Perl, 2008; Chang et al., 2013).

1.4.3 Explosive Autofrettage

Explosive autofrettage is a dynamic way of introducing beneficial compressive residual stresses in the vicinity of the inner wall of a thick-walled cylinder. In this process, an explosive charge at the inside of the cylinder is detonated to cause the plastic deformation of the inner wall, extending to a certain intermediate radial position. A typical setup for achieving explosive autofrettage is shown in Figure 1.5. A rapid burning explosive charge is placed along the longitudinal axis of a ductile metal tube called a radial piston, which is then placed along the longitudinal axis of the thick-walled cylinder to be autofrettaged (Mote et al., 1971; Ezra et al., 1973). The space between the radial piston and the thick-walled cylinder is filled with an energy transmitting medium such as air or water. The energy transmitting medium is sealed axially with the help of end plugs and restraining fixtures. When the explosive charge is detonated by means of a blasting cap, the radial piston gets expanded due to explosion gases. This pressurizes the energy transmitting medium confined between the radial piston and the inner wall of the cylinder. With the proper choice of the explosive charge and the correct dimension of the radial piston, the pressure can be increased up to such a high value that the inner wall of the cylinder gets plastically deformed. Upon dissipation of the explosively created pressure, substantial residual compressive hoop stresses are generated at the inner side of the cylinder, propagating to a certain thickness of the cylinder wall.

The capital cost involved in explosive autofrettage is less expensive than that of the hydraulic and swage autofrettage processes. The method does not require expensive equipment, such as hydraulic pumps or hydraulic presses. However, due to the involvement of explosives, the control of the process requires skill. Also, in general, the residual compressive hoop stress level is smaller than that achieved by the hydraulic or swage autofrettage

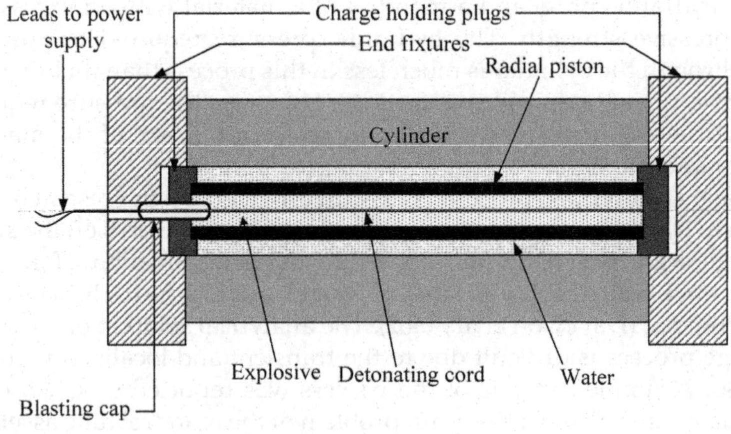

FIGURE 1.5
Schematic diagram of explosive autofrettage (Kamal, 2016).

processes. In order to produce the residual hoop stresses of the order of the hydraulic and swage autofrettage processes, the explosive autofrettage process requires a more sophisticated configuration of explosive tube.

That the explosive autofrettage process has not drawn the considerable attention of researchers is perhaps due to the involvement of explosives, or the requirement of legal permission for its experimental implementation. Also, the modeling of the process involves dynamic elasto-plasticity, which makes the analysis complex. Nevertheless, the earliest analytical study for the feasibility of the explosive autofrettage process was carried out by Mote et al., (1971). His analytical model was based on a strain rate formulation of plasticity using the von Mises yield criterion and its associated flow rule without strain hardening. They also conducted experiments on cannon barrel model M81 made up of steel alloy 4340 of bore diameter 152 mm to validate their analytical model. Other contributions on the experimental front were made by (Ren-rui et al., 1999; Taylor et al., 2012). Ren-rui et al., (1999) also carried out a non-linear finite element method (FEM) analysis of the process in addition to the experimental study.

1.4.4 Thermal Autofrettage

The generation of residual stresses due to thermal loading is well-known among the research community. The original idea of generating compressive residual stresses in a thick-walled tube by means of thermal stresses is attributed to General Thomas Rodman of the Ordnance Department of the United States Army. In 1847, Rodman fabricated gun barrels from direct casting using the principle of differential cooling. The outer surface of the cylinder was heated, and the bore was cooled with cold water. Based on a similar principle, the concept of thermal autofrettage for thick-walled cylinders was conceived in 2015 by Kamal and Dixit (2015a,b). In this process, the thick-walled cylinder or sphere is subjected to a temperature gradient across the wall thickness in the loading stage. In the unloading stage, the cylinder is cooled down to room temperature where the elastic unloading of thermal stresses occurs and compressive residual stresses are induced at the inner side. The thermal autofrettage process and its experimental setup is shown schematically in Figure 1.6. As shown in Figure 1.6, a thick-walled cylinder with inner radius a and outer radius b is considered. The outer wall of the cylinder is subjected to a temperature T_b by gradual external heating using an electrical heating element. A regular flow of cold fluid *e.g.*, cold water is made through the bore of the cylinder, which keeps the inner surface at a lower temperature, T_a. Thus, the desired temperature difference, $(T_b–T_a)$ is created across the cylinder wall thickness, which will deform the inner wall plastically. At this stage, the cylinder wall becomes plastic up to some intermediate radius, whilst the outer portion of the cylinder remains in the elastic state. This is so long as the temperature difference is not large enough to cause the yielding of the outer wall. This stage is called the first stage of

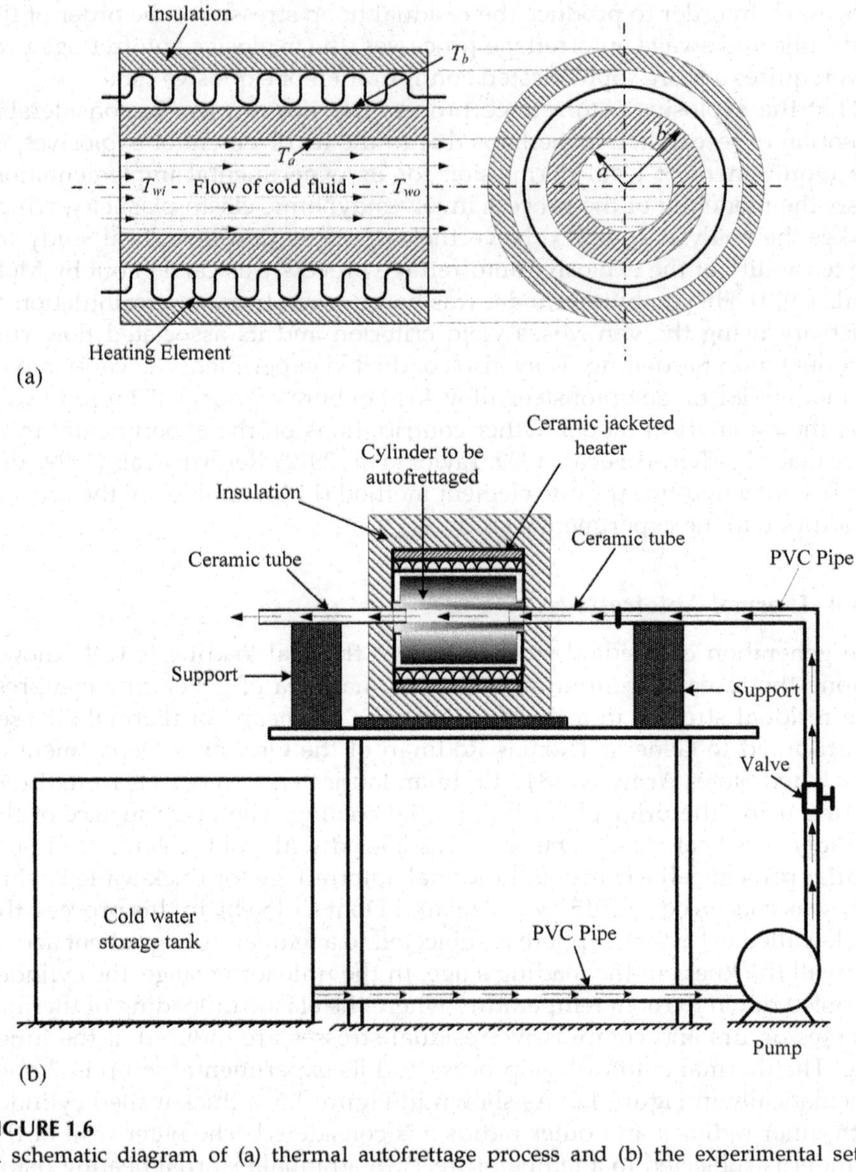

FIGURE 1.6
A schematic diagram of (a) thermal autofrettage process and (b) the experimental setup (Kamal, 2016).

plastic deformation. As the temperature difference is increased further, after crossing a certain threshold of (T_b–T_a), the outer wall of the cylinder begins to yield and the simultaneous plastic deformation of both the inner and outer portion of the cylinder takes place. At this stage the cylinder will have a plastic zone at the inner side and a plastic zone at the outer side with an intermediate elastic zone within the wall thickness. This stage is called the second stage of plastic deformation. When the temperature difference across the wall thickness is released, *i.e.*, when the entire cylinder is cooled to room temperature, compressive residual stresses are generated at and around the inner wall of the cylinder. The compressive residual stresses significantly reduce the magnitude of tensile stress field at and around the inner wall of the cylinder when it is subjected to loading by internal pressure or temperature gradient. This enhances the pressure-carrying capacity as well as the working thermal gradient of the cylinder.

Barbero and Wen (2002) were the first to coin the term "thermal autofrettage" as an alternative method for the autofrettage of composite metal-lined cryogenic pressure vessels by cooling the vessel to a temperature lower than that of the operating temperature using cryogenic fluids. However, the behavior of the process was not analyzed in detail either theoretically or experimentally. Kargarnovin et al., (2005) carried out an analysis of thermal autofrettage for thick-walled spherical vessels made of an elastic, perfectly plastic material. In the study, the inner wall of the sphere was maintained at a higher temperature than that of the outer surface. The analysis of Kargarnovin et al., (2005) was extended for a case of linear hardening material by Darijani et al., (2009). Based on the principle depicted in Figure 1.6, Kamal and Dixit (2015a,b) developed a thermal autofrettage process for the first time by creating a radial thermal gradient across the wall thickness. Kamal and Dixit (2015a, 2015b) presented analytical models for the thermal autofrettage of a thick-walled cylinder assuming plane–stress and a generalized plane–strain condition. A comparison of the analytical models was also carried out by Kamal et al., (2017) with 3D FEM analysis results. It was shown that the generalized plane–strain analytical model was valid for length to wall thickness ratios greater than 6 and the plane–stress analytical model was valid for the length to wall thickness ratios less than 0.5. A comparative study of the thermal autofrettage process with the hydraulic autofrettage process was carried out by Kamal and Dixit (2016a); this analysis showed that the achievable increase in pressure-carrying capacity in the thermal autofrettage process was lesser than that possible in the hydraulic autofrettage process. This was due to a limitation on the maximum allowable temperature on the outer surface of the cylinder to avoid recrystallization. Kamal et al., (2016) also carried out a rigorous experimental study of the thermal autofrettage process. Further, Kamal and Dixit (2016b) analyzed the thermal autofrettage combined with shrink-fit in order to enhance the performance of thermally autofrettaged cylinders.

The arrangement for the thermal autofrettage process is simple and easy to handle compared to the existing methods of hydraulic, swage, or explosive autofrettage. The process does not involve any ultra-high hydraulic pressure or explosives. Thus, it does not require a costly hydraulic power pack or an expensive arrangement for detonating explosives. This makes the process comparatively inexpensive. Moreover, due to absence of any moving part in the arrangement for thermal autofrettage, friction is avoided. The process utilizes heating elements and cold fluid (such as cold water) to create the desired thermal gradient across the wall thickness of the cylinder, which is not harmful to the environment. Thus, the process appears to be a greener manufacturing procedure.

1.4.5 Rotational Autofrettage

The concept of rotational autofrettage is very recent. It is based on the principle of plastically deforming the inner wall of the cylinder by rotating the cylinder about its axis at a certain angular velocity. The process was first proposed by Zare and Darijani (2016). In the rotational autofrettage process, the cylinder to be autofrettaged is subjected to a gradually increasing angular velocity. Due to rotation, a centrifugal force acts at the inner surface. When the angular velocity attains a certain value, the cylinder experiences yielding at the inner surface. After crossing a certain threshold value of angular velocity, the cylinder wall becomes plastic up to a certain radial depth propagating from the inner surface, keeping the material near the outer surface in an elastic state. After achieving the desired elastic–plastic deformation across the cylinder wall, the plastically deforming angular velocity is gradually reduced to zero. This unloading of the centrifugal force generates beneficial compressive residual stresses in the vicinity of the inner surface of the cylinder in a similar fashion as in the case of other autofrettage processes. The process is schematically represented in Figure 1.7. The required angular velocity for achieving plastic deformation in the cylinder wall may be obtained by using an electric motor and a transmission system may be employed for controlling the operation of the process.

The feasibility analysis of the rotational autofrettage process was first carried out by Zare and Darijani (2016), theoretically considering plane–strain assumption with the incorporation of the Tresca yield criterion for non-hardening material. Zare and Darijani (2017) extended their theoretical model in order to incorporate strain-hardening using linear kinematic hardening. A plane–stress model analyzing the rotational autofrettage of a thick-walled short hollow cylinder or thin hollow disk is also available considering the

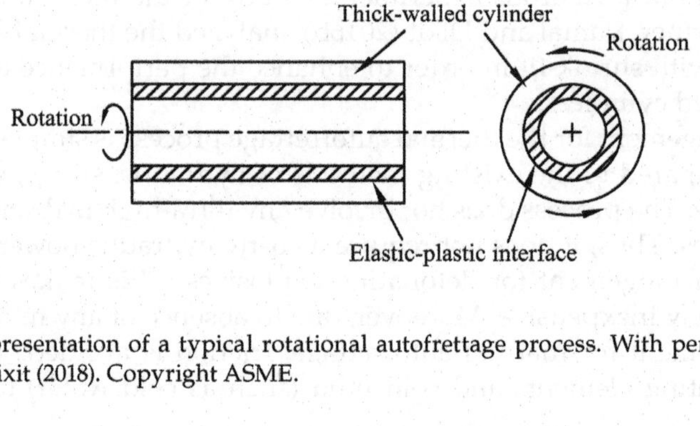

FIGURE 1.7
Schematic representation of a typical rotational autofrettage process. With permission from Shufen and Dixit (2018). Copyright ASME.

Tresca yield criterion with strain-hardening (Kamal, 2018). The experimental verification of the theoretical models is yet to be carried out by the researchers to confirm the soundness and value of the process.

1.5 Conclusion

In this chapter, a very general introduction of the autofrettage process is presented. The principle of the autofrettage technology is introduced, followed by a brief history of the development of the process. Different types of autofrettage processes, namely hydraulic, swage, explosive, thermal, and rotational autofrettage are briefly described. Due to the effectiveness and advancement in the analysis of the processes on both the theoretical and experimental fronts by researchers, the hydraulic and swage autofrettage processes remain the most widely practiced processes in industry, despite certain disadvantages. The explosive autofrettage process is less popular and has not been explored by researchers in detail. The thermal autofrettage process is a potential procedure for generating beneficial compressive residual stresses in thick-walled cylindrical vessels. The process has been validated both theoretically and experimentally. The process appears to be simple, inexpensive, and environmentally friendly when compared to the hydraulic, swage, and explosive autofrettage processes. However, the process has yet to find its place in industry. Further research is needed on the development of equipment for carrying out thermal autofrettage processes in industry. The concept of rotational autofrettage has been conceived of recently. A few theoretical analyses are available for the feasibility analysis of the process. There is no experimental verification of the process to date. In future, the process may be established as a potential autofrettage procedure through accurate modeling, development of experimental setup, and experimental verification of the residual stresses due to rotation.

References

Allen, D.N. and Sopwith, D.G., (1951), The stresses and strains in a partially thick tube under internal pressure and end load, *Proceedings of the Royal Society, London, Series A*, **205**, pp. 69–82.

Barbero, E.J. and Wen, E.W., (2002), Coefficient of thermal expansion compatibility through mechanical and thermal autofrettage in metal lined composite pipes, ASTM STP 1436, edited by C.E. Bakis, American society for Testing and Materials, West Conshohocken, PA.

Bihamta, R., Movahhedy, M.R. and Mashreghi, A.R., (2007), A numerical study of swage autofrettage of thick-walled tubes, *Materials and Design*, **28**, pp. 804–815.

Chang, L., Pan, Y. and Ma, X., (2013), Residual stress calculation of swage autofrettage gun barrel, *IJCSI International Journal of Computer Science Issues*, **10**(2), pp. 52–59.

Chen, P.C.T., (1988), A simple analysis of the swage autofrettage process, Technical Report ARCCB-TR-88030, US Army Armament Research, Development and Engineering Center, Close Combat Armaments Center, Benét Laboratories, Watervliet, NY.

Darijani, H., Kargarnovin, M.H. and Naghdabadi, R., (2009), Design of thick-walled cylindrical vessels under internal pressure based on elasto-plastic approach, *Materials and Design*, **30**, pp. 3537–3544.

Davidson, T.E., Barton, C.S., Reiner, A.N. and Kendall, D.P., (1962), New approach to the autofrettage of high-strength cylinders, *Experimental Mechanics*, **2**, pp. 33–40.

Dirmoser, O., (1931), *Design of High Strength Gun Tubes (Built Up and Cold Worked) Having Normal and Artificially Raised Elastic Limit*, Vienna Technical College, Vienna, Austria.

Ezra, A., Glick, H., Howell, W. and Kaplan, M., (1973), Method and apparatus for explosive autofrettage. U.S. Patent No. 3,751,954.

Gao, X.-L., (1992), An exact elasto-plastic solution for an open-ended thick-walled cylinder of a strain hardening material, *International Journal of Pressure Vessel and Piping*, **52**, pp. 129–144.

Gao, X.-L., (1993), An exact elasto-plastic solution for a closed-end thick-walled cylinder of elastic linear-hardening material with large strains, *International Journal of Pressure Vessel and Piping*, **56**, pp. 331–350.

Gao, X.-L., (2003) Elasto-plastic analysis of an internally pressurized thick-walled cylinder using a strain gradient plasticity theory, *International Journal of Solids and Structures*, **40**, pp. 6445–6455.

Gao, X.-L., (2007), Strain gradient plasticity solution for an internally pressurized thick-walled cylinder of an elastic linear-hardening material, *Zeitschrift für angewandte Mathematik und Physik*, **58**(1), pp. 161–173.

Gao, X.-L., Wen, J.-F., Xuan, F.-Z. and Tu, S.-T., (2015), Autofrettage and shakedown analyses of an internally pressurized thick-walled cylinder based on strain gradient plasticity solutions, *ASME Journal of Pressure Vessel Technology*, **82**, pp. 041010-1–041010-12.

Gibson, M.C., (2008), Determination of residual stress distributions in autofrettaged thick-walled cylinders, Ph.D. Thesis, Defence College of Management and Technology, Cranfield University, UK.

Hill, R., Lee, E.H. and Tupper, S.J., (1947), The theory of combined plastic and elastic deformation with particular reference to a thick tube under internal pressure, *Proceedings of the Royal Society of London, Series A, Mathematical and Physical Sciences*, **191**, pp. 278–303.

Huang, X.P., (2005), A general autofrettage model of a thick-walled cylinder based on tensile compressive stress–strain curve of a material, *Journal of Strain Analysis for Engineering Design*, **40**(6), pp. 599–608.

Iremonger, M.J. and Kalsi, G.S., (2003), A numerical study of swage autofrettage, *ASME Journal of Pressure Vessel Technology*, **125**, pp. 347–351.

Jacob, L., (1907), La Résistance et L'équilibre Elastique des Tubes Frettés, *Mémoire de L'artillerie Navale*, **1**(5), pp. 43–155 (in French).

Jahed, H. and Dubey, R.N., (1997), An axisymmetric method of elastic–plastic analysis capable of predicting residual stress field, *ASME Journal of Pressure Vessel Technology*, **119**, pp. 264–273.

Kamal, S.M., (2016), A theoretical and experimental study of thermal autofrettage process, Ph.D. Thesis, IIT Guwahati.

Kamal, S.M., (2018), Analysis of residual stress in the rotational autofrettage of thick-walled disks, *ASME Journal of Pressure Vessel Technology*, **140**(6), 061402-1–061402-10.

Kamal, S.M., Borsaikia, A.C. and Dixit, U.S., (2016), Experimental assessment of residual stresses induced by the thermal autofrettage of thick-walled cylinders, *The Journal of Strain Analysis for Engineering Design*, **51**(2), pp. 144–160.

Kamal, S.M. and Dixit, U.S., (2015a), Feasibility study of thermal autofrettage of thick-walled cylinders, *ASME Journal of Pressure Vessel Technology*, **137**(6), pp. 061207-1–061207-18.

Kamal, S.M. and Dixit, U.S., (2015b), Feasibility study of thermal autofrettage process, in *Advances in Material Forming and Joining, 5th International and 26th All India Manufacturing Technology, Design and Research Conference, AIMTDR 2014*, edited by R.G. Narayanan and U.S. Dixit, Springer, New Delhi, India.

Kamal, S.M. and Dixit, U.S., (2016a), A comparative study of thermal and hydraulic autofrettage, *Journal of Mechanical Science and Technology*, **30**(6), pp. 2483–2496.

Kamal, S.M. and Dixit, U.S., (2016b), A study on enhancing the performance of thermally autofrettaged cylinder through shrink-fitting, *ASME Journal of Manufacturing Science and Engineering*, **138**(9), pp. 094501–094501-5.

Kamal, S.M., Dixit, U.S., Roy, A., Liu, Q., and Silberschmidt, V.V., (2017), Comparison of plane–stress, generalized-plane–strain, and 3D FEM elastic–plastic analyses of thick-walled cylinders subjected to radial thermal gradient, *International Journal of Mechanical Sciences*, **131–132**, pp. 744–752.

Kargarnovin, M.H., Zarei, A.R. and Darijani, H., (2005), Wall thickness optimization of thick-walled spherical vessel using thermo-elasto-plastic concept, *International Journal of Pressure Vessel and Piping*, **82**, pp. 379–385.

MacGregor, C.W., Coffin Jr., L.F. and Fisher, J.C., (1948), Partially plastic thick-tubes, *Journal of the Franklin Institute*, **245**(1), pp. 135–158.

Macrae, A.E., (1930), *Overstrain in Metals*, H.M. Stationery Office, London, UK.

May, I.L., (2015), 60 years of pressure in retrospect, *Procedia Engineering*, **130**, pp. 3–16.

Mote, J.D., Ching, L.K.W., Knight, R.E., Fay, R.J. and Kaplan, M.A., (1971), Explosive autofrettage of cannon barrels, AMMRC CR 70-25, Army Materials and Mechanics Research Center, Watertown, MA.

Nádai, A., (1931), *Plasticity*, McGraw-Hill Book Company, Inc., New York.

O'Hara, G.P., (1992), Analysis of the swage autofrettage process, Technical report ARCCB-TR-92016, U.S. Army Armament Research Development and Engineering Center, Benét Laboratories, Waterville, NY.

Parker, A.P., (2001a), Autofrettage of open-end tubes—pressures, stresses, strains, and code comparisons, *ASME Journal of Pressure Vessel Technology*, **123**, pp. 271–281.

Parker, A.P., (2001b), Bauschinger effect design procedures for compound tubes containing an autofrettaged layer, *ASME Journal of Pressure Vessel Technology*, **123**, pp. 203–206.

Parker, A.P., (2005), Assessment and extension of an analytical formulation for prediction of residual stress in autofrettaged thick cylinders, ASME Pressure Vessels and Piping Conference, Denver, CO, July 17–21.

Parker, A.P. and Kendall, D.P., (2003), Residual stresses and lifetimes of tubes subjected to shrink-fit prior to autofrettage, *ASME Journal of Pressure Vessel Technology*, **125**(3), pp. 282–286.

Parker, A.P., Underwood, J.H., Throop, J.F. and Andrasic, C.P., (1983), Stress intensity and fatigue crack growth in a pressurized, autofrettaged thick cylinder, American Society for Testing and Materials 14th National Symposium on Fracture Mechanics, UCLA, ASTM STP 791, pp. 216–237.

Parker, A.P., Underwood, J.H. and Kendall, D.P., (1999), Bauschinger effect design procedures for autofrettaged tubes including material removal and Sachs method, *ASME Journal of Pressure Vessel Technology*, **121**, pp. 430–437.

Perl, M. and Aroné, R., (1988), Stress intensity factors for a radial multi-jacketed partially autofrettaged pressurized thick-walled cylinder, *ASME Journal of Pressure Vessel Technology*, **110**, pp. 147–154.

Perl, M., (1998), An improved split-ring method for measuring the level of autofrettage in thick-walled cylinders, *Journal of Pressure Vessel Technology*, **120**(1), pp. 69–73.

Perl, M. and Perry, J., (2006), An experimental-numerical determination of the three-dimensional autofrettage residual stress field incorporating Bauschinger effects, *ASME Journal of Pressure Vessel Technology*, **128**, pp. 173–178.

Perry, J. and Perl, M. (2008), A 3-D model for evaluating the residual stress field due to swage autofrettage, *ASME Journal of Pressure Vessel Technology*, **130**, pp. 041211-1–041211-6.

Rees, D.W.A., (1987), A theory of autofrettage with applications to creep and fatigue, *International Journal of Pressure Vessel and Piping*, **30**, pp. 57–76.

Rees, D.W.A., (1990), Autofrettage theory and fatigue life of open-ended cylinders, *Journal of Strain Analysis for Engineering Design*, **25**, pp. 109–121.

Ren-rui, Z., Chun-da, T. and Guo-zhen, Z., (1999) Elasto-plastical dynamic analysis of explosive autofrettage, *Southwest Petroleum University (Natural Science)*, **21**(4), pp. 82–85.

Retrieved on December 26, 2018 from https://www.flickr.com/photos/17968829@N03/22743260562)

Retrieved on December 26, 2018 from https://www.flickr.com/photos/bilfinger/14074154115

Retrieved on December 26, 2018 from https://www.flickr.com/photos/iaea_imagebank/3441138290

Retrieved on December 26, 2018 from https://pxhere.com/en/photo/712201

Rodman, T.J., (1847), Improvement in casting ordnance, U. S. Patent No. 5236, August 14.

Shufen, R. and Dixit, U.S., (2018), A review of theoretical and experimental research on various autofrettage processes, *ASME Journal of Pressure Vessel Technology*, **140**, pp. 050802–050802-15.

Stacey, A. and Webster, G.A., (1988), Determination of residual stress distributions in autofrettaged tubing, *International Journal of Pressure Vessel and Piping*, **31**, pp. 205–220.

Steele, M.C., (1952), Partially plastic thick-walled cylinder theory, *ASME Journal of Applied Mechanics*, **19**, pp. 133–140.

Taylor, D.J., Watkins, T.R., Hubbard, C.R., Hill, M.R. and Meith, W.A., (2012), Residual stress measurements of explosive clad cylindrical pressure vessels, *ASME Journal of Pressure Vessel Technology*, **134**, pp. 011501-1–011501-8.

Wilder, B.G., (1865), Five hundred and fifty-eighth meeting, November 14, 1865. Monthly meeting; on the Nephila Plumipes, or Silk Spider, *Proceedings of the American Academy of Arts and Sciences*, **7**, pp. 44–57.

Zare, H.R. and Darijani, H., (2016), A novel autofrettage method for strengthening and design of thick-walled cylinders, *Materials and Design*, **105**, pp. 366–374.

Zare, H.R. and Darijani, H., (2017), Strengthening and design of the linear hardening thick-walled cylinders using the new method of rotational autofrettage, *International Journal of Mechanical Sciences*, **124–125**, pp. 1–8.

2

A Review on Plasticity and the Finite Element Method

2.1 Introduction

When a metal subjected to an external load is deformed beyond the elastic limit, it undergoes plastic deformation. Once plastic deformation has occurred, the metal does not return to its original configuration upon removal of the load. In the elastic state, deformation can be obtained as a function of load by solving the following sets of equations: (1) equations of motion, which reduce to stress equilibrium equations in the absence of inertia forces, (2) strain-displacement relations, and (3) stress–strain relations called constitutive equations. These equations need to be solved along with certain boundary conditions. During plastic deformation, the first set of equations, i.e., equations of motion, remain unchanged, but the other two sets of equations are modified. Usually, a new measure of strain is used to account for large deformation. Unlike in elastic deformation, stress is not a unique function of strain in plastic deformation. Hence, stress–strain relation is expressed either as stress and incremental strain relation or as stress and strain rate relation. One also needs a yield criterion to know the onset of plastic deformation based on some combination of stress components. A model for strain hardening (and sometimes strain softening) is also required.

Problems involving plasticity are generally difficult to solve by using a conventional approach. Most of the time, engineers have to rely on numerical techniques such as the finite difference method (FDM) and the finite element method (FEM). The present chapter discusses these fundamentals of classical plasticity and FEM. In this chapter, first index notations are described. After which a brief overview of the theory of plasticity is provided. Finally, a simplified introduction to FEM is presented.

2.2 Index Notation

The field variables in the governing equations of elastic–plastic deformation use operations on scalars, vectors, and tensors. For the convenience of representation of various equations, the index notation along with Einstein's summation convention is used. In index notation, a tensor component is represented by using indices. An index is designated by a small letter, say *i*, which can take on the values 1, 2, and 3 for the *x*, *y*, and *z* coordinates of a Cartesian system, respectively. For example, the components of a typical vector ***a*** (which is a first-order tensor) are represented by a_i that denotes 3 components formed by the index *i* (1, 2, 3). Thus,

$$\{a_i\} \equiv \begin{Bmatrix} a_1 \\ a_2 \\ a_3 \end{Bmatrix}. \tag{2.1}$$

The components of a second-order tensor ***A*** are represented by a_{ij}, which may be arranged in the form of a matrix with nine components formed by varying each of the indices *i* and *j* from 1 to 3. Thus,

$$\left[a_{ij}\right] = \begin{bmatrix} a_{11} & a_{12} & a_{13} \\ a_{21} & a_{22} & a_{23} \\ a_{31} & a_{32} & a_{33} \end{bmatrix}. \tag{2.2}$$

If an index *i* appears without any repetition (as in the case of Equation (2.1) or Equation (2.2)) it is called a free index and represents an independent feature of the variable. Thus, a vector has one free index that represents its direction coordinate. A second-order tensor has two free indices: one of them represents the direction coordinate and the other represents the plane on which the force per unit area is measured.

If in any term an index is repeated, it indicates summation by varying the index over its range and adding all resulting components. The repeating index is called the dummy index, and this convention is called the Einstein summation convention. For example, in the three-dimensional (3D) space, a_{ii} means the summation of three components by varying the dummy index *i* from 1 to 3, i.e.,

$$a_{ii} = a_{11} + a_{22} + a_{33}. \tag{2.3}$$

Similarly, $a_i b_i$ with a repeating index *i* represents the summation of the products of the components of vectors *a* and *b*, i.e.,

$$a_i b_i = a_1 b_1 + a_2 b_2 + a_3 b_3 \tag{2.4}$$

This indicates the dot product of vectors ***a*** and ***b***. Another frequently used symbol in tensor algebra is the Kronecker delta function represented by δ_{ij}. It is defined as:

$$\delta_{ij} = \begin{cases} 1 & \text{if } i = j, \\ 0 & \text{if } i \neq j. \end{cases} \tag{2.5}$$

If its components are arranged in a matrix form, it represents an identity matrix. For a second-order tensor $\boldsymbol{A}$ with its components as a_{ij}, the following relation holds good:

$$a_{ij}\delta_{ij} = a_{ii}. \tag{2.6}$$

For the sake of brevity, the comma notation is also in vogue. As per this, "$_{,j}$" indicates partial derivative with respect to coordinate x_j. Thus,

$$u_{i,j} = \frac{\partial u_i}{\partial x_j}. \tag{2.7}$$

2.3 Measures of Strain and Stress

Elementary solid mechanics deals with mostly small deformations with elastic strains. For small deformations, infinitesimal linear strain tensor may be defined as:

$$\varepsilon_{ij} = \frac{1}{2}\left(u_{i,j} + u_{j,i}\right), \tag{2.8}$$

where u_i represents a displacement component as a function of spatial coordinates. One-dimensional (1D) constitutive stress–strains equations are generally expressed in terms of the engineering stresses and strains. The engineering strain denoted by e is defined as:

$$e = \left(\frac{\Delta l}{l_o}\right), \tag{2.9}$$

where Δl is the change in length and l_o is the original length of the specimen in a uniaxial-tensile test. For infinitesimal small change in length, e reduces to one component of infinitesimal strain tensor. When the deformation is large, it is more convenient to use an alternate measure of strain, which is the logarithmic strain. The logarithmic (often called true) strain denoted by ε is defined as:

$$\varepsilon = \ln\left(\frac{l_f}{l_o}\right), \tag{2.10}$$

where l_f is the final length of the specimen. An important advantage of using the logarithmic strain is demonstrated by the following example.

Consider a uniaxial-tensile test specimen of original length l_o subjected to an elongation of Δl to stretch it up to length l_f. If the specimen is subsequently subjected to a compression Δl back to its original length l_o, the effective final strain in the specimen should be zero because there is no net change in length. The engineering strains for the elongation and compression can be expressed as $e_1 = \Delta l / l_o$ and $e_2 = -\Delta l / l_f$, respectively, which upon addition does not give zero as:

$$\left(\frac{\Delta l}{l_o}\right)+\left(\frac{-\Delta l}{l_f}\right)\neq 0. \tag{2.11}$$

On the other hand, the final strain based on the logarithmic measure of strain is:

$$\varepsilon = \ln\left(\frac{l_f}{l_o}\right)+\ln\left(\frac{l_o}{l_f}\right) = \ln\left(\frac{l_f}{l_o}\times\frac{l_o}{l_f}\right) = 0. \tag{2.12}$$

Hence, the logarithmic strain measure provides the realistic picture. The relationship between the engineering strain and the logarithmic strain can be readily derived as:

$$\varepsilon = \ln\left(\frac{l_f}{l_o}\right) = \ln\left(\frac{l_0+\Delta l}{l_o}\right) = \ln\left(1+\frac{\Delta l}{l_o}\right) = \ln(1+e). \tag{2.13}$$

One can obtain a logarithmic strain tensor by using the definition of 1D logarithmic strain along three principal directions. Logarithmic strain tensor expressed along these principal directions can be transformed to any other convenient coordinate system using tensor transformation relations.

The engineering stress denoted by S is defined as:

$$S = \frac{F}{A_o}, \tag{2.14}$$

where A_o is the original cross-sectional area of the test specimen and F is the component of the force. When F is normal to the surface of area A_o (as in a uniaxial-tensile test), the corresponding stress S is designated as normal or direct stress; when F is tangential to the surface, the corresponding stress is called shear stress. A true stress measure denoted by σ is defined as:

$$\sigma = \frac{F}{A_f}, \tag{2.15}$$

where A_f is the current (final) cross-sectional area of the specimen. The relationship between engineering stress and true stress can be derived as follows. Volume change during an elastic deformation is very small; during the plastic deformation of most metals, volume remains constant. Thus, assuming overall volume constancy:

$$A_f l_f = A_o l_o. \tag{2.16}$$

Substituting A_f from Equation (2.16) into Equation (2.15), provides:

$$\sigma = \frac{F l_f}{A_o l_o}. \tag{2.17}$$

Substituting Equation (2.14) and Equation (2.9) into Equation (2.17) provides:

$$\sigma = \frac{S l_f}{l_o} = \frac{S(l_o + \Delta l)}{l_o} = S\left(1 + \frac{\Delta l}{l_o}\right) = S(1 + e). \tag{2.18}$$

In general, on any surface, there may be tangential as well as normal force components. Moreover, these force components may vary from point to point. Hence, a traction vector $\boldsymbol{t}_n$ at a point on the surface with normal $\boldsymbol{n}$ is defined as:

$$\boldsymbol{t}_n = \lim_{\Delta A \to 0} \frac{\Delta \boldsymbol{F}}{\Delta A}, \tag{2.19}$$

where ΔF is the resultant of distributed internal forces on small area ΔA of the surface. As $\boldsymbol{t}_n$ is a vector, it has three components in 3D space. If we know the traction vectors on three orthogonal planes passing through a point, we can find out the traction vector on any plane passing through that point. Hence, the state of stress at a point is fully defined by nine components. A typical component is written as σ_{ij}, where i designates the plane (more specifically the direction of normal to the plane) and j denotes the direction of the force component. Stress defined in the aforesaid manner is called the Cauchy stress tensor $\boldsymbol{\sigma}$. Note that σ_{ij} represents the components of stress tensor $\boldsymbol{\sigma}$ and not the tensor itself. When each component is associated with a plane and a direction of force, then only the stress tensor is fully described. It is just like u_i represents the displacement component u_x, u_y, and u_z. Complete vector will be represented as:

$$\boldsymbol{u} = u_x \mathbf{i} + u_y \mathbf{j} + u_z \mathbf{k}, \tag{2.20}$$

where **i**, **j**, and **k** are unit vectors along the x, y, and z directions in 3D space.

2.4 Equations of Motion

Applying Newton's laws or balancing linear momentum in three orthogonal directions in a Cartesian coordinate system, three equations of motion are obtained, which are valid for any continuous matter, irrespective of the type of deformation. In index notation, these equations are represented as:

$$\sigma_{ji,j} + b_i = \rho a_i, \tag{2.21}$$

where ρ is the density of the material and b_i is the i-th component of the body force per unit volume. Unlike surface forces, body forces act on the particles of surface as well as on the insides of the body. An example of body force is gravity force, and for a unit volume it is ρg in the downward vertical direction, where g is the acceleration due to gravity. The components of acceleration vector are represented by a_i. In the absence of inertia effects, Equation (2.21) reduces to:

$$\sigma_{ji,j} + b_i = 0, \tag{2.22}$$

which is called stress equilibrium equation.

Balance of angular momentum (in the absence of body moments) provides:

$$\sigma_{ji} = \sigma_{ij}. \tag{2.23}$$

Hence, stress equilibrium equation (Equation 2.22) may also be written as:

$$\sigma_{ij,j} + b_i = 0. \tag{2.24}$$

Often the boundary conditions are expressed in terms of the tractions. The stress and traction components in a plane with normal vector $\boldsymbol{n}$ are related as:

$$\sigma_{ji} n_j = (t_n)_i, \tag{2.25}$$

where $(t_n)_i$ is the i-th traction component and n_j is the j-th component of the normal vector. In view of symmetry of the stress tensor $\boldsymbol{\sigma}$ as shown in Equation (2.23), Equation (2.25) is popularly written as:

$$\sigma_{ij} n_j = (t_n)_i. \tag{2.26}$$

2.5 Strain-Displacement Relations

The definition of infinitesimal linear strain tensor (Equation 2.8) provides strain-displacement relation. Using Equation (2.8) with the three displacement field components known, nine components of the strain tensor can be obtained. This tensor is symmetric by the very definition of strain. Thus, effectively there are only six components of this strain tensor. Given six components of strain tensor, one can obtain the displacement field by integration. However, as there are six equations to obtain three unknown components of the displacement field, an arbitrary strain field may not provide a unique displacement field. Hence, the strain components need to be related, providing three independent compatibility conditions. In most of the books on continuum mechanics, six compatibility conditions are presented, out of which only three are independent.

2.6 Incremental Strain and Strain Rate Measures of Plastic Deformation

The infinitesimal linear strain tensor defined in Equation (2.8) cannot be used as a measure of plastic deformation where the deformation is large. Also in plastic deformation, stress and strain do not have a one to one relation, i.e., for a particular level of strain, there may be several possible values of stress. During plastic deformation, the final stress at the current strain depends on the history of the deformation. Hence, it is more convenient to use incremental strain or strain rate as a measure of deformation to express the constitutive equations. The components of incremental strain tensor are expressed as:

$$d\varepsilon_{ij} = \frac{1}{2}\left(du_{i,j} + du_{j,i}\right), \tag{2.27}$$

where du_i represents a component of incremental displacement as a function of spatial coordinates. In particular, updated Lagrangian formulation uses incremental strain as a measure of deformation.

Another measure of deformation can be strain rate, whose components are expressed as:

$$\dot{\varepsilon}_{ij} = \frac{1}{2}\left(v_{i,j} + v_{j,i}\right) \tag{2.28}$$

where v_i represents a component of velocity as a function of spatial coordinates. Eulerian formulation uses strain rate as the measure of deformation. In the textbooks on continuum mechanics, strain rate tensor is called rate of deformation tensor.

2.7 Yield Criterion

The yield criterion defines the elastic limit of a material at which it begins to yield. Mathematically, the yield criterion is expressed as a scalar function of the components of stresses as:

$$f(\sigma_{ij}) = 0, \tag{2.29}$$

where σ_{ij} are the components of the Cauchy stress tensor. The function f is called the yield function. For the purpose of developing the yield criterion, it is convenient to additively decompose the stress tensor into hydrostatic and deviatoric parts. Thus,

$$\sigma = \left(\frac{1}{3} tr\sigma\right)1 + \sigma', \tag{2.30}$$

where $tr\boldsymbol{\sigma}$ is the trace (sum of diagonal components) of stress tensor, **1** is a unit tensor whose components are represented by an identity matrix and $\boldsymbol{\sigma}'$ is the deviatoric stress tensor with zero trace. In index notation, Equation (2.30) is expressed as:

$$\sigma_{ij} = \left(\frac{1}{3}\sigma_{kk}\right)\delta_{ij} + \sigma'_{ij}. \tag{2.31}$$

It has been observed that, in most metals, only the deviatoric part of the stress tensor causes plastic deformation. Two popularly used yield criteria are explained here.

2.7.1 *Tresca Yield Criterion*

The Tresca yield criterion is also called the maximum shear stress criterion. It states that yielding of the material commences when the state of stress at any point in the material satisfies the following function (Dixit and Dixit, 2008):

$$f(\sigma_{ij}) = \max\{|\sigma_1 - \sigma_2|, |\sigma_2 - \sigma_3|, |\sigma_3 - \sigma_1|\} - \sigma_Y = 0, \tag{2.32}$$

where σ_Y is the yield stress of the material and σ_1, σ_2, and σ_3 are the principal stresses. Principal stresses are the eigenvalues of the stress tensor. Many ductile materials follow the Tresca criterion very closely. It is easily seen that hydrostatic part does not affect the yield criterion.

2.7.2 *Von Mises Yield Criterion*

In the von Mises yield criterion, the yield function is defined as a scalar function of an invariant (called a second invariant) of the deviatoric part of the Cauchy stress tensor. Mathematically, it is expressed as (Dixit and Dixit, 2008)

$$f(\sigma_{ij}) = 0 \equiv 3J_2 - \sigma_Y^2 = 0, \tag{2.33}$$

where J_2 is the second invariant of the deviatoric part of the Cauchy stress tensor. It is defined as

$$J_2 = \frac{1}{2}\sigma'_{ij}\sigma'_{ij}, \tag{2.34}$$

where σ'_{ij} is the deviatoric part of the Cauchy stress tensor σ_{ij}.

It is also sometimes convenient to reduce the von Mises yield criterion to the following form:

$$f(\sigma_{ij}) = \sigma_{eq} - \sigma_Y = 0, \tag{2.35}$$

where σ_{eq} called the equivalent stress is expressed as:

$$\sigma_{eq} = \sqrt{3J_2}. \tag{2.35}$$

The equivalent stress represents a measure that is equivalent to that of the stress in a uniaxial tensile/compression test. In terms of the principal stresses, the von Mises yield criterion reduces to the following form:

$$f(\sigma_{ij}) = (\sigma_1 - \sigma_2)^2 + (\sigma_2 - \sigma_3)^2 + (\sigma_3 - \sigma_1)^2 - 2\sigma_Y^2 = 0. \tag{2.36}$$

The yield function $f(\sigma_{ij})$ when plotted in the three-dimensional stress space of σ_1, σ_2, and σ_3 traces the yield locus. The Tresca yield locus is a regular hexagonal prism, whereas the von Mises yield locus is a right circular cylinder of radius $2\sigma_Y/\sqrt{3}$, whose axis is oriented along the line $\sigma_1 = \sigma_2 = \sigma_3 = 0$. The Tresca yield locus is inscribed inside the von Mises cylinder.

2.8 Criterion for Subsequent Yielding

When a material has commenced initial yielding, the material develops a resistance to further deformation and more load needs to be applied for subsequent plastic deformation. When a plastically deformed material is reloaded, the material yields at a stress larger than the initial yield stress on further loading. This phenomenon is called strain hardening. In mathematical modeling, strain hardening is incorporated using a hardening function in the constitutive equations. Strain hardening depends only on the plastic part of the strain. Hence, the plastic part of the strain must be separated from the total strain. For small rotations, incremental strain and strain rate can be decomposed into elastic and plastic parts. Thus,

$$d\varepsilon_{ij} = d\varepsilon_{ij}^{e} + d\varepsilon_{ij}^{p}, \tag{2.37}$$

$$\dot{\varepsilon}_{ij} = \dot{\varepsilon}_{ij}^{e} + \dot{\varepsilon}_{ij}^{p}. \tag{2.38}$$

Two popular types of hardening models used in elastoplastic modeling are isotropic hardening and kinematic hardening. In isotropic hardening, the yield locus increases in size without any change in the position of the center or shape of the yield locus as more strain is applied in the plastically deformed material. In uniaxial tension, isotropic strain-hardening can be expressed as a scalar function of the plastic strain as follows:

$$\sigma = H(\varepsilon^{p})^{n}, \tag{2.39}$$

where H is called the hardening function and σ is the true stress due to the applied strain in the uniaxial tension. Some hardening functions that have been proposed and popularly adopted for the numerical modeling of elastoplastic problems are as follows (Dixit and Dixit, 2014):

(i) Holloman's law:

$$\sigma = K(\varepsilon^{p})^{n}, \tag{2.40}$$

(ii) Ludwik's law:

$$\sigma = \sigma_Y + K(\varepsilon^{p})^{n}, \tag{2.41}$$

(iii) Ramberg–Osgood equation:

$$\varepsilon^{p} = \frac{\sigma}{E}\left\{1 + K\left(\frac{\sigma}{\sigma_Y}\right)^{n-1}\right\}, \tag{2.42}$$

(iv) Swift's law:

$$\sigma = \sigma_Y\left(1 + K\varepsilon^{p}\right)^{n}, \tag{2.43}$$

(v) Prager's law:

$$\sigma = \tanh\left(\frac{E\varepsilon^{p}}{\sigma_Y}\right), \tag{2.44}$$

(vi) Johnson–Cook model:

$$\sigma = \left(\alpha + \beta\varepsilon_p^n\right)\left(1 + \gamma \ln \frac{\dot{\varepsilon}^p}{\dot{\varepsilon}_o}\right)\left\{1 - \left(\frac{T - T_o}{T_{\text{melt}} - T_o}\right)\right\} \tag{2.45}$$

where E is Young's modulus of elasticity, $\dot{\varepsilon}_o$ is a reference strain rate, $\dot{\varepsilon}^p$ is the plastic strain rate, ε_p^n is the plastic strain, T is temperature, T_o is the room temperature, T_{melt} is the melting temperature, and K, n, α, β, and γ are hardening parameters that are determined by fitting the hardening function in the experimental true stress versus plastic strain. These relations are used for any general 3D loading, by replacing the uniaxial stress with equivalent stress and strain rate by equivalent plastic strain rate defined by:

$$\dot{\varepsilon}_{eq}^p = \sqrt{\frac{2}{3}\dot{\varepsilon}_{ij}^p \dot{\varepsilon}_{ij}^p}. \tag{2.46}$$

Similarly, strain in 1D is replaced by equivalent strain in 3D that is obtained by path line integration of equivalent strain rate. In kinematic hardening, the position of the center of the yield locus changes without any change in the size of the yield locus with increasing plastic strain. This model is used to incorporate the Bauschinger effect, wherein a strain hardening material that has already commenced tensile yielding yields at a smaller magnitude of stress when subsequently subjected to a compressive load. The first kinematic hardening law was proposed by Prager (1955) for the rigid translation of the yield locus without any change in its shape or size. The yield criterion for subsequent yielding in the kinematic hardening model is governed by the expression:

$$f(\sigma_{ij} - \alpha_{ij}) = 0, \tag{2.47}$$

where α_{ij} is called the back stress and represents the incremental translation of the yield surface in the stress space. Based on the von Mises yield function, the criterion for subsequent yielding in a kinematic hardening material becomes:

$$\left(\sigma'_{ij} - \mathrm{d}\alpha_{ij}\right)\left(\sigma'_{ij} - \mathrm{d}\alpha_{ij}\right) - \frac{2}{3}\sigma_Y^2 = 0. \tag{2.48}$$

where $\mathrm{d}\alpha_{ij}$ is the incremental back stress and σ'_{ij} is the deviatoric part of the stress tensor. Prager's hardening law states that the incremental translation of the yield locus takes place along the direction of the incremental plastic strain $\mathrm{d}\varepsilon_{ij}^p$ as:

$$\mathrm{d}\alpha_{ij} = c\mathrm{d}\varepsilon_{ij}^p, \tag{2.49}$$

where $d\alpha_{ij}$ is the incremental translation of the yield locus due to the applied incremental plastic strain $d\varepsilon_{ij}^p$ and c is the material hardening parameter. If c is a constant, then the hardening is called linear kinematic hardening, whereas if c is dependent on deformation history, it is called non-linear kinematic hardening. Here, only the linear kinematic hardening is discussed. For linear kinematic hardening in Prager's model, c is equal to 2/3 times the slope of the plastic part of the bilinear stress strain curve. In a uniaxial tension test, Prager's law predicts hardening along the direction of applied stress but transverse softening. This is a drawback of Prager's hardening law since transverse softening is not observed experimentally. This drawback was corrected by Ziegler (Ziegler, 1959). According to Ziegler's hardening law, the incremental translation of the yield locus takes place along the direction of the vector connecting the center of the yield locus to the point of the current stress tensor. The linear form of Ziegler's hardening law is defined as (Dixit and Dixit, 2008):

$$d\alpha_{ij} = \frac{H'}{\sigma_Y}\left(\sigma_{ij} - \alpha_{ij}\right)d\varepsilon_{eq}^p, \tag{2.50}$$

where H' is the constant slope of the plastic part of the bilinear stress–strain curve and $(\sigma_{ij} - \alpha_{ij})$ represents the vector along the direction of the line connecting the center of the yield locus to the point of the current stress tensor.

2.9 Stress-Incremental Strain and Stress–Strain Rate Relation During Plastic Deformation

It has been explained in Section 2.6 that the constitutive stress strain relations for plastic deformation should either be in the form of incremental strain or strain rate. Once the material has reached the plastic state, the 3D stress strain relation can no longer be governed by the generalized Hooke's law. For plastic deformation, the stress strain constitutive equations are developed based on the *flow rule*. According to this, the incremental plastic strain is obtained from a scalar function $g(\sigma_{ij})$ of the stress tensor, which is called the plastic potential. Thus,

$$d\varepsilon_{ij}^p = d\lambda\frac{\partial g}{\partial \sigma_{ij}}, \tag{2.51}$$

where $d\lambda$ is a positive scalar called the plastic multiplier. The plastic potential does not have any physical meaning; it has only a mathematical significance. It is generally assumed that the plastic potential is identical to the yield function $f(\sigma_{ij})$. When the flow rule is based on the yield function as the plastic potential, it is called the *associated flow rule*, which is expressed as follows:

$$d\varepsilon_{ij}^{p} = d\lambda \frac{\partial f}{\partial \sigma_{ij}}. \tag{2.52}$$

2.10 Introduction to the Finite Element Method (FEM)

In general, a mechanical system may be of two types—discrete and continuous. The defining characteristics of a discrete system include a finite number of degrees of freedom, which may be modeled by a system of linear equations. A discrete system is often represented as spring-mass assembly, in which the lumped masses are treated as nodes. Displacements of the nodes are functions of the spring stiffness and loading. In contrast, a continuous system has an infinite number of degrees of freedom and may be modeled by a differential equation along with certain boundary conditions.

The FEM may be considered as a technique of transforming a continuous system into a discrete system. The continuous domain is treated as a collection of discrete elements. Due to this, instead of solving differential equations, one needs to solve only algebraic equations in FEM. Today, the scope of FEM has expanded into a number of fields including engineering, science, and mathematics, and it can be treated as a general numerical technique for solving differential equations.

In FEM, for each element, algebraic equations are developed in terms of the unknown variables at specific points of the elements called nodes. A global system of equations is constructed by assembling elemental equations. The steps of FEM are explained in the following subsections.

2.10.1 *Preprocessing*

This step comprises discretizing the continuous domain into a finite number of elements. Elements can be of different forms and sizes. As the computational time and cost increases with an increasing number of elements, the size of each element is usually decided on the basis of a mesh sensitivity analysis. Some commonly used elements are depicted in Figures 2.1– 2.3.

In each element, the unknown variable is approximated by a function (generally polynomial) in terms of the known values of the variables at the nodes. At the common node, suitable continuity should be satisfied. At some boundary nodes, the value of the variables may be known due to so-called essential boundary conditions.

FIGURE 2.1
One-dimensional (a) 2-noded and (b) 3-noded line element.

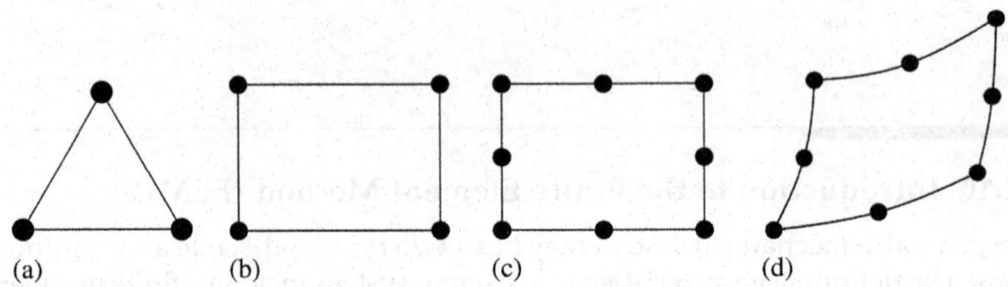

FIGURE 2.2
Two-dimensional (a) 3-noded triangular, (b) 4-noded rectangular, (c) 8-noded rectangular element, and (d) 8-noded curvillinear element.

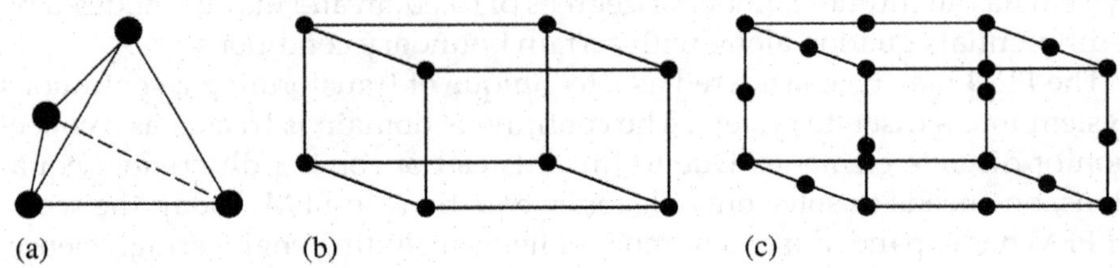

FIGURE 2.3
Three-dimensional (a) 4-noded tetrahedral, (b) 8-noded brick, and (c) 20-noded brick element.

As an illustration, consider a two-node line element shown in Figure 2.4. Assume that it is used for predicting the displacement field. Let the approximating polynomial for displacement be:

$$u = ax + b, \tag{2.53}$$

where x is the spatial coordinate and a and b are the element-specific constants. Substituting the nodal values, $u(x_1) = u_1$, and $u(x_2) = u_2$, in Equation (2.53) provides the following system of equations:

$$\begin{aligned} u_1 &= ax_1 + b, \\ u_2 &= ax_2 + b. \end{aligned} \tag{2.54}$$

Solving it, we get:

$$a = \frac{u_1 - u_2}{x_1 - x_2} \quad \text{and} \quad b = \frac{u_2x_1 - u_1x_2}{x_1 - x_2}.$$

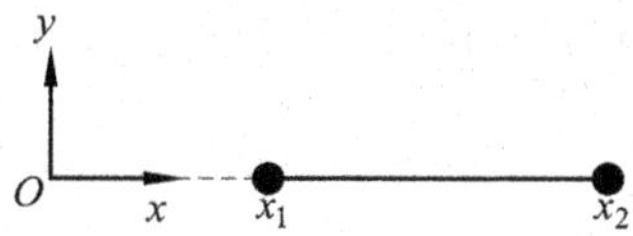

FIGURE 2.4
A 2-noded line element with spatial node coordinates.

Substituting these expressions of a and b in Equation (2.53) and rearranging, the following expression is obtained:

$$u = N_1 u_1 + N_2 u_2, \tag{2.55}$$

where:

$$N_1 = \left(\frac{x - x_2}{x_1 - x_2}\right) \quad \text{and} \quad N_2 = \left(\frac{x - x_1}{x_2 - x_1}\right) \tag{2.56}$$

are called shape functions because they convey an idea of the shape of the function that is being used for approximation. The functions of Equation (2.56) are Lagrangian polynomial interpolation functions. This form of expression in terms of the nodal values offers the advantage that the continuity of the function at each common node between two adjacent elements is automatically ensured. Similarly, for a three-node line element, the approximating function for u can be derived and written as follows:

$$u = N_1 u_1 + N_2 u_2 + N_3 u_3, \tag{2.57}$$

where:

$$N_1 = \frac{(x - x_2)(x - x_3)}{(x_1 - x_2)(x_1 - x_3)}, \; N_2 = \frac{(x - x_1)(x - x_3)}{(x_2 - x_1)(x_2 - x_3)} \text{ and } N_3 = \frac{(x - x_1)(x - x_2)}{(x_3 - x_1)(x_3 - x_2)}. \tag{2.58}$$

These shape functions have the following characteristics:

- The sum of all of the shape functions is always one.
- The value of a shape function at the assigned node is one, while other shape functions are zero at that node.

2.10.2 Developing Elemental Equations

There are various ways of obtaining the elemental equations. Some common approaches are the direct stiffness method, virtual work method, principle of minimum energy method, and the Galerkin method. For the sake of brevity, only the direct stiffness method and the Galerkin method are explained in this chapter.

2.10.2.1 Direct Stiffness Method

The direct stiffness method can be easily employed for finding out the elemental equations for one-dimensional elements such as bars and beams. Consider a bar subjected to uniaxial force F as shown in Figure 2.5. The bar is discretized into three equal elements as shown in Figure 2.6(a). The free body diagram in a particular element with forces F_i and F_j acting at the designated end nodes i and j, respectively, are shown in Figure 2.6(b).

From elementary solid mechanics, the compressive strain ε_i at the ith node is given by:

$$\varepsilon_i = \frac{du}{dx} = \frac{u_i - u_j}{l}, \tag{2.59}$$

where l is the length of the element. Corresponding compressive stress σ_i is given by:

$$\sigma_i = E\varepsilon_i, \tag{2.60}$$

where E is the Young's modulus of elasticity. Hence, the compressive force F_i can be expressed as:

$$F_i = \sigma_i A, \tag{2.61}$$

where A is the cross-sectional area of the element. Substituting Equations (2.59) and (2.60) in Equation (2.61) provides:

$$F_i = \frac{EA}{l}\left(u_i - u_j\right). \tag{2.63}$$

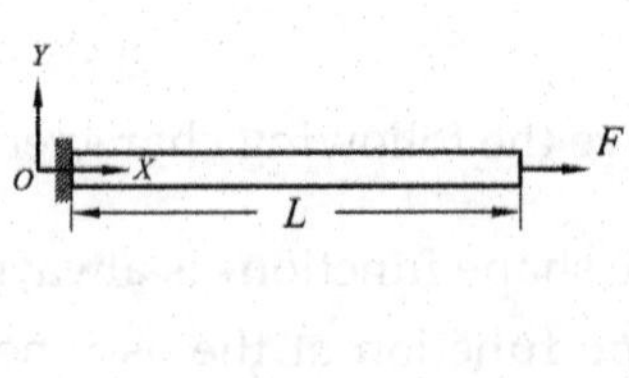

FIGURE 2.5
Axially loaded bar.

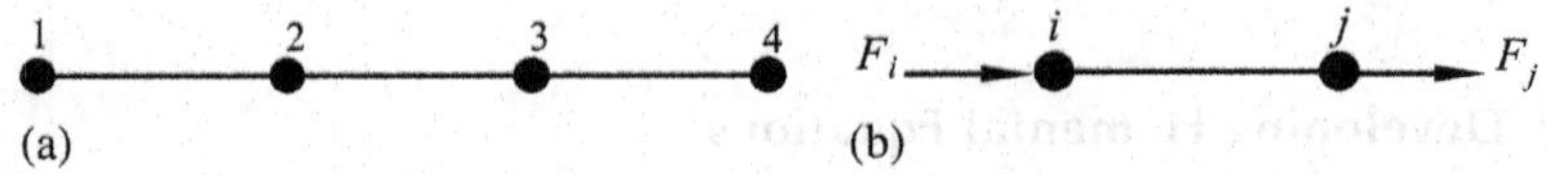

FIGURE 2.6
(a) Three element discretization of the bar and (b) free body diagram of a typical 2-node element in the bar.

Similarly, nodal tensile force F_j can be expressed as:

$$F_j = \frac{EA}{l}\left(u_j - u_i\right). \tag{2.64}$$

Equations (2.63) and (2.64) can be arranged in the following form:

$$\frac{EA}{l}\begin{bmatrix} 1 & -1 \\ -1 & 1 \end{bmatrix}\begin{Bmatrix} u_i \\ u_j \end{Bmatrix} = \begin{Bmatrix} F_i \\ F_j \end{Bmatrix}. \tag{2.65}$$

In solid mechanics, the coefficient matrix of Equation (2.65) is called the elemental stiffness matrix and the right-hand side vector is called the elemental internal load vector.

2.10.2.2 Weighted Residual Method

The weighted residual method is a method where the governing differential equations of a problem are solved approximately based on an assumed solution. The assumed solution is also called the trial solution. Consider a general differential equation:

$$D\phi + q = 0, \tag{2.66}$$

where D is a differential operator, ϕ is the primary variable function that is to be solved and q is a known function. If a trial function $\tilde{\phi}$ is chosen as the approximate solution, then Equation (2.66) becomes:

$$D\tilde{\phi} + q = R, \tag{2.67}$$

where R is called residual. It will be zero if the trial function is the exact solution.

Besides satisfying the boundary conditions, the chosen trial function $\tilde{\phi}$ must be differentiable up to the order of the differential equation. Hence, its requirement for differentiability is strong and Equation (2.67) is called strong form. The trick is to choose trial function such that it minimizes some integral of the weighted residual. Depending on the chosen weight function, different types of weighted residual methods are obtained. In the collocation method, the residual is zero at certain specific points. In the sub-domain method, the integral of the residual is zero at certain selected intervals. In the least square method, the integral of R^2 is minimized over the whole domain. In the Galerkin method, a weighted residual is minimized over the whole domain; however, the form of weight function is the same as the trial function. In this chapter, only the Galerkin method is discussed in detail. In this method, the residual is multiplied by a weight function and the integral is equated to zero. Thus,

$$\int_V wR\,\mathrm{d}V = \int_V w\left(D\tilde{\phi}+q\right)\mathrm{d}V = 0, \tag{2.68}$$

where V is the domain and w is the weight function. Equation (2.68) is integrated by parts to reduce the order of derivatives, so that trial function with a lesser requirement for differentiability can be selected. The form of the equation with reduced differentiability requirement is called weak form. In Galerkin method, the weight function is the arbitrary linear combination of the shape functions of the trial function. The arbitrary coefficients of the shape function in the weight function are often called nodal weights and are akin to nodal variables. The procedure yields a system of algebraic equations.

The Galerkin method is demonstrated here with a simple example. Consider a uniaxial bar of length L and a uniform cross-sectional area A loaded with an axial load with an intensity of q as shown in Figure 2.7. The problem is governed by the following differential equation:

$$\frac{\mathrm{d}}{\mathrm{d}x}\left(EA\frac{\mathrm{d}u}{\mathrm{d}x}\right)+q=0, \tag{2.69}$$

with boundary conditions:

$$u(0)=0,\quad \left.\frac{\mathrm{d}u}{\mathrm{d}x}\right|_{x=L}=0. \tag{2.70}$$

The bar is discretized into three two-node elements each of length $l = L/3$ as shown in Figure 2.8.

To derive the elemental equation for a particular element with first and second node coordinates as x_e and x_{e+1}, respectively, the following procedure is adopted. Multiplying the differential equation (Equation 2.70) by a weight function w and integrating by parts yields:

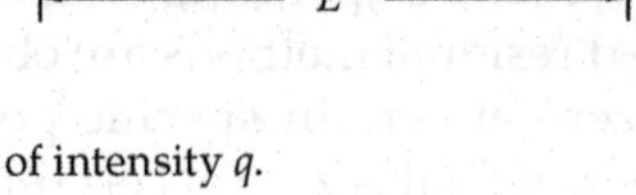

FIGURE 2.7
Bar loaded with uniaxial load of intensity q.

FIGURE 2.8
Finite element mesh for the bar.

$$\left.\begin{aligned} &\int_{x_e}^{x_e+1} w\left\{\frac{\mathrm{d}}{\mathrm{d}x}\left(EA\frac{\mathrm{d}u^e}{\mathrm{d}x}\right)+q\right\}\mathrm{d}x=0 \\ \Rightarrow &\int_{x_e}^{x_e+1}\left(-EA\frac{\mathrm{d}w}{\mathrm{d}x}\frac{\mathrm{d}u^e}{\mathrm{d}x}+qw\right)\mathrm{d}x+wEA\frac{\mathrm{d}u^e}{\mathrm{d}x}\Bigg|_{x_e}^{x_e+1}=0 \end{aligned}\right\}, \tag{2.71}$$

where u^e is the displacement for the element. The linear approximation of the displacement is:

$$u_e=\lfloor N_1 \quad N_2\rfloor\begin{Bmatrix}u_1\\u_2\end{Bmatrix}, \tag{2.72}$$

where the shape functions N_1 and N_2 are defined in terms of the node coordinates as follows:

$$\left.\begin{aligned} N_1 &= \frac{(x-x_{e+1})}{(-x_{e+1}+x_e)}, \\ N_2 &= \frac{(x-x_e)}{(x_{e+1}-x_e)}. \end{aligned}\right\} \tag{2.73}$$

From Equation (2.72), we have:

$$\frac{\mathrm{d}u^e}{\mathrm{d}x}=\left\lfloor\frac{\mathrm{d}N_1}{\mathrm{d}x} \quad \frac{\mathrm{d}N_2}{\mathrm{d}x}\right\rfloor\begin{Bmatrix}u_1\\u_2\end{Bmatrix}=\lfloor N_{1,x} \quad N_{2,x}\rfloor\begin{Bmatrix}u_1\\u_2\end{Bmatrix}, \tag{2.73}$$

where the comma and x in the suffices denotes differentiation with respect to x.

In the Galerkin method, the weight function is similar to the approximating function. Hence, here it is taking as arbitrary the linear combination of the shape functions and is represented as:

$$w=\lfloor w_1 \quad w_2\rfloor\begin{Bmatrix}N_1\\N_2\end{Bmatrix}, \tag{2.74}$$

where w_1 and w_2 are arbitrary nodal weights. Without the loss of generality, the coordinates of the nodes in the element can be taken as $x_e=0$ and $x_{e+1}=l$. Therefore,

$$\left.\begin{aligned} N_1 &= \left(1-\frac{x}{l}\right)\Rightarrow\frac{\mathrm{d}N_1}{\mathrm{d}x}=\frac{-1}{l} \\ N_2 &= \frac{x}{l}\Rightarrow\frac{\mathrm{d}N_2}{\mathrm{d}x}=\frac{1}{l} \end{aligned}\right\} \tag{2.75}$$

Substituting Equations (2.72), (2.73), and (2.74) in Equation (2.71) provides:

$$
-\int_0^l \begin{bmatrix} w_1 & w_2 \end{bmatrix} EA \begin{Bmatrix} N_{1,x} \\ N_{2,x} \end{Bmatrix} \begin{bmatrix} N_{1,x} & N_{2,x} \end{bmatrix} \begin{Bmatrix} u_1 \\ u_2 \end{Bmatrix} \mathrm{d}x + \int_0^l q \begin{bmatrix} w_1 & w_2 \end{bmatrix} \begin{Bmatrix} N_1 \\ N_2 \end{Bmatrix} \mathrm{d}x
$$

$$
+ \begin{bmatrix} w_1 & w_2 \end{bmatrix} \begin{Bmatrix} -E_1A_1 \left(\dfrac{\mathrm{d}u^e}{\mathrm{d}x} \right)_1 \\ E_2A_2 \left(\dfrac{\mathrm{d}u^e}{\mathrm{d}x} \right)_2 \end{Bmatrix} = 0 \tag{2.76}
$$

Since the weights are arbitrary, Equation (2.76) reduces to:

$$
\int_0^l EA \begin{Bmatrix} N_{1,x} \\ N_{2,x} \end{Bmatrix} \begin{bmatrix} N_{1,x} & N_{2,x} \end{bmatrix} \begin{Bmatrix} u_1 \\ u_2 \end{Bmatrix} \mathrm{d}x = \int_0^l q \begin{Bmatrix} N_1 \\ N_2 \end{Bmatrix} \mathrm{d}x + \begin{Bmatrix} -E_1A_1 \left(\dfrac{\mathrm{d}u^e}{\mathrm{d}x} \right)_1 \\ E_2A_2 \left(\dfrac{\mathrm{d}u^e}{\mathrm{d}x} \right)_2 \end{Bmatrix} \tag{2.77}
$$

Substituting Equation (2.75) in (2.77) provides:

$$
\int_0^l EA \begin{Bmatrix} \dfrac{-1}{l} \\ \dfrac{1}{l} \end{Bmatrix} \begin{bmatrix} \dfrac{-1}{l} & \dfrac{1}{l} \end{bmatrix} \begin{Bmatrix} u_1 \\ u_2 \end{Bmatrix} \mathrm{d}x = \int_0^l q \begin{Bmatrix} 1 - \dfrac{x}{l} \\ \dfrac{x}{l} \end{Bmatrix} \mathrm{d}x + \begin{Bmatrix} -E_1A_1 \left(\dfrac{\mathrm{d}u^e}{\mathrm{d}x} \right)_1 \\ E_2A_2 \left(\dfrac{\mathrm{d}u^e}{\mathrm{d}x} \right)_2 \end{Bmatrix} \tag{2.78}
$$

$$
\Rightarrow \frac{EA}{l} \begin{bmatrix} 1 & -1 \\ -1 & 1 \end{bmatrix} \begin{Bmatrix} u_1 \\ u_2 \end{Bmatrix} = \frac{q}{l} \begin{Bmatrix} -\dfrac{(x-l)^2}{2} \Bigg|_0^l \\ \dfrac{x^2}{2} \Bigg|_0^l \end{Bmatrix} + \begin{Bmatrix} -E_1A_1 \left(\dfrac{\mathrm{d}u^e}{\mathrm{d}x} \right)_1 \\ E_2A_2 \left(\dfrac{\mathrm{d}u^e}{\mathrm{d}x} \right)_2 \end{Bmatrix} \tag{2.79}
$$

$$
\Rightarrow \frac{EA}{l} \begin{bmatrix} 1 & -1 \\ -1 & 1 \end{bmatrix} \begin{Bmatrix} u_1 \\ u_2 \end{Bmatrix} = \begin{Bmatrix} \dfrac{ql}{2} \\ \dfrac{ql}{2} \end{Bmatrix} + \begin{Bmatrix} -E_1A_1 \left(\dfrac{\mathrm{d}u^e}{\mathrm{d}x} \right)_1 \\ E_2A_2 \left(\dfrac{\mathrm{d}u^e}{\mathrm{d}x} \right)_2 \end{Bmatrix} \tag{2.80}
$$

Equation (2.80) is the final elemental equation of the problem. The next step is to assemble the elemental equations of all of the elements. This is explained in Subsection 2.10.3.

2.10.3 Assembling the Elemental Equations

The global elemental equation is the equation of the entire domain. The size of the stiffness matrix in the global equation is equal to the total number of degrees of freedom at all of the nodes. For example, if a bar is discretized into three two-node elements, then for one degree of freedom per node, the order of the global stiffness matrix should be 4×4. In a particular element, the global node number can be related to the local node number using the connectivity matrix. The connectivity matrix is a matrix that shows the relationship between the locations of the global and the local nodes, where the columns represent the element number and the rows represent the global node numbers. The connectivity matrix of the example from Section 2.10.2 with the mesh shown in Figure 2.8 is given in Figure 2.9.

For four nodes and one degree of freedom per node, the size of the global stiffness matrix will be 4×4. Using the relevant global node locations, the elemental equations for each element can be expressed as follows:

$$\text{Element 1}: \frac{EA}{l}\begin{bmatrix} 1 & -1 & 0 & 0 \\ -1 & 1 & 0 & 0 \\ 0 & 0 & 0 & 0 \\ 0 & 0 & 0 & 0 \end{bmatrix}\begin{Bmatrix} u_1 \\ u_2 \\ u_3 \\ u_4 \end{Bmatrix} = \begin{Bmatrix} \frac{ql}{2} \\ \frac{ql}{2} \\ 0 \\ 0 \end{Bmatrix} + \begin{Bmatrix} -E_1A_1\left(\frac{du^e}{dx}\right)_1 \\ E_2A_2\left(\frac{du^e}{dx}\right)_2 \\ 0 \\ 0 \end{Bmatrix}, \quad (2.81)$$

$$\text{Element 2}: \frac{EA}{l}\begin{bmatrix} 0 & 0 & 0 & 0 \\ 0 & 1 & -1 & 0 \\ 0 & -1 & 1 & 0 \\ 0 & 0 & 0 & 0 \end{bmatrix}\begin{Bmatrix} u_1 \\ u_2 \\ u_3 \\ u_4 \end{Bmatrix} = \begin{Bmatrix} 0 \\ \frac{ql}{2} \\ \frac{ql}{2} \\ 0 \end{Bmatrix} + \begin{Bmatrix} 0 \\ -E_2A_2\left(\frac{du^e}{dx}\right)_2 \\ E_3A_3\left(\frac{du^e}{dx}\right)_3 \\ 0 \end{Bmatrix}, \quad (2.82)$$

$$\begin{array}{lcc} & \text{Local node 1} & \text{Local node 2} \\ \text{Element 1} & \left[\begin{matrix} 1 \\ 2 \\ 3 \end{matrix}\right. & \left.\begin{matrix} 2 \\ 3 \\ 4 \end{matrix}\right] \\ \text{Element 2} & & \\ \text{Element 3} & & \end{array}$$

FIGURE 2.9
An illustration of a connectivity matrix.

$$\text{Element 3}: \frac{EA}{l}\begin{bmatrix} 0 & 0 & 0 & 0 \\ 0 & 0 & 0 & 0 \\ 0 & 0 & 1 & -1 \\ 0 & 0 & -1 & 1 \end{bmatrix}\begin{Bmatrix} u_1 \\ u_2 \\ u_3 \\ u_4 \end{Bmatrix} = \begin{Bmatrix} 0 \\ 0 \\ \frac{ql}{2} \\ \frac{ql}{2} \end{Bmatrix} + \begin{Bmatrix} 0 \\ 0 \\ -E_3A_3\left(\frac{du^e}{dx}\right)_3 \\ E_4A_4\left(\frac{du^e}{dx}\right)_4 \end{Bmatrix}, \tag{2.83}$$

Adding Equations (2.81), (2.82), and (2.83) provides the following system of global equations:

$$\frac{EA}{l}\begin{bmatrix} 1 & -1 & 0 & 0 \\ -1 & 2 & -1 & 0 \\ 0 & -1 & 2 & -1 \\ 0 & 0 & -1 & 1 \end{bmatrix}\begin{Bmatrix} u_1 \\ u_2 \\ u_3 \\ u_4 \end{Bmatrix} = \begin{Bmatrix} \frac{ql}{2} \\ ql \\ ql \\ \frac{ql}{2} \end{Bmatrix} + \begin{Bmatrix} -E_1A_1\left(\frac{du^e}{dx}\right)_1 \\ 0 \\ 0 \\ E_1A_1\left(\frac{du^e}{dx}\right)_4 \end{Bmatrix}. \tag{2.84}$$

2.10.4 Applying Boundary Conditions and Solving the System of Equations

The boundary conditions for the example of Section 2.10.2 with the mesh shown in Figure 2.8 are given in Equation (2.70). In nodal form, these can be written as:

$$u_1 = 0, \quad \left(\frac{du^e}{dx}\right)_4 = 0. \tag{2.85}$$

The first boundary condition with a prescribed value of the primary variable is called the essential boundary condition. The second boundary condition providing traction-free condition is called the natural or force boundary condition. The boundary conditions can be inserted in Equation (2.84) as follows:

$$\frac{EA}{l}\begin{bmatrix} 1 & 0 & 0 & 0 \\ -1 & 2 & -1 & 0 \\ 0 & -1 & 2 & -1 \\ 0 & 0 & -1 & 1 \end{bmatrix}\begin{Bmatrix} u_1 \\ u_2 \\ u_3 \\ u_4 \end{Bmatrix} = \begin{Bmatrix} 0 \\ ql \\ ql \\ \frac{ql}{2} \end{Bmatrix}. \tag{2.86}$$

Note that the first equation in the system of equations (Equation 2.86) is the essential boundary condition equation, which has replaced the previous equation. Solving Equation (2.86) yields:

$$u_1 = 0,\ u_2 = \frac{5ql^2}{2EA},\ u_3 = \frac{4ql^2}{EA} \text{ and } u_4 = \frac{9ql^2}{2EA}. \tag{2.87}$$

2.10.5 Post-Processing

In this step, the nodal displacements obtained in Equation (2.87) are used to determine strain and stress in the bar. This is called post processing. Strain in element 1 shown in Figure 2.8 is calculated as follows:

$$\varepsilon^{(1)} = \frac{u_2 - u_1}{l} = \frac{1}{l}\left(\frac{5ql^2}{2EA} - 0\right) = \frac{5ql}{2EA}. \tag{2.88}$$

Knowing the strain, one can also determine the stresses. Post-processing also includes the representation of the results in the form of tables, graphs, and contours, etc.

All types of FEM procedures basically follow the abovementioned five steps. Nowadays, there are many commercial as well as free packages that can be used for solving various physical problems.

2.11 An Example of Hydraulic Autofrettage of a Thick Cylinder

A 60 mm long thick-walled cylinder with inner radius 10 mm and outer radius 30 mm is considered as a typical specimen of hydraulic autofrettage. The cylinder is open-ended and the ends are free. The material of the cylinder is SS305 steel having a Young's modulus of 193 GPa, Poisson's ratio of 0.3, an initial yield stress of 205 MPa, and a linear hardening modulus of 10505 MPa. The cylinder is pressurized using various magnitudes of internal pressure. The elastic–plastic analysis of the cylinder is carried out using FEM. The yield pressure of the cylinder is taken as 105 MPa.

Figure 2.10 depicts the distribution of the radial and hoop stresses in the cylinder after pressurization with various magnitudes of pressure—50 MPa, 70 MPa, and 100 MPa, where the cylinder remains elastic. The axial stress remains negligible in the case of an open-ended cylinder and hence it is not depicted. The hoop stress is tensile whereas the radial stress is compressive throughout the thickness of the cylinder. Both are the maximum at the inner wall and decrease gradually with increasing radial position. Hence, the initial plastic deformation of the cylinder will occur at the inner wall of the cylinder.

As the magnitude of applied pressure is increased gradually, the cylinder undergoes initial yielding at 105 MPa. With further increase a plastic zone spreads from the inner wall. Two cases of material behavior are

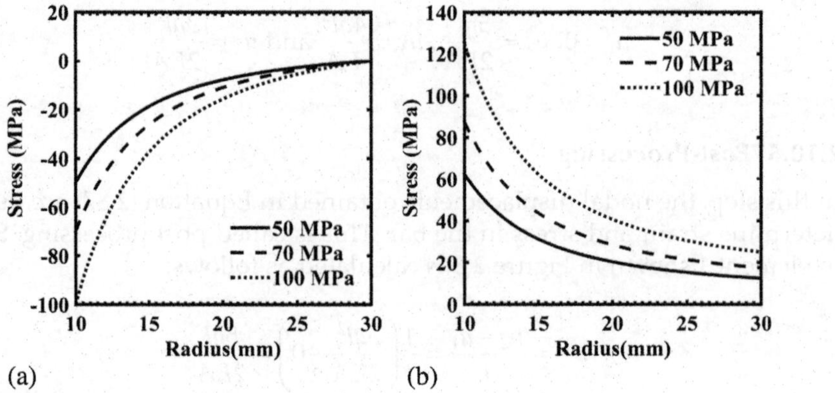

FIGURE 2.10
Distribution of (a) radial stresses and (b) hoop stresses due to elastic deformation of the cylinder for three internal pressures.

considered, namely one where the material hardens isotropically without the Bauschinger effect, and another where there is kinematic hardening with the Bauschinger effect.

Figure 2.11 depicts the distribution of the radial and hoop stresses after pressurization with pressures 110 MPa, 160 MPa, and 220 MPa where the cylinder undergoes plastic deformation with isotropic hardening behavior. Figure 2.12 depicts the distribution of the residual hoop stress after unloading the cylinder from the aforementioned pressure loading conditions. Referring to Figure 2.11, the radial stress remains compressive throughout whose magnitude is the maximum at the inner wall. The hoop stress remains tensile throughout, but it is the maximum at the elastic–plastic interface. After

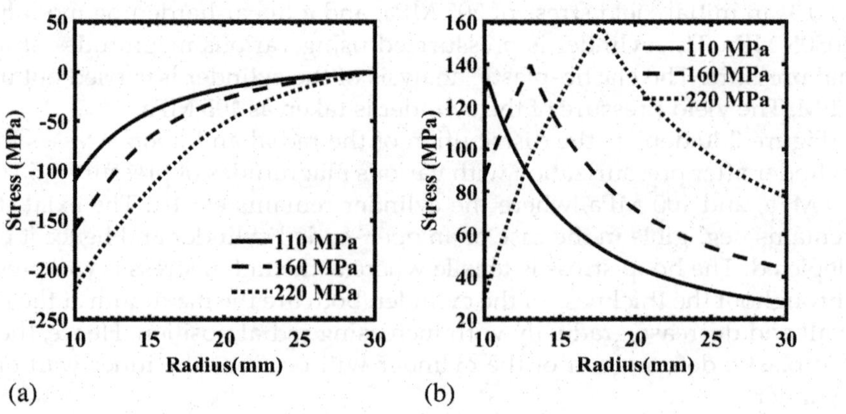

FIGURE 2.11
Distribution of (a) radial stresses and (b) hoop stresses due to elastic–plastic deformation of the cylinder for three internal pressures considering isotropic hardening.

unloading the cylinder from each case of loading, compressive residual stress is induced in the cylinder. The magnitude of the compressive hoop residual stress is maximum at the inner wall as shown in Figure 2.12 and decreases gradually toward the outer wall. Also, as the magnitude of the applied pressure increases, the magnitude of the maximum compressive hoop residual stress at the inner wall also increases. With further increase in the applied pressure, there comes a point where the cylinder yields due to compressive stresses at the inner wall during the unloading process. Hence, the autofrettage pressure is limited to a magnitude where the cylinder avoids reverse yielding due to compressive residual stresses during the unloading process. In the case considering isotropic hardening, this is 220 MPa for the current case of the material and cylinder dimensions.

Figure 2.13 depicts the distribution of the radial and hoop stresses after pressurization with pressures 110 MPa, 160 MPa, and 210 MPa where the

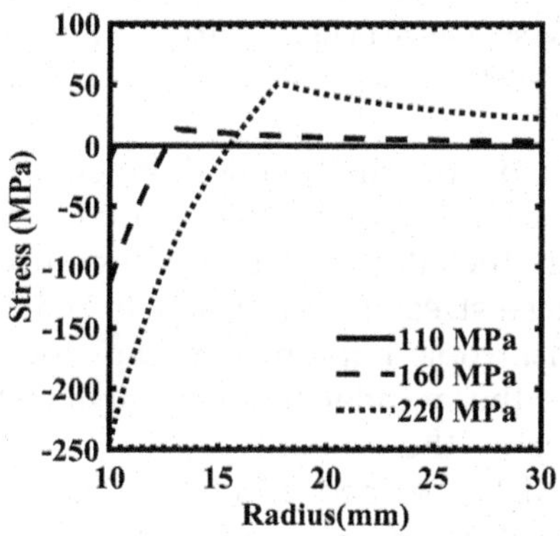

FIGURE 2.12
Distribution of residual hoop stresses in the autofrettaged cylinder for three internal pressures considering isotropic hardening.

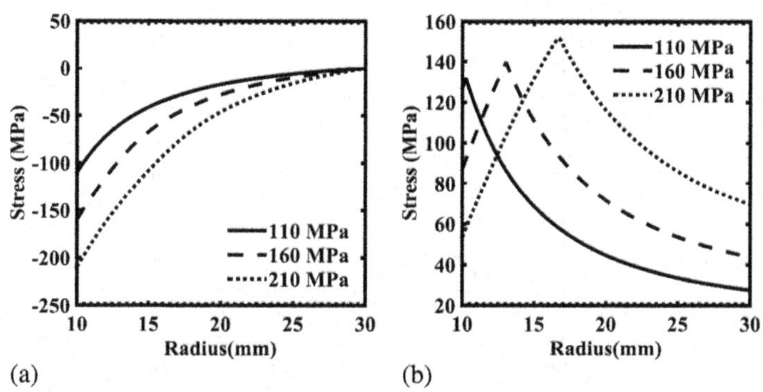

FIGURE 2.13
Distribution of (a) radial stresses and (b) hoop stresses due to elastic–plastic deformation of the cylinder for three internal pressures considering kinematic hardening.

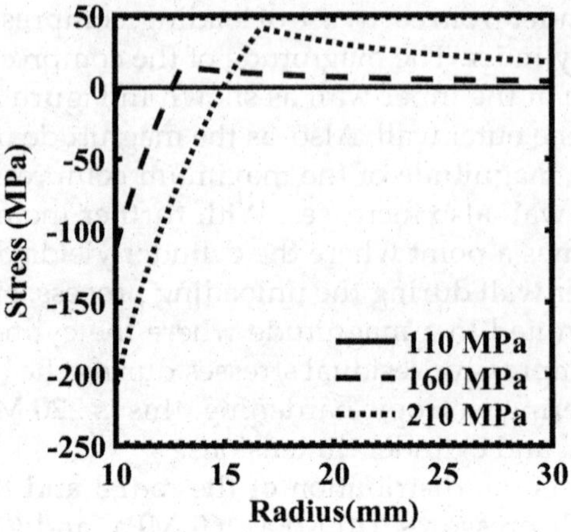

FIGURE 2.14
Distribution of residual hoop stresses in the autofrettaged cylinder for three internal pressures considering kinematic hardening.

cylinder undergoes plastic deformation with kinematic hardening behavior. Figure 2.14 depicts the distribution of the residual hoop stress after unloading the cylinder from the aforementioned pressure loading conditions. The distribution of these stresses are similar to that already explained for the case considering isotropic hardening except for the magnitude of the applied pressure where the cylinder avoids reverse yielding due to compressive residual stress during the unloading process, which is 210 MPa. This is lesser than the 220 MPa obtained in the case of isotropic hardening. It is due to the Bauschinger effect in the case of kinematic hardening. Due to the

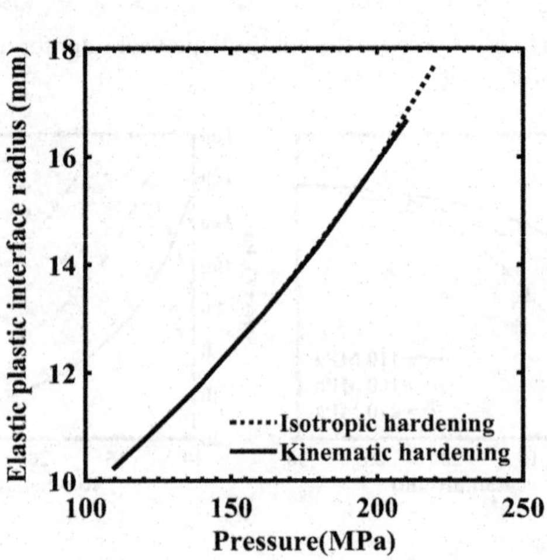

FIGURE 2.15
Variation of the elastic–plastic interface radius with internal pressure.

Bauschinger effect, the cylinder has a lesser compressive yield strength and the cylinder yields at a lesser stress.

The variation of the elastic–plastic interface radius with internal pressure is depicted in Figure 2.15. It is almost the same for the isotropic and the kinematic hardening cases. However, considering isotropic hardening assumption, it is possible to apply a greater pressure without reverse yielding. This is evident from the figure.

2.12 Conclusion

In this chapter, reviews of classical plasticity and the FEM have been presented. First, the system of index notation and tensor algebra is briefly described. This is followed by a discussion of the fundamentals of classical plasticity, which include the measure of plastic strain, the criterion for initial yielding, and the criterion for subsequent yielding of the material. The reader is then introduced to the finite element method, which is a numerical technique for solving differential equations. The procedure for adopting the method is presented. This comprises preprocessing, formulating the elemental equations, assembling these to obtain global equations, applying the boundary conditions, solving the equations, and finally post-processing. This background/recapitulation will help in understanding the modeling of autofrettage processes.

References

Dixit, P.M. and Dixit, U.S., (2008), *Modeling of Metal Forming and Machining Processes: By Finite Element and Soft Computing Methods*, 2008 edition. Springer, London, UK.

Dixit, P.M. and Dixit, U.S., (2014), *Plasticity: Fundamentals and Applications, 1st edition.* CRC Press, Boca Raton, FL.

Prager, W., (1955), The theory of plasticity: a survey of recent achievements, *Proceedings of the Institution of Mechanical Engineers*, **169**, pp. 41–57. https://doi.org/10.1243/PIME_PROC_1955_169_015_02

Ziegler, H., (1959), A modification of Prager's hardening rule, *Quarterly of Applied Mathematics*, **17**, pp. 55–65.

3

Hydraulic Autofrettage

3.1 Introduction

Hydraulic autofrettage is the most widely practiced autofrettage process. It is the earliest concept of autofrettage that was originally conceived by Jacob (1907). Here, the thick-walled cylinder/sphere is filled with hydraulic oil and then pressurized with ultra-high pressure. The hydraulic pressure causes the initial yielding of the inner wall at a threshold of pressure called the yield pressure. The yield pressure represents the original pressure-carrying capacity of the cylinder with the specific material and dimensions. As the pressure is gradually increased, further yielding takes place and a plastic zone propagates toward the outer wall. This creates two deformed zones in the cylinder, namely, an inner plastic zone extending from the inner surface to an intermediate radius and an outer elastic zone extending from the intermediate radius to the outer surface. At this point, if the cylinder is depressurized, the outer elastic zone recovers and compressive residual stresses are induced in the vicinity of the inner wall of the cylinder/sphere.

3.2 A Typical Hydraulic Autofrettage Process

A schematic of a typical hydraulic autofrettage setup was shown in Figure 1.3. That was for the autofrettage of a cylindrical pressure vessel. Hydraulic autofrettage is also carried out for spherical pressure vessels of the type shown in Figure 3.1. The photograph is taken from reference (Sedmak et al., 2016). For the hydraulic autofrettage of a cylinder, the ultra-high autofrettage pressure is generated by a pump and intensifier arrangement and delivered to the cylinder. Typically, the pressure is of the order of 10^3-10^4 bar. Seal plugs are provided at the ends of the cylinder to prevent leakage. To conserve the volume of the oil used in the process, a solid spacer is inserted inside the cylinder. A typical hydraulic oil is made up of castor oil with an anti-freezing

FIGURE 3.1
Typical pressure vessels that can be subjected to hydraulic autofrettage. From (Sedmak et al., 2016) under the CC BY-NC-ND license.

additive in order to avoid the freezing of the oil due to high pressure. Detailed description of one experimental setup of hydraulic autofrettage is available in (Franklin and Morrison, 1960).

3.3 Aspects in Modeling Hydraulic Autofrettage

The section of a thick-walled cylinder/sphere with inner and outer radii *a* and *b*, respectively, is shown in Figure 3.2. The bore is pressurized with a hydraulic pressure of magnitude *p*, which is greater than the yield pressure p_Y (to start yielding) of the cylinder. The pressure causes partial yielding up to an intermediate radius *c* of the cylinder resulting in two deformed zones—an inner plastic zone and an outer elastic zone.

Like most problems in plasticity, the general basis for the mathematical modeling of hydraulic autofrettage processes consists of formulating a set of governing differential equations, which are essentially the stress–equilibrium equations, stress–strain constitutive equations, and the strain–displacement compatibility equations that must be solved by invoking certain boundary conditions. The elastic limit of the cylinder is governed by a yield criterion. The stress response in the cylinder due to the evolution of plastic strain post-yielding is described by a strain-hardening function. Since the loading cycle in an autofrettage process includes an unloading stage, the Bauschinger effect also becomes an important factor. Therefore, three common aspects are important in the modeling of hydraulic autofrettage

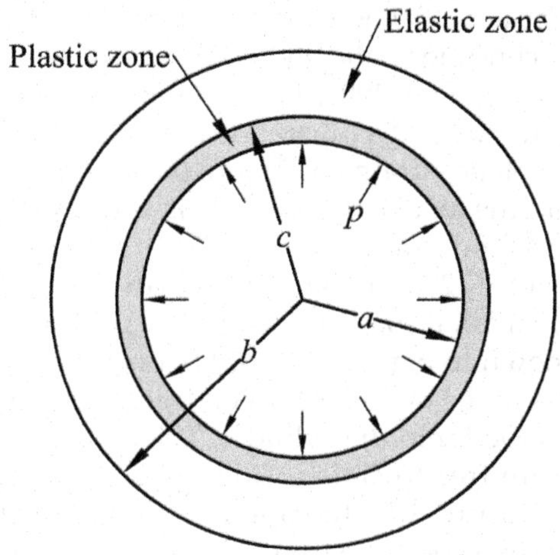

FIGURE 3.2
Section of the thick-walled cylinder/sphere subjected to hydraulic autofrettage.

processes: (i) yield criterion, (ii) incorporation of the material behavior for subsequent plastic deformation, and (iii) boundary conditions. The first two have been already discussed in Sections 2.7 and 2.8, respectively, and will not be repeated in this chapter. The third aspect is discussed as follows.

The mechanical boundary conditions depend on the type of vessel that is being studied. For a sphere, the boundary condition depends on whether the outer wall remains traction-free or constrained. For cylinders, the formulation may be based on either a plane–stress or a generalized plane–strain assumption. In the plane–stress assumption, the axial stress is taken as zero and is applicable in the analysis of thin rings or discs (Avitzur, 1994; Chen, 1973; Lu and Hsu, 1977; Perry and Aboudi, 2003). In the generalized plane–strain assumption, the axial displacement of the cylinder is assumed to remain constant throughout the length of the cylinder and is applicable in the analysis of long tubes (Alexandrov et al., 2015; Allen and Sopwith, 1951; Bland, 1956; Chen, 1980, 1986; Chu, 1972; Davidson et al., 2013; Durban and Kubi, 1992; Elder et al., 1975; Gao, 1992, 1993, 2003, 2007; Gao et al., 2015; Hill et al., 1947; Hosseinian et al., 2009; Huang and Moan, 2009; Huang, 2005; Huang and Cui, 2004, 2005; Koiter, 1953; Lazzarin and Livieri, 1997; Li et al., 1988; Livieri and Lazzarin, 2001; Loghman and Wahab, 1994; MacGregor et al., 1948; Marcal, 1965a, 1965b; Mendelson, 1968; Nádai, 1950; Parker, 2000; Perl and Perry, 2005; Rees, 1990; Thomas, 1953; Wang, 1988; Zeng et al., 1993). Again, based on the type of the axial constraints, the long cylinder satisfying the generalized plane–strain assumption may have one of the three types of end conditions—open-end, closed-end, and plane–strain. In the open-end condition, the ends of the cylinder remain free (Allen and Sopwith, 1951; Bland, 1956; Chen, 1980, 1986; Chu, 1972; Davidson et al., 2013; Elder et al., 1975; Gao, 1992; Koiter, 1953; Li et al., 1988; MacGregor et al., 1948; Marcal, 1965a;

Mendelson, 1968; Parker, 2000; Rees, 1990; Thomas, 1953; Zeng et al., 1993). In the closed-end condition, the cylinder has structurally sealed ends (Alexandrov et al., 2015; Gao, 1993; Lazzarin and Livieri, 1997; Livieri and Lazzarin, 2001; Loghman and Wahab, 1994; Marcal, 1965b; Perl and Perry, 2005). Finally, in the plane–strain condition, the ends of the cylinder are rigidly constrained and the axial displacement is zero (Alexandrov et al., 2015; Avitzur, 1994; Durban and Kubi, 1992; Gao, 2003, 2007; Gao et al., 2015; Hill et al., 1947; Hosseinian et al., 2009; Huang and Moan, 2009; Huang, 2005; Huang and Cui, 2004, 2005; Nádai, 1950; Wang, 1988). The schematic of the three types of conditions are shown in Figure 3.3. For the open-end condition shown in Figure 3.3a, the ends of the cylinder remain free, when pressurized oil from the intensifier is pumped into the cylinder. The entire axial load is taken by the pistons, which are free to slide inside the cylinder. For the closed-end condition shown in Figure 3.3b, the applied pressure is transmitted to the wall as well as to the ends of the cylinder. Finally, for the plane–strain condition shown in Figure 3.3c, the ends of the cylinder are constrained while the bore of the cylinder is pressurized.

The analyses of the plane–stress and the open-end conditions have been considered equivalent by some researchers (Avitzur, 1994; Rees, 1990) because the ends of the cylinder remain traction-free in both of the cases. This was also shown by Gibson et al. (2005) using results based on a finite element method (FEM) study. Gibson et al. also showed that the results given by closed-end and plane–strain conditions were similar. In the case of a thick-walled sphere, the mechanical boundary conditions only consist of whether the outer wall remains traction-free or constrained.

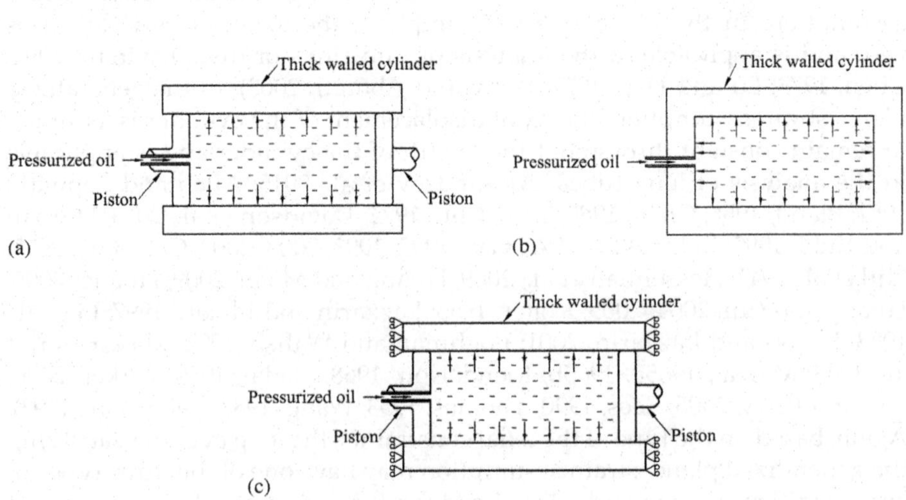

FIGURE 3.3
Cylindrical autofrettage specimen with (a) open-end, (b) closed-end, and (c) plane–strain condition.

The various modeling aspects in hydraulic autofrettage affect the complexity of the problem formulation and the particular solution method. This leads to a classification of the mathematical models into two broad types—analytical modeling and FEM modeling. These are explained in the following subsections of this chapter. As a basis for subsequent elasto-plastic analysis, a solution for the elastic analysis of an internally pressurized thick-walled cylinder with various end conditions is first presented in Section 3.4. Then, closed-form models of a hydraulically autofrettaged cylinder based on the Tresca and von Mises yield criterion are explained in Sections 3.5 and 3.6, respectively. Likewise, Section 3.7 presents the closed-form model for the elastic analysis of an internally pressurized thick-walled sphere. A closed-form model for a hydraulically autofrettaged thick-walled sphere is presented in Section 3.8.

3.4 Elastic Analysis of an Internally Pressurized Thick-Walled Cylinder

As a basis for subsequent elasto-plastic analysis, the solution for the elastic analysis of an internally pressurized thick-walled cylinder with various end conditions is presented first. This analysis is available in several books of solid mechanics, e.g., Chakrabarty (2012). The solution of stresses for the elastic analysis of an internally/externally pressurized thick-walled cylinder has been provided by the French mathematician, Gabriel Lamé (1795–1870). According to his 1852 paper, the distribution of stresses as a function of the radial position r is:

$$\sigma_r = A + \frac{B}{r^2}, \tag{3.1}$$

$$\sigma_\theta = A - \frac{B}{r^2}, \tag{3.2}$$

where σ_r and σ_θ are the radial and circumferential stress components. The constants A and B are obtained by invoking the traction boundary conditions in the cylinder. For internal pressurization, a negative pressure $-p$ acts at the inner wall while the outer wall remains free. Substituting $(\sigma_r)_{r=a} = -p$ and $(\sigma_r)_{r=b} = 0$ in Equations (3.1) and (3.2) provides:

$$\sigma_r = -p\left(\frac{\dfrac{b^2}{r^2} - 1}{\dfrac{b^2}{a^2} - 1}\right), \tag{3.3}$$

$$\sigma_\theta = p\left(\frac{\frac{b^2}{r^2}+1}{\frac{b^2}{a^2}-1}\right). \tag{3.4}$$

The above set of expressions shows that for internal pressurization, the radial stress σ_r is always compressive whereas the circumferential stress σ_θ is always tensile. The axial stress, σ_z depends on the type of end condition of the cylinder, which are derived as follows for three cases:

(1) Cylinder with closed ends:

By equilibrium of forces in the longitudinal direction,

$$P = p\pi a^2 = \sigma_z \pi\left(b^2 - a^2\right), \tag{3.5}$$

where P is the resultant force acting at the ends of the cylinder. Thus,

$$\sigma_z = \frac{pa^2}{b^2 - a^2}. \tag{3.6}$$

(2) Cylinder with open ends/disc with plane–stress condition:

Since the cylinder ends are traction-free (*i.e.*, $P = 0$) the axial stress remains zero. Hence,

$$\sigma_z = 0. \tag{3.7}$$

(3) Cylinder with constrained ends/plane–strain condition:

From a generalized Hook's law, the axial stress is:

$$\sigma_z = E\varepsilon_z + \nu\left(\sigma_r + \sigma_\theta\right). \tag{3.8}$$

Substituting $\varepsilon_z = 0$ in Equation (3.8) due to plane–strain condition leads to:

$$\sigma_z = \nu\left(\sigma_r + \sigma_\theta\right). \tag{3.9}$$

Using Equations (3.3) and (3.4) in Equation (3.9) provides the expression for the distribution of axial stress as:

$$\sigma_z = \frac{2\nu pa^2}{b^2 - a^2}. \tag{3.10}$$

The closed-form model of a hydraulically autofrettaged thick-walled cylinder is formulated as a statically determinate problem because the solution can be obtained using only the force–balance relationship between the elastic and plastic zone. The solution of stresses is obtained from the equilibrium equation that is to be solved by using the following conditions—continuity of radial stress at the elastic–plastic interface and the satisfaction of the yield criterion at the elastic–plastic interface. Two types of solutions, namely one based on Tresca yield criterion and another based on von Mises yield criterion, are derived in the following Sections.

3.5 Closed-Form Model of Hydraulically Autofrettaged Cylinder Based on the Tresca Yield Criterion

For all the three end conditions (closed-end, open-end, and plane–strain), the axial stress, σ_z is the intermediate principal stress. As per the Tresca yield criterion, at the onset of yielding in the cylinder,

$$\sigma_\theta - \sigma_r = \sigma_Y, \tag{3.11}$$

where σ_Y is the yield stress in uniaxial tension or compression. From Equations (3.3) and (3.4),

$$\sigma_\theta - \sigma_r = \frac{2p\dfrac{b^2}{r^2}}{\dfrac{b^2}{a^2}-1}. \tag{3.12}$$

it can be inferred from Equation (3.12) that the magnitude of the stress difference is the maximum at the inner wall where initial yielding takes place. Evaluating $(\sigma_\theta - \sigma_r)$ at $r = a$ using Equation (3.12) and then substituting it in the Tresca yield criterion, the minimum pressure for the initial yielding is obtained as:

$$p_Y = \frac{\sigma_Y}{2}\left(1 - \frac{a^2}{b^2}\right), \tag{3.13}$$

which is valid for all end conditions.

3.5.1 Stress Distribution after Loading

Refer to Figure 3.2. In the elastic zone $c \le r \le b$, the radial and hoop stresses are given by Lamé's equations (Equations (3.1) and (3.2)). Applying the boundary conditions $(\sigma_r)_{r=b} = 0$ and $(\sigma_\theta - \sigma_r)_{r=c} = \sigma_Y$, the radial and hoop stress distributions in the elastic zone are given by:

$$\sigma_r = -\frac{\sigma_Y c^2}{2b^2}\left(\frac{b^2}{r^2} - 1\right), \tag{3.14}$$

$$\sigma_\theta = \frac{\sigma_Y c^2}{2b^2}\left(\frac{b^2}{r^2} + 1\right). \tag{3.15}$$

In the plastic zone ($a \le r \le c$), the solution for the stresses is derived from the stress–equilibrium equation that remains satisfied through all stages of deformation and expressed as:

$$\sigma_\theta - \sigma_r = r\frac{d\sigma_r}{dr}. \tag{3.16}$$

It is assumed that the material of the cylinder does not strain harden. It is already discussed that σ_z is the intermediate principal stress. It is assumed that the material is perfectly plastic with no strain-hardening. Applying the Tresca yield criterion (Equation 3.11) in the equilibrium equation (Equation 3.16) and solving provides:

$$\sigma_r = \sigma_Y \ln r + C, \tag{3.17}$$

where C is an integration constant and can be evaluated using the continuity of radial stress at $r = c$. Evaluating the radial stress component for elastic zone from Equation (3.14) and equating it with the plastic radial stress component from Equation (3.17) at $r = c$ leads to:

$$C = -\frac{\sigma_Y}{2}\left(1 - \frac{c^2}{b^2}\right) - \sigma_Y \ln c. \tag{3.18}$$

Substituting C in Equation (3.17), the radial stress distribution in the plastic zone, $a \le r \le c$, is given by:

$$\sigma_r = -\frac{\sigma_Y}{2}\left(1 - \frac{c^2}{b^2} + \ln\frac{c^2}{r^2}\right), \tag{3.19}$$

From the Tresca yield criterion, the hoop stress distribution is given by:

$$\sigma_\theta = \frac{\sigma_Y}{2}\left(1 + \frac{c^2}{b^2} - \ln\frac{c^2}{r^2}\right). \tag{3.20}$$

The unknown radius of the elastic–plastic interface c can be obtained by using the boundary condition at the inner radius a. For an autofrettage pressure, $p > p_Y$, at $r = a$, $\sigma_r = -p$. Thus, from Equation (3.19):

$$p = \frac{\sigma_Y}{2}\left(1 - \frac{c^2}{b^2} + \ln\frac{c^2}{a^2}\right). \tag{3.21}$$

Equation (3.21) can be numerically solved for the unknown radius c using a bisection method.

3.5.2 Residual Stress Distribution after Unloading

When the cylinder is unloaded by releasing the autofrettage pressure, residual stresses are generated within the wall of the cylinder. Assuming the unloading process to be completely elastic, the residual stresses in the elastic zone ($c \le r \le b$) are obtained by subtracting Equations (3.3) and (3.4) from Equations (3.14) and (3.15), respectively. The resulting stresses are given by:

$$\sigma_r = -\frac{\sigma_Y}{2}\left(\frac{c^2}{a^2} - \frac{p}{p_Y}\right)\left(\frac{a^2}{r^2} - \frac{a^2}{b^2}\right), \tag{3.22}$$

$$\sigma_\theta = \frac{\sigma_Y}{2}\left(\frac{c^2}{a^2} - \frac{p}{p_Y}\right)\left(\frac{a^2}{r^2} + \frac{a^2}{b^2}\right). \tag{3.23}$$

The residual stresses in the plastic zone ($a \le r \le c$) are also obtained by subtracting Equations (3.3) and (3.4) from Equations (3.19) and (3.20) and are given as:

$$\sigma_r = -\frac{\sigma_Y}{2}\left\{\frac{p}{p_Y}\left(1 - \frac{a^2}{r^2}\right) - \ln\frac{r^2}{a^2}\right\}, \tag{3.24}$$

$$\sigma_\theta = -\frac{\sigma_Y}{2}\left\{\frac{p}{p_Y}\left(1 + \frac{a^2}{r^2}\right) - \left(2 + \ln\frac{r^2}{a^2}\right)\right\}. \tag{3.25}$$

Induced compressive residual stresses in an autofrettaged cylinder may cause reyielding in the cylinder depending on the degree of overstrain. The condition for compressive reyielding is determined by applying the yield criterion. In the elastic zone, the application of Tresca yield criterion in the Equations (3.22) and (3.23) provides:

$$\sigma_\theta - \sigma_r = \sigma_Y \frac{a^2}{r^2}\left(\frac{c^2}{a^2} - \frac{p}{p_Y}\right). \tag{3.26}$$

Similarly, for the plastic zone, Equations (3.23) and (3.24) provide:

$$\sigma_\theta - \sigma_r = \sigma_Y\left(1 - \frac{pa^2}{p_Y r^2}\right). \tag{3.27}$$

As can be inferred from Equation (3.27), the magnitude of $(\sigma_\theta - \sigma_r)$ is the maximum at the inner wall. Hence, reyielding will take place at $r = a$ if the applied pressure is at least twice the yield pressure of the cylinder, i.e.,

$$p \geq 2p_Y \tag{3.28}$$

Substituting, p_Y from Equation (3.13) provides:

$$p \geq \sigma_Y\left(1 - \frac{a^2}{b^2}\right) \tag{3.29}$$

For complete yielding of the cylinder, the required autofrettage pressure p_o can be determined by integrating the equilibrium equation (Equation 3.16) between the limits a and b, which provides:

$$p_o = \sigma_Y \ln\left(\frac{b}{a}\right) \tag{3.30}$$

However, Equation (3.30) is valid for a perfectly plastic material as it is assumed that $(\sigma_\theta - \sigma_r)$ is equal to σ_Y throughout the plastic zone. Substituting Equation (3.30) in Equation (3.29), the smallest wall ratio (b/a) for which the secondary yielding can occur on unloading can be obtained from the following expression:

$$\ln\left(\frac{b}{a}\right) \geq \left(1 - \frac{a^2}{b^2}\right), \tag{3.31}$$

which upon solving gives $(b/a) \geq 2.22$ and the fully plastic pressure corresponding to $(b/a) = 2.22$ is approximately 0.8 σ_Y. Thus, for a cylinder with $(b/a) \geq 2.22$ and an autofrettage pressure in the range $2p_Y \leq p \leq p_o$, reyielding will take place in the cylinder during the unloading.

3.6 Closed-Form Model of Hydraulically Autofrettaged Cylinder Based on von Mises Yield Criterion

According to the von Mises yield criterion, at the onset of yielding,

$$\sqrt{\frac{1}{2}\left\{(\sigma_r - \sigma_\theta)^2 + (\sigma_\theta - \sigma_z)^2 + (\sigma_z - \sigma_r)^2\right\}} = \sigma_Y, \tag{3.32}$$

where σ_r, σ_θ, and σ_z are the principal stresses as the shear stress components are zero. In the following subsections, the models of a hydraulically autofrettaged cylinder are presented based on the von Mises criterion for the plane–stress and plane–strain conditions. It is to be mentioned that solutions for open-end and plane–stress are identical. Solutions for plane–strain and constrained ends are identical. The results for a closed-end case are very close to that of the plane–strain case, and will be identical for the material having a Poisson's ratio of 0.5. The solutions using the von Mises criterion are available in the literature (Avitzur, 1994). However, in the following subsections, the style of presentation and some final expressions are different from (Avitzur, 1994).

3.6.1 Cylinder with Open Ends or Disc with Plane–Stress Condition

Here, $\sigma_z = 0$, and the von Mises yield criterion (Equation 3.32) reduces to:

$$\sqrt{\sigma_r^2 - \sigma_r\sigma_\theta + \sigma_\theta^2} = \sigma_Y. \tag{3.33}$$

Using Equations (3.3) and (3.4) in Equation (3.33), the yield pressure can be expressed as:

$$p_Y = \frac{\dfrac{\sigma_Y}{\sqrt{3}}\left(1 - \dfrac{a^2}{b^2}\right)}{\sqrt{1 + \dfrac{a^4}{3b^4}}} \tag{3.34}$$

3.6.1.1 Stress Distribution after Loading

In the elastic zone ($c \le r \le b$), the radial and hoop stresses are given by Lamé's equations. Applying the boundary conditions $(\sigma_r)_{r=b} = 0$ and $\left(\sqrt{\sigma_r^2 - \sigma_r\sigma_\theta + \sigma_\theta^2}\right)_{r=c} = \sigma_Y$ in Equations (3.8) and (3.9), the radial and hoop stress distributions in the elastic zone are given by:

$$\sigma_r = -\frac{\left(\dfrac{b^2}{r^2} - 1\right)\sigma_Y}{\sqrt{3\left(\dfrac{b^4}{c^4}\right) + 1}}, \tag{3.35}$$

$$\sigma_\theta = \frac{\left(\dfrac{b^2}{r^2} + 1\right)\sigma_Y}{\sqrt{3\left(\dfrac{b^4}{c^4}\right) + 1}}, \tag{3.36}$$

In the plastic zone ($a \le r \le c$), quadratic Equation (3.33) provides:

$$\sigma_\theta = \begin{cases} \dfrac{\sigma_r + \sqrt{4\sigma_Y^2 - 3\sigma_r^2}}{2} \\ \text{or} \\ \dfrac{\sigma_r - \sqrt{4\sigma_Y^2 - 3\sigma_r^2}}{2} \end{cases} \tag{3.37}$$

The term $(4\sigma_Y^2 - 3\sigma_r^2)$ is always positive. For internal pressurization in a thick-walled cylinder, hoop stress is always tensile. This is possible only in the first case. Therefore,

$$\sigma_\theta = \frac{\sigma_r + \sqrt{4\sigma_Y^2 - 3\sigma_r^2}}{2}. \tag{3.38}$$

Hence,

$$\sigma_\theta - \sigma_r = \frac{-\sigma_r + \sqrt{4\sigma_Y^2 - 3\sigma_r^2}}{2}. \tag{3.39}$$

Substituting Equation (3.39) in the equilibrium Equation (3.16) provides:

$$\frac{\mathrm{d}\sigma_r}{-\sigma_r + \sqrt{4\sigma_Y^2 - 3\sigma_r^2}} = \frac{1}{2}\frac{\mathrm{d}r}{r}, \tag{3.40}$$

or:

$$\int \frac{\mathrm{d}\sigma_r}{-\sigma_r + \sqrt{4\sigma_Y^2 - 3\sigma_r^2}} = \frac{1}{2}\int \frac{\mathrm{d}r}{r} \tag{3.41}$$

To integrate the left-hand side, let:

$$\sqrt{3}\sigma_r = 2\sigma_Y \sin\theta. \tag{3.42}$$

Hence,

$$\sqrt{3}\,\mathrm{d}\sigma_r = 2\sigma_Y \cos\theta\,\mathrm{d}\theta, \tag{3.43}$$

or:

$$\mathrm{d}\sigma_r = \frac{2\sigma_Y \cos\theta\,\mathrm{d}\theta}{\sqrt{3}}. \tag{3.44}$$

Substituting Equation (3.44) in Equation (3.41) and integrating yields the following steps:

$$\frac{\ln|r|}{2}=\int \frac{2}{\sqrt{3}} \frac{\sigma_Y \cos\theta}{\left(-\frac{2}{\sqrt{3}}\sigma_Y \sin\theta + 2\sigma_Y \cos\theta\right)} \mathrm{d}\theta, \tag{3.45}$$

$$\frac{\ln|r|}{2}=\int \frac{\cos\theta}{\left(-\sin\theta+\sqrt{3}\cos\theta\right)} \mathrm{d}\theta, \tag{3.46}$$

$$\frac{\ln|r|}{2}=\int \frac{\frac{\cos\theta}{2}}{\left(-\frac{\sin\theta}{2}+\frac{\sqrt{3}\cos\theta}{2}\right)} \mathrm{d}\theta, \tag{3.47}$$

$$\frac{\ln|r|}{2}=\int \frac{\frac{\cos\theta}{2}}{\left(-\sin\theta\sin\frac{\pi}{6}+\cos\theta\cos\frac{\pi}{6}\right)} \mathrm{d}\theta, \tag{3.48}$$

$$\frac{\ln|r|}{2}=\frac{1}{2}\int \frac{\cos\theta}{\cos\left(\theta+\frac{\pi}{6}\right)} \mathrm{d}\theta, \tag{3.49}$$

$$\frac{\ln|r|}{2}=\frac{1}{2}\int \frac{\cos\left(\theta+\frac{\pi}{6}-\frac{\pi}{6}\right)}{\cos\left(\theta+\frac{\pi}{6}\right)} \mathrm{d}\theta, \tag{3.50}$$

$$\frac{\ln|r|}{2}=\frac{1}{2}\int \frac{\frac{\sqrt{3}}{2}\cos\left(\theta+\frac{\pi}{6}\right)+\frac{1}{2}\sin\left(\theta+\frac{\pi}{6}\right)}{\cos\left(\theta+\frac{\pi}{6}\right)} \mathrm{d}\theta, \tag{3.51}$$

$$\frac{\ln|r|}{2}=\frac{1}{2}\int \left\{\frac{\sqrt{3}}{2}+\frac{1}{2}\tan\left(\theta+\frac{\pi}{6}\right)\right\} \mathrm{d}\theta, \tag{3.52}$$

$$\frac{\ln|r|}{2}=\frac{1}{4}\left\{\sqrt{3}\,\theta-\ln\left|\cos\left(\theta+\frac{\pi}{6}\right)\right|\right\}+C, \tag{3.53}$$

$$\ln r^2=\sqrt{3}\sin^{-1}\left(\frac{\sqrt{3}}{2}\frac{\sigma_r}{\sigma_Y}\right)-\ln\left|\cos\left\{\sin^{-1}\left(\frac{\sqrt{3}}{2}\frac{\sigma_r}{\sigma_Y}\right)+\frac{\pi}{6}\right\}\right|+D, \tag{3.54}$$

where D is a constant of integration.

From the expression for the radial stress component in the elastic region, at $r=c$,

$$\sigma_r\Big|_{r=c} = \frac{\left(1-\dfrac{b^2}{c^2}\right)\sigma_Y}{\sqrt{3\left(\dfrac{b^4}{c^4}\right)+1}}. \tag{3.55}$$

Due to continuity of the radial stress component at the elastic–plastic interface, substitution of Equation (3.55) in Equation (3.54) at $r=c$ provides:

$$\ln c^2 = \sqrt{3}\sin^{-1}\left\{\frac{\sqrt{3}}{2}\frac{\left(1-\dfrac{b^2}{c^2}\right)}{\sqrt{3\left(\dfrac{b^4}{c^4}\right)+1}}\right\} - \ln\left|\cos\left\{\sin^{-1}\left(\frac{\sqrt{3}}{2}\frac{\left(1-\dfrac{b^2}{c^2}\right)}{\sqrt{3\left(\dfrac{b^4}{c^4}\right)+1}}\right)+\frac{\pi}{6}\right\}\right| + D. \tag{3.56}$$

Therefore,

$$D = \ln c^2 - \sqrt{3}\sin^{-1}\left\{\frac{\sqrt{3}}{2}\frac{\left(1-\dfrac{b^2}{c^2}\right)}{\sqrt{3\left(\dfrac{b^4}{c^4}\right)+1}}\right\} + \ln\left|\cos\left[\sin^{-1}\left\{\frac{\sqrt{3}}{2}\frac{\left(1-\dfrac{b^2}{c^2}\right)}{\sqrt{3\left(\dfrac{b^4}{c^4}\right)+1}}\right\}+\frac{\pi}{6}\right]\right|. \tag{3.57}$$

Substituting Equation (3.57) in Equation (3.54) provides the following expression for the radial stress component in the plastic zone:

$$\ln\frac{r^2}{c^2} = \sqrt{3}\sin^{-1}\left(\frac{\sqrt{3}}{2}\frac{\sigma_r}{\sigma_Y}\right) - \ln\left|\cos\left\{\sin^{-1}\left(\frac{\sqrt{3}}{2}\frac{\sigma_r}{\sigma_Y}\right)+\frac{\pi}{6}\right\}\right|$$
$$-\sqrt{3}\sin^{-1}\left\{\frac{\sqrt{3}}{2}\frac{\left(1-\dfrac{b^2}{c^2}\right)}{\sqrt{3\left(\dfrac{b^4}{c^4}\right)+1}}\right\} + \ln\left|\cos\left[\sin^{-1}\left\{\frac{\sqrt{3}}{2}\frac{\left(1-\dfrac{b^2}{c^2}\right)}{\sqrt{3\left(\dfrac{b^4}{c^4}\right)+1}}\right\}+\frac{\pi}{6}\right]\right|. \tag{3.58}$$

Although Equation (3.58) is an implicit function, the σ_r can be calculated at any r, using a suitable numerical method such as the bisection method. The distribution of the hoop stress component in the plastic zone is obtained from Equation (3.38) once the radial stress is known from Equation (3.58). The unknown radius of the elastic–plastic interface c can be obtained by using the boundary condition at the inner radius a. At $r=a$, $\sigma_r=-p$. From Equation (3.58):

$$\ln\frac{a^2}{c^2} = -\sqrt{3}\sin^{-1}\left(\frac{\sqrt{3}}{2}\frac{p}{\sigma_Y}\right) - \ln\left|\cos\left\{-\sin^{-1}\left(\frac{\sqrt{3}}{2}\frac{p}{\sigma_Y}\right)+\frac{\pi}{6}\right\}\right|$$
$$-\sqrt{3}\sin^{-1}\left\{\frac{\sqrt{3}}{2}\frac{\left(1-\frac{b^2}{c^2}\right)}{\sqrt{3\left(\frac{b^4}{c^4}\right)+1}}\right\} + \ln\left|\cos\left[\sin^{-1}\left\{\frac{\sqrt{3}}{2}\frac{\left(1-\frac{b^2}{c^2}\right)}{\sqrt{3\left(\frac{b^4}{c^4}\right)+1}}\right\}+\frac{\pi}{6}\right]\right| \tag{3.60}$$

Equation (3.60) may also be solved numerically to obtain c for a particular p.

3.6.1.2 Residual Stress Distribution after Unloading

The residual stresses in the elastic zone ($c \le r \le b$) are obtained by subtracting Equations (3.3) and (3.4) from Equations (3.35) and (3.36), respectively. The resulting residual stress components are given by:

$$\sigma_r = -\frac{\left(\frac{b^2}{r^2}-1\right)\sigma_Y}{\sqrt{3\left(\frac{b^4}{c^4}\right)+1}} + p\left(\frac{\frac{b^2}{r^2}-1}{\frac{b^2}{a^2}-1}\right), \tag{3.61}$$

$$\sigma_\theta = \frac{\left(\frac{b^2}{r^2}+1\right)\sigma_Y}{\sqrt{3\left(\frac{b^4}{c^4}\right)+1}} - p\left(\frac{\frac{b^2}{r^2}+1}{\frac{b^2}{a^2}-1}\right). \tag{3.62}$$

The residual stresses in the plastic zone ($a \le r \le c$) are obtained by subtracting Equations (3.3) and (3.4) from numerically evaluated radial and hoop stresses, respectively.

3.6.2 Cylinder with Constrained Ends/Plane–Strain Condition

For the plane–strain case, $\varepsilon_z=0$. Since the axial strain is always zero, there is no plastic deformation in the axial direction. Hence, as per the associated flow rule with the von Mises yield criterion, the deviatoric part of the axial component of the principal stress is zero:

$$\sigma_z - \frac{\sigma_r + \sigma_\theta + \sigma_z}{3} = 0, \tag{3.63}$$

or:

$$\sigma_z = \frac{(\sigma_r + \sigma_\theta)}{2}. \tag{3.64}$$

Using Equation (3.64), the von Mises yield criterion becomes:

$$\sqrt{\frac{1}{2}\left\{(\sigma_r - \sigma_\theta)^2 + \left(\sigma_\theta - \frac{(\sigma_r + \sigma_\theta)}{2}\right)^2 + \left(\frac{(\sigma_r + \sigma_\theta)}{2} - \sigma_r\right)^2\right\}} = \sigma_Y, \tag{3.65}$$

which simplifies to:

$$(\sigma_\theta - \sigma_r) = \frac{2}{\sqrt{3}}\sigma_Y. \tag{3.66}$$

Equation (3.66) is identical to the Tresca yield criterion except with σ_Y multiplied by a factor of $2/\sqrt{3}\sigma_Y$. Thus, for the plane–strain condition with the von Mises yield criterion, the stresses after loading and the residual stresses after unloading can be readily obtained from the expressions derived in Section 3.5 on the basis of the Tresca yield criterion by replacing σ_Y with $2\sigma_Y/\sqrt{3}$. Referring to Equation (3.13), the yield pressure for the initial yielding in a cylinder with constrained ends derived on the basis of the von Mises yield criterion is:

$$p_Y = \frac{\sigma_Y}{\sqrt{3}}\left(1 - \frac{a^2}{b^2}\right). \tag{3.67}$$

3.6.2.1 Stress Distribution after Loading

The radial and hoop stress distributions in the elastic zone ($c \le r \le b$) are obtained by modifying Equations (3.14) and (3.15), respectively, as:

$$\sigma_r = -\frac{\sigma_Y c^2}{\sqrt{3}b^2}\left(\frac{b^2}{r^2} - 1\right), \tag{3.68}$$

$$\sigma_\theta = \frac{\sigma_Y c^2}{\sqrt{3}b^2}\left(\frac{b^2}{r^2} + 1\right). \tag{3.69}$$

The radial and hoop stress distributions in the plastic zone ($a \le r \le c$) are obtained by modifying Equations (3.19) and (3.20), respectively, as:

$$\sigma_r = -\frac{\sigma_Y}{\sqrt{3}}\left(1-\frac{c^2}{b^2}+\ln\frac{c^2}{r^2}\right), \tag{3.70}$$

$$\sigma_\theta = \frac{\sigma_Y}{\sqrt{3}}\left(1+\frac{c^2}{b^2}-\ln\frac{c^2}{r^2}\right). \tag{3.71}$$

Using the boundary condition, $\sigma_r = -p$ at $r = a$ in Equation (3.70):

$$p = \frac{\sigma_Y}{\sqrt{3}}\left(1-\frac{c^2}{b^2}+\ln\frac{c^2}{a^2}\right). \tag{3.72}$$

The radius of the elastic–plastic interface, c, is obtained by numerically solving Equation (3.72).

3.6.2.2 Residual Stress Distribution after Unloading

The residual stress components in the elastic zone ($c \le r \le b$) are obtained by modifying Equations (3.22) and (3.23), respectively. They are expressed as:

$$\sigma_r = -\frac{\sigma_Y}{\sqrt{3}}\left(\frac{c^2}{a^2}-\frac{p}{p_Y}\right)\left(\frac{a^2}{r^2}-\frac{a^2}{b^2}\right), \tag{3.73}$$

$$\sigma_\theta = \frac{\sigma_Y}{\sqrt{3}}\left(\frac{c^2}{a^2}-\frac{p}{p_Y}\right)\left(\frac{a^2}{r^2}+\frac{a^2}{b^2}\right). \tag{3.74}$$

The residual stress distributions in the plastic zone ($a \le r \le c$) are obtained by modifying Equations (3.24) and (3.25), respectively. They are:

$$\sigma_r = -\frac{\sigma_Y}{\sqrt{3}}\left\{\frac{p}{p_Y}\left(1-\frac{a^2}{r^2}\right)-\ln\frac{r^2}{a^2}\right\}, \tag{3.75}$$

$$\sigma_\theta = -\frac{\sigma_Y}{\sqrt{3}}\left\{\frac{p}{p_Y}\left(1+\frac{a^2}{r^2}\right)-\left(2+\ln\frac{r^2}{a^2}\right)\right\}. \tag{3.76}$$

3.7 Elastic Analysis of an Internally Pressurized Thick-Walled Sphere

For a thick-walled sphere subjected to pressure, Lamé provided the following expressions for radial and hoop stress components, respectively, (Chakrabarty, 2012):

$$\sigma_r = A + \frac{B}{r^3}, \quad (3.77)$$

$$\sigma_\theta = A - \frac{B}{2r^3}, \quad (3.78)$$

where A and B are constants. Refer to Figure 3.2 as a representative section of the sphere. For an internally pressurized sphere, the constants A and B can be obtained from the boundary conditions of vanishing radial stress at the outer radius and $(\sigma_r)_{r=a} = -p$. Thus,

$$\sigma_r = -p\left(\frac{\frac{b^3}{r^3} - 1}{\frac{b^3}{a^3} - 1}\right), \quad (3.79)$$

$$\sigma_\theta = p\left(\frac{\frac{b^3}{2r^3} + 1}{\frac{b^3}{a^3} - 1}\right). \quad (3.80)$$

3.8 Closed-Form Model of a Hydraulically Autofrettaged Sphere

Closed-form models for a hydraulically autofrettaged sphere are well-established (Chakrabarty, 2012). For a sphere, both the Tresca and von Mises yield criteria reduce to the identical form given by:

$$\sigma_\theta - \sigma_r = \sigma_Y. \quad (3.81)$$

From Equations (3.79) and (3.80),

$$\sigma_\theta - \sigma_r = \frac{2p\frac{b^3}{r^3}}{\frac{b^3}{a^3} - 1}. \quad (3.82)$$

When the autofrettage pressure reaches the yield pressure, yielding initiates at the inner wall at $r = a$, and the yield criterion is satisfied. Therefore, substituting Equation (3.82) in Equation (3.81) provides the following expression for the yield pressure of the sphere:

$$p_Y = \frac{2\sigma_Y}{3}\left(1-\frac{a^3}{b^3}\right). \tag{3.83}$$

3.8.1 Stress Distribution after Loading

In the elastic zone ($c \le r \le b$) the radial and hoop stresses are given by Lame's equations (Equations (3.79) and (3.80)). Applying the boundary conditions $(\sigma_r)_{r=b}=0$ and $(\sigma_\theta-\sigma_r)_{r=c}=\sigma_Y$, the radial and hoop stress distributions in the elastic zone are:

$$\sigma_r = -\frac{2\sigma_Y c^3}{3b^3}\left(\frac{b^3}{r^3}-1\right), \tag{3.84}$$

$$\sigma_\theta = \frac{2\sigma_Y c^3}{3b^3}\left(\frac{b^3}{2r^3}+1\right). \tag{3.85}$$

In the plastic zone ($a \le r \le c$), the solution for the stresses is derived from the stress–equilibrium equation that remains satisfied through all stages of deformation and is expressed as:

$$\frac{2}{r}(\sigma_\theta-\sigma_r) = \frac{d\sigma_r}{dr}. \tag{3.86}$$

It is assumed that the material of the cylinder does not strain harden. Applying the von Mises (or Tresca) yield criterion (Equation 3.81) in the equilibrium equation (Equation 3.86):

$$\frac{2}{r}\sigma_Y = \frac{d\sigma_r}{dr}. \tag{3.87}$$

Solving Equation (3.87), the radial stress component is expressed as:

$$\sigma_r = 2\sigma_Y \ln r + C, \tag{3.88}$$

where C is a constant of integration; it can be evaluated using the continuity of radial stress at $r=c$. Evaluating the radial stress component for the elastic zone from Equation (3.79) and equating it with the plastic radial stress component from Equation (3.88) at $r=c$ leads to:

$$C = -\frac{2\sigma_Y}{3}\left(1-\frac{c^3}{b^3}\right)-2\sigma_Y \ln c. \tag{3.89}$$

Substituting C in Equation (3.88), the radial stress distribution in the plastic zone, $a \le r \le c$, is given by:

$$\sigma_r = -\frac{2\sigma_Y}{3}\left(1-\frac{c^3}{b^3}+\ln\frac{c^3}{r^3}\right), \tag{3.90}$$

From yield criterion, the hoop stress distribution is given by:

$$\sigma_\theta = \frac{2\sigma_Y}{3}\left(\frac{1}{2}+\frac{c^3}{b^3}-\ln\frac{c^3}{r^3}\right). \tag{3.91}$$

The unknown radius of the elastic–plastic interface c can be obtained by using the boundary condition at the inner radius, a. At $r=a$, $\sigma_r=-p$. Thus, from Equation (3.90):

$$p = \frac{2\sigma_Y}{3}\left(1-\frac{c^3}{b^3}+\ln\frac{c^3}{a^3}\right), \tag{3.92}$$

Equation (3.92) can be numerically solved for the unknown radius c using a numerical method such as the bisection method.

3.8.2 Residual Stress Distribution after Unloading

Residual stress distribution in elastic zone ($c \le r \le b$) is obtained by subtracting Equations (3.79) and (3.80) from Equations (3.84) and (3.85), respectively. The residual stress components are:

$$\sigma_r = -\frac{2\sigma_Y}{3}\left(\frac{c^3}{a^3}-\frac{p}{p_Y}\right)\left(\frac{a^3}{r^3}-\frac{a^3}{b^3}\right), \tag{3.93}$$

$$\sigma_\theta = \frac{2\sigma_Y}{3}\left(\frac{c^3}{a^3}-\frac{p}{p_Y}\right)\left(\frac{a^3}{2r^3}+\frac{a^3}{b^3}\right). \tag{3.94}$$

Residual stress distribution in plastic zone ($a \le r \le c$) is obtained by subtracting Equations (3.79) and (3.80) from Equations (3.90) and (3.91), respectively. The residual stress components are:

$$\sigma_r = -\frac{2\sigma_Y}{3}\left\{\frac{p}{p_Y}\left(1-\frac{a^3}{r^3}\right)-\ln\frac{r^3}{a^3}\right\}, \tag{3.95}$$

$$\sigma_\theta = -\frac{2\sigma_Y}{3}\left\{\frac{p}{p_Y}\left(1+\frac{a^2}{r^2}\right)-\left(\frac{3}{2}+\ln\frac{r^3}{a^3}\right)\right\}. \tag{3.96}$$

The condition for compressive reyielding is determined by applying the yield criterion. In the elastic zone, Equations (3.93) and (3.94) provide:

$$\sigma_r - \sigma_\theta = \sigma_Y\frac{2a^2}{3r^2}\left(\frac{p}{p_Y}-\frac{c^3}{a^3}\right). \tag{3.97}$$

Similarly, for the plastic zone, Equations (3.95) and (3.96) provide:

$$\sigma_r - \sigma_\theta = \sigma_Y \left(\frac{pa^3}{p_Y r^3} - 1 \right). \tag{3.98}$$

As can be inferred from Equation (3.98), the magnitude of $(\sigma_\theta - \sigma_r)$ is the maximum at the inner wall. Hence, reyielding will take place at $r = a$ if the applied pressure is at least twice the yield pressure of the cylinder, i.e.,

$$p \geq 2p_Y. \tag{3.99}$$

Substitution of p_Y from Equation (3.83) provides:

$$p \geq \frac{4\sigma_Y}{3} \left(1 - \frac{a^3}{b^3} \right). \tag{3.100}$$

For the complete yielding of the cylinder, the required autofrettage pressure p_o can be determined by integrating both sides of the equilibrium Equation (3.87), after putting the yield criterion, between the limits a and b. Thus, the pressure for complete yielding of the sphere is given by:

$$p_o = \sigma_Y \ln \left(\frac{b^2}{a^2} \right). \tag{3.101}$$

Substituting Equation (3.101) in Equation (3.100), the smallest wall ratio (b/a), for which the secondary yielding can occur on unloading, can be obtained from the following expression:

$$\ln \left(\frac{b^2}{a^2} \right) \geq \frac{4}{3} \left(1 - \frac{a^3}{b^3} \right). \tag{3.102}$$

Equation (3.102) gives $(b/a) \geq 1.7$, and the fully plastic pressure for $(b/a) = 1.7$ is approximately $1.06\sigma_Y$. For a sphere with $(b/a) \geq 1.7$ and an autofrettage pressure in the range $2p_Y \leq p \leq p_o$, reyielding will take place in the sphere during the unloading.

3.9 Results and Discussion

This section presents the typical results of the hydraulic autofrettage of cylinders and spheres. Radial dimensions with inner radius 15 mm and outer radius 60 mm are considered for both the cylinder and the sphere. The material of the autofrettage specimen is high strength, low alloy steel 4333 M4, having Young's modulus 207 GPa, and yield strength 1070 MPa.

The yield pressure of the cylinder according to the Tresca yield criterion using Equation (3.13) is 501.56 MPa. Similarly, for the von Mises yield criterion the yield pressure calculated using Equation (3.34) for open-end condition is 578.78 MPa; and that calculated using Equation (3.67) for the constrained end condition is 579.15 MPa. The Tresca yield criterion predicts a lower yield pressure compared to that predicted by the von Mises yield criterion for all types of end conditions. Hence, the Tresca yield criterion may be adopted for a more conservative design. For all the end conditions, the cylinder is subjected to a hydraulic autofrettage process using a pressure of 700 MPa.

For the open-end condition, the distribution of stresses after loading and residual stresses after unloading are shown in Figure 3.4a,b, respectively. Figure 3.4a also shows equivalent stresses calculated based on both the von Mises and Tresca yield criteria. The equivalent stress based on the von Mises yield criterion is given by Equation (2.35), whereas the equivalent stress based on the Tresca yield criterion is equal to ($\sigma_\theta - \sigma_r$). The equivalent stress equal to the yield strength (1070 MPa, in this case) indicates the plastic deformation in the cylinder. The elastic–plastic interface radius based on the von Mises yield criterion is 16.67 mm, and that based on the Tresca yield criterion is 18.34 mm. The plastic zone is larger in the case of the Tresca yield criterion due to the earlier commencement of the initial yielding. The maximum compressive residual hoop stress at the inner wall is 261.63 MPa, based on the von Mises yield criterion, whereas that based on the Tresca yield criterion is 423.33 MPa. This is a significant difference. Compared to residual hoop stress, residual radial stress is very small.

For the constrained end condition, the distribution of stresses after loading and residual stresses after unloading are shown in Figure 3.5a,b, respectively. Figure 3.5a also shows corresponding equivalent stresses calculated on the

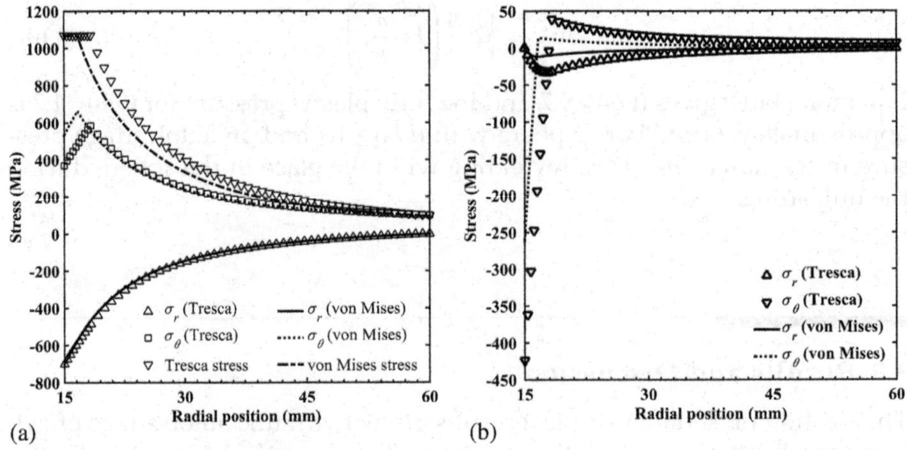

FIGURE 3.4
Distribution of (a) stresses after loading and (b) residual stresses after unloading of a cylinder with open ends.

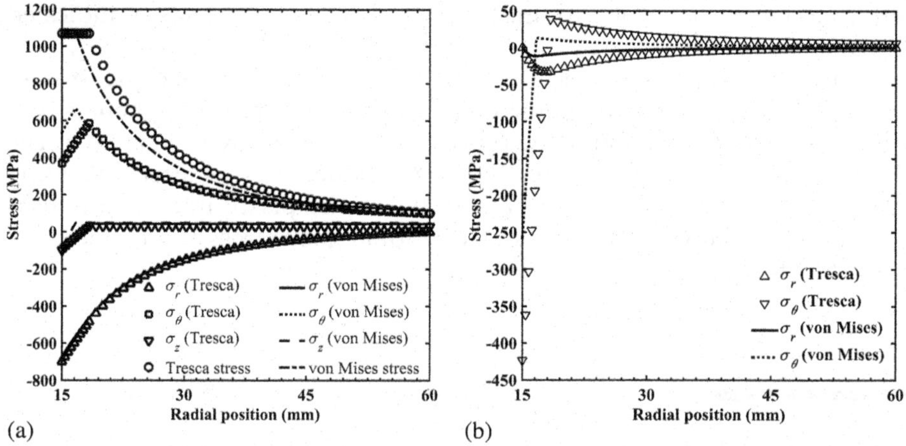

FIGURE 3.5
Distribution of (a) stresses after loading and (b) residual stresses after unloading of a cylinder with constrained ends.

basis of each respective yield criterion. The elastic–plastic interface radius based on the von Mises yield criterion is 16.66 mm and that based on the Tresca yield criterion is 18.34 mm. The maximum compressive hoop residual at the inner wall is 257.70 MPa whereas that based on the Tresca yield criterion is 423.33 MPa. Here too, the difference is significant. Hence, although the Tresca yield criterion gives a more conservative design for a non-autofrettaged cylinder with any end condition, it tends to overestimate the maximum compressive hoop residual stress after being subjected to autofrettage. This in turn underestimates the safe limit pressure in the autofrettaged cylinder when it is reloaded with the working pressure. In Figure 3.5b, axial residual

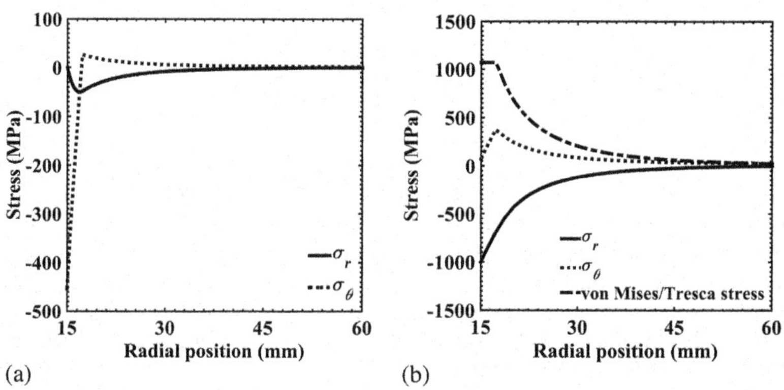

FIGURE 3.6
Distribution of (a) stresses after loading and (b) residual stresses after unloading of a sphere.

stress has not be ploted; it is the mean of radial and hoop stresses, as indicated in Equation (3.64).

Figure 3.6 depicts the typical results for a sphere. The yield pressure of the sphere calculated using Equation (3.13) is 702.19 MPa. The sphere is subjected to a hydraulic autofrettage process using a pressure of 1000 MPa. The distribution of stresses after loading and residual stresses after unloading are shown in Figure 3.6a,b, respectively. The elastic–plastic interface radius is 17.29 mm. The maximum compressive residual hoop stress at the inner wall is 453.81 MPa.

3.10 Conclusion

This chapter presents a detailed description of the hydraulic autofrettage process in a thick-walled cylinder and sphere. A typical hydraulic autofrettage process is first explained. This is followed by a discussion of the various aspects involved in the mathematical modeling of the process. Three main aspects are of importance: the yield criterion, the incorporation of the material behavior, and the boundary condition in the autofrettage specimen. Based on these aspects, closed-form models of hydraulic autofrettage in cylinders and sphere are presented. As an initial basis, the elastic analyses are presented first. In cylinders, two types of models—one based on the Tresca yield criterion and another based on the von Mises yield criterion are derived. Typical results are then presented for the hydraulic autofrettage in cylinders and spheres. It is seen that the choice of yield criterion has a significant effect on the results of the autofrettage of cylinders. However, both the Tresca and von Mises yield criteria provide the same results for the autofrettage of a sphere.

References

Alexandrov, S., Jeong, W. and Chung, K., (2015), Descriptions of reversed yielding in internally pressurized tubes, *Journal of Pressure Vessel Technology*, **138**, 011204-1–011204-10. https://doi.org/10.1115/1.4031029

Allen, D.N. de G. and Sopwith, D.G., (1951), The stresses and strains in a partly plastic thick tube under internal pressure and end-load, *Proceedings of the Royal Society of London A: Mathematical, Physical and Engineering Sciences*, **205**, 69–83. https://doi.org/10.1098/rspa.1951.0018

Avitzur, B., (1994), Autofrettage – stress distribution under load and retained stresses after depressurization, *International Journal of Pressure Vessels and Piping*, **57**, 271–287.

Bland, D.R., (1956), Elasto-plastic thick-walled tubes of work-hardening material subject to internal and external pressures and to temperature gradients, *Journal of the Mechanics and Physics of Solids*, **4**, 209–229. https://doi.org/10.1016/0022-5096(56)90030-8

Chakrabarty, J., (2012), *Theory of Plasticity*, 3rd edition. Butterworth-Heinemann.

Chen, P., (1980), Generalized plane-strain problems in an elastic-plastic thick-walled cylinder (No. ARLCB-TR-80028), Army Armament Research and Development Center, Watervliet, NY.

Chen, P.C.T., (1973), A comparison of flow and deformation theories in a radially stressed annular plate, *Journal of Applied Mechanics*, **40**, 283–287. https://doi.org/10.1115/1.3422941

Chen, P.C.T., (1986), The bauschinger and hardening effect on residual stresses in an autofrettaged thick-walled cylinder, *Journal of Pressure Vessel Technology*, **108**, 108–112. https://doi.org/10.1115/1.3264743

Chu, S.-C., (1972), A more rational approach to the problem of an elasto-plastic thick-walled cylinder, *Journal of the Franklin Institute*, **294**, 57–65. https://doi.org/10.1016/0016-0032(72)90113-5

Davidson, T., Barton, C., Reiner, A. and Kendall, D., (2013), Overstrain of high strength, open end cylinders of intermediate diameter ratio, Presented at the Proceedings of the 1st International Congress on Experimental Mechanics, pp. 335–352.

Durban, D. and Kubi, M., (1992), A general solution for the pressurized elasto-plastic tube, *Journal of Applied Mechanics*, **59**, 20–26. https://doi.org/10.1115/1.2899431

Elder, A.S., Tomkins, R. and Mann, T.L., (1975), Generalized plane–strain in an elastic, perfectly plastic cylinder, with reference to the hydraulic autofrettage process, Presented at the Tran. 21st Conference of Army Mathematicians, pp. 623–659.

Franklin, G.J. and Morrison, J.L.M., (1960), Autofrettage of cylinders: prediction of pressure/external expansion curves and calculation of residual stresses, *Proceedings of the Institution of Mechanical Engineers*, **174**, 947–974. https://doi.org/10.1243/PIME_PROC_1960_174_069_02

Gao, X., (1992), An exact elasto-plastic solution for an open-ended thick-walled cylinder of a strain-hardening material, *International Journal of Pressure Vessels and Piping*, **52**, 129–144. https://doi.org/10.1016/0308-0161(92)90064-M

Gao, X.-L., (1993), An exact elasto-plastic solution for a closed-end thick-walled cylinder of elastic linear-hardening material with large strains, *International Journal of Pressure Vessels and Piping*, **56**, 331–350. https://doi.org/10.1016/0308-0161(93)90004-D

Gao, X.-L., (2003), Elasto-plastic analysis of an internally pressurized thick-walled cylinder using a strain gradient plasticity theory, *International Journal of Solids and Structures*, **40**, 6445–6455. https://doi.org/10.1016/S0020-7683(03)00424-4

Gao, X.-L., (2007), Strain gradient plasticity solution for an internally pressurized thick-walled cylinder of an elastic linear-hardening material, *Zeitschrift für Angewandte Mathematik und Physik*, **58**, 161–173. https://doi.org/10.1007/s00033-006-0083-4

Gao, X.-L., Wen, J.-F., Xuan, F.-Z. and Tu, S.-T., (2015), Autofrettage and shakedown analyses of an internally pressurized thick-walled cylinder based on strain gradient plasticity solutions, *Journal of Applied Mechanics*, **82**, 041010–041010-12. https://doi.org/10.1115/1.4029798

Gibson, M.C., Hameed, A., Parker, A.P. and Hetherington, J.G., (2005), A comparison of methods for predicting residual stresses in strain-hardening, autofrettaged thick cylinders, including the bauschinger effect, *Journal of Pressure Vessel Technology*, **128**, 217–222. https://doi.org/10.1115/1.2172964

Hill, R., Lee, E.H. and Tupper, S.J., (1947), The theory of combined plastic and elastic deformation with particular reference to a thick tube under internal pressure, *Proceedings of the Royal Society of London A: Mathematical, Physical and Engineering Sciences*, **191**, 278–303. https://doi.org/10.1098/rspa.1947.0116

Hosseinian, E., Farrahi, G.H. and Movahhedy, M.R., (2009), An analytical framework for the solution of autofrettaged tubes under constant axial strain condition, *Journal of Pressure Vessel Technology*, **131**, 061201–061201-8. https://doi.org/10.1115/1.3148082

Huang, X. and Moan, T., (2009), Residual stress in an autofrettaged tube taking bauschinger effect as a function of the prior plastic strain, *Journal of Pressure Vessel Technology*, **131**, 021207–021207-7. https://doi.org/10.1115/1.3062937

Huang, X.P., (2005), A general autofrettage model of a thick-walled cylinder based on tensile-compressive stress-strain curve of a material, *The Journal of Strain Analysis for Engineering Design*, **40**, 599–607. https://doi.org/10.1243/030932405X16070

Huang, X.P. and Cui, W., (2004), Autofrettage analysis of thick-walled cylinder based on tensile-compressive curve of material, *Key Engineering Materials*, **274–276**, 1035–1040. https://doi.org/10.4028/www.scientific.net/KEM.274-276.1035

Huang, X.P. and Cui, W.C., (2005), Effect of bauschinger effect and yield criterion on residual stress distribution of autofrettaged tube, *Journal of Pressure Vessel Technology*, **128**, 212–216. https://doi.org/10.1115/1.2172621

Jacob, L., (1907), La Résistance et L'équilibre Élastique des Tubes Frettés, *Memorial de L'artillerie Navale*, **1**, 43–155.

Koiter, W., (1953), On partially plastic thick-walled tubes, *Biezeno Anniversary Volume* 232–251.

Lazzarin, P. and Livieri, P., (1997), Different solutions for stress and strain fields in autofrettaged thick-walled cylinders, *International Journal of Pressure Vessels and Piping*, **71**, 231–238. https://doi.org/10.1016/S0308-0161(97)00002-1

Li, G., Zeng, X., Li, J. and Huang, L., (1988), Elasto-plastic analysis of an open-ended cylinder from the twelve polygonal yield condition, *International Journal of Pressure Vessels and Piping*, **33**, 143–152. https://doi.org/10.1016/0308-0161(88)90067-1

Livieri, P. and Lazzarin, P., (2001), Autofrettaged cylindrical vessels and bauschinger effect: an analytical frame for evaluating residual stress distributions, *Journal of Pressure Vessel Technology*, **124**, 38–46. https://doi.org/10.1115/1.1425809

Loghman, A. and Wahab, M.A., (1994), Loading and unloading of thick-walled cylindrical pressure vessels of strain-hardening material, *Journal of Pressure Vessel Technology*, **116**, 105–109. https://doi.org/10.1115/1.2929562

Lu, W.Y. and Hsu, Y.C., (1977), Elastic-plastic analysis of a flat ring subject to internal pressure, *Acta Mechanica*, **27**, 155–172. https://doi.org/10.1007/BF01180083

MacGregor, C.W., L.F. Coffin and Fisher, J.C., (1948), Partially plastic thick-walled tubes, *Journal of the Franklin Institute*, **245**, 135–158. https://doi.org/10.1016/0016-0032(48)90679-6

Marcal, P.V., (1965a), A note on the elastic-plastic thick cylinder with internal pressure in the open and closed-end condition, *International Journal of Mechanical Sciences*, **7**, 841–845. https://doi.org/10.1016/0020-7403(65)90036-6

Marcal, P.V., (1965b), A stiffness method for elastic-plastic problems, *International Journal of Mechanical Sciences*, **7**, 229–238. https://doi.org/10.1016/0020-7403(65)90040-8

Mendelson, A., (1968), *Plasticity, Theory and Application*, Macmillan, New York.

Nádai, A., (1950), *Theory of Flow and Fracture of Solids* Vol-1, McGraw-Hill, New York.

Parker, A.P., (2000), Autofrettage of open-end tubes – pressures, stresses, strains, and code comparisons, *Journal of Pressure Vessel Technology*, **123**, 271–281. https://doi.org/10.1115/1.1359209

Perl, M. and Perry, J., (2005), An experimental-numerical determination of the three-dimensional autofrettage residual stress field incorporating bauschinger effects, *Journal of Pressure Vessel Technology*, **128**, 173–178. https://doi.org/10.1115/1.2172959

Perry, J. and Aboudi, J., (2003), Elasto-plastic stresses in thick walled cylinders, *Journal of Pressure Vessel Technology*, **125**, 248–252. https://doi.org/10.1115/1.1593078

Rees, D.W.A., (1990), Autofrettage theory and fatigue life of open-ended cylinders. *The Journal of Strain Analysis for Engineering Design*, **25**, 109–121. https://doi.org/10.1243/03093247V252109

Sedmak, A., Kirin, S., Golubovic, T., Mitrovic, S. and Stanojevic, P., (2016), Risk based approach to integrity assessment of a large spherical pressure vessel, *Procedia Structural Integrity*, 21st European Conference on Fracture, ECF21, 20–24 June 2016, Catania, Italy, **2**, 3654–3659. https://doi.org/10.1016/j.prostr.2016.06.454

Thomas, D.G.B., (1953), The autofrettage of thick tubes with free ends, *Journal of the Mechanics and Physics of Solids*, **1**, 124–133. https://doi.org/10.1016/0022-5096(53)90016-7

Wang, G.S., (1988), An elastic-plastic solution for a normally loaded center hole in a finite circular body, *International Journal of Pressure Vessels and Piping*, **33**, 269–284. https://doi.org/10.1016/0308-0161(88)90075-0

Zeng, X., Li, Jinxiang, Li, Jinwei, Li, G., Yang, Z. and Li, Jinyu, (1993), The application of the twelve-angled polygonal yield criterion to pressure vessel problems, *International Journal of Pressure Vessels and Piping*, **55**, 385–393. https://doi.org/10.1016/0308-0161(93)90059-3

4

Swage and Explosive Autofrettage

4.1 Introduction

The swage autofrettage process was proposed by Davidson et al. (1962) as an economical alternative to the conventional hydraulic autofrettage process. Here, the autofrettage process is carried out by forcing a specially shaped, oversized mandrel into the tube. The oversized mandrel plastically deforms the vicinity of the inner wall during the forced pass along the cylinder. The schematic of a typical swage autofrettage process and mandrel geometry are shown in Figure 4.1a,b, respectively. The mandrel has three main geometrical specifications—a front taper of θ_F, a parallel section, and a rear taper. The front taper provides the angle of approach for the gradual application of the autofrettage load, the parallel section ensures the uniform plastic deformation of the inner wall of the cylinder, and the rear taper enables the elastic unloading of the cylinder. The method is only suitable for tube specimens, and it is generally used in the autofrettage of gun barrels. Spherical vessels cannot be autofrettaged by this process.

Explosive autofrettage is carried out by detonating an explosive charge inside the thick-walled cylinder filled with a pressure propagating medium like air or water. This process has not been actively pursued by researchers (as is the case with other autofrettage processes) perhaps because of the involvement of explosives and the fact that the process may require legal permission for its practical implementation. For this reason, the process is only reviewed based on the literature that is available. This chapter is organized as follows: Section 4.2 gives a brief description of a typical swage autofrettage process. This is followed by the various issues in modeling the swage autofrettage process in Section 4.3. Section 4.4 presents the closed-form model of swage autofrettage. Section 4.5 presents the typical results of the swage autofrettage process. Section 4.6 is devoted to the description and review of the explosive autofrettage process. Finally, the chapter is concluded in Section 4.7.

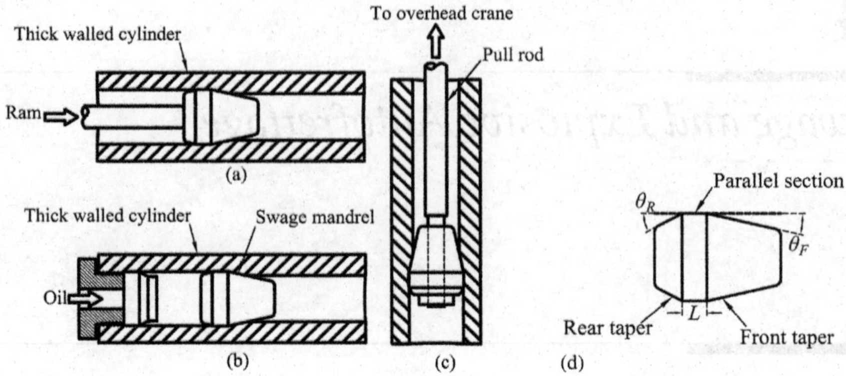

FIGURE 4.1
Schematic of (a) mechanical push, (b) hydraulic push, and (c) mechanical pull swage autofrettage processes; (d) a typical swage mandrel.

4.2 A Typical Swage Autofrettage Process

In the original proposal, Davidson et al. (1962) presented three types of experimental setups: mechanical push swage autofrettage, mechanical pull swage autofrettage, and hydraulic push autofrettage. The first arrangement used a hydraulic ram to push the mandrel, the second used an overhead crane to vertically pull up the mandrel, and the third used hydraulic pressure applied at the end of the mandrel. The three types of arrangements, along with a typical swage mandrel geometry, are shown in Figure 4.1. The mandrel with a front and rear taper of 1.5° and 3°, respectively, with a 19 mm long parallel section was used in all three of the arrangements. The mandrel had a surface finish of about 0.05–0.4 μm Ra. Three types of lubricants—a molybdenum disulfide suspension in oil, the copper plating of the inner cylinder wall, and a combination of the two were tried, out of which the third gave the best performance. Unless mentioned, all subsequent discussions on swage autofrettage in this chapter will mean the mechanical push swage autofrettage process.

4.3 Issues in Modeling the Swage Autofrettage Process

In general, researchers have identified and focused on two main aspects in the practical implementation of swage autofrettage, namely, the mandrel driving force, and the maximum compressive hoop-residual stress at the inner wall of the autofrettaged cylinder. The mandrel driving force is the forming load that is required to push the mandrel along the tube. As explained by Perry and Perl (2008), the mandrel driving force in a swage autofrettage process is calculated by multiplying the measured axial strain by Young's modulus and the cross-sectional area of the parallel section of the mandrel. The mandrel driving force and the maximum compressive residual stress at the inner wall depends on four important factors. These are the mandrel geometry, the mandrel material behavior, friction between the mandrel and the cylinder, and the number of passes of the swage mandrel. Researchers have carried out parametric studies based on various numerical models. These are discussed in detail in the following subsections.

4.3.1 Effect of Mandrel Geometry

The mandrel, which has three main geometrical specifications—angle of front taper θ_F, angle of rear taper θ_R, and length of parallel section as shown in Figure 4.1(d) has a significant effect on the mandrel driving force. The angle of front taper is sometimes taken as being smaller (Davidson et al., 1962), greater (Iremonger and Kalsi, 2003), or equal (Till and Rammerstorfer, 1983) compared to the angle of rear taper. Iremonger and Kalsi (2003) carried out a rigorous analysis based on a finite difference method (FDM). The autofrettage specimen was a barrel with an inner diameter of 107.95 mm. The study showed that an increase in the angle of front taper from 3° to 6° resulted in a 24% increase in the predicted driving force. This was due to the increase in the interference between the tube and the mandrel per unit axial displacement of the mandrel, leading to larger contact stresses between the barrel and the mandrel. On the other hand, increasing the rear angle from 1.5 to 4.5 resulted in an 18.4% decrease in the mandrel driving force. This was due to an increased rate of unloading that led to a more rapid reduction of contact stresses between the mandrel and the barrel. Similarly, reducing the length of the parallel section from 9.9 mm to 6.6 mm decreased the area of contact between the barrel and the mandrel and resulted in a 6.4% decrease in the mandrel driving force. Bihamta et al. (2007) and Gibson et al. (2012) observed the same by incorporating kinematic hardening and the effect of friction in a FEM. Bihamta et al. (2007) suggested and showed that the mandrel driving force could be reduced by using a mandrel with straight or helical grooves. The mandrel was attached to the ram in such a way that it could rotate freely about its longitudinal axis as it was pushed along the tube.

For the variation of the maximum compressive hoop-residual stress, Gibson et al. (2014) showed that it increased with the increase in both of the angles, although the variation due to the rear angle was more sensitive. An increase in the length of the parallel section beyond an optimum value equal to 0.12 times of the inner radius led to a drop in the hoop-residual stresses.

4.3.2 Effect of the Mandrel Material Behavior

The rigidity of the mandrel plays a significant role in the induced compressive residual stresses. The material behavior of the mandrel may be assumed as either rigid or elastic. Studies based on the rigid assumption include (Barbachano et al., 2011.; Chen, 1988a), whereas those based on the elastic assumption include (Barbachano et al., n.d.; Chang et al., 2013; Chen, 1988b, 1988a, Gibson et al., 2012, 2014; Hua and Penumarthy, 2014; O'Hara, 1992; Perry and Perl, 2008). Davidson et al. (1962) originally used a mandrel made of steel alloy 4340, but researchers have since also used tungsten carbide as the material of the mandrel; a material whose elastic modulus is three times greater than that of a typical steel alloy. Based on an finite element method (FEM(without considering material hardening and the effect of friction, Chen (1988a) showed that the assumption of elasticity in the mandrel reduced the overstrain from 70% to 60%, which subsequently led to the reduction of the maximum compressive hoop-residual stress in the swage autofrettaged cylinder. A more rigorous FEM-based analysis was carried out by Barbáchano et al. (2011) by incorporating Chaboche's combined isotropic–kinematic hardening model (Chaboche, 2008). It was shown that the effect of considering the elasticity of the mandrel on the maximum radial and hoop-residual stresses in the cylinder was insignificant. On the other hand, the maximum compressive axial residual stress in the cylinder decreased by about 19% due to the elasticity of the mandrel.

4.3.3 Effect of Friction between Mandrel and Cylinder

The swage autofrettage process involves significant friction at the interface of the parallel section of the mandrel and the inner wall of the cylinder. Friction is incorporated by using the coefficient of friction μ governed by Coulomb's law. Gibson et al., (2012) carried out an FEM-based parametric study by varying the coefficient of friction from 0 to 0.09 in steps of 0.015, and showed that the driving force increased in almost the same proportion as the friction coefficient. An empirical equation for the quick prediction of the mandrel driving force as a function of the coefficient of friction was suggested, which is as follows:

$$F = F_{\mu=0} + k\mu, \tag{4.1}$$

where F is the mandrel driving force, $F_{\mu=0}$ is the mandrel driving force at zero friction, and k is the constant of proportionality. A similar result was shown in the experimentally validated FEM by Bihamta et al. (2007).

Gibson et al. (2014) and Chang et al. (2013) carried out FEM-based parametric studies to investigate the effect of friction on the residual stresses. Both research groups reported that the effect of friction on the maximum compressive residual stress was insignificant. Since the frictional force did not exert any normal force on the inner wall of the cylinder, increasing the friction did not result in much change in the overstrain of the autofrettaged cylinder. Thus, the maximum compressive hoop-residual stress was not significantly affected.

4.3.4 Effect of Number of Passes

For a very large interference, it may be suitable to perform the swage autofrettage process in more than one pass to reduce the overall requirement of the mandrel driving force (Gibson et al., 2012). Usually, the first pass uses a lesser interference and the final pass uses the desired maximum interference. Iremonger and Kalsi (2003) observed that a double-pass swage autofrettage process could reduce the mandrel driving force by 22% compared to that in a single-pass process, when the interference used in the first pass was 70% of the desired maximum interference. Gibson et al. (2012) carried out a similar study by considering the kinematic hardening in the cylinder material. The role of the Bauschinger effect for the first pass was insignificant. However, after the second pass, the mandrel driving force increased slightly due to the strain hardening of the inner wall of the tube resulting from the first pass. There was also a reduction in the compressive yield stress at the inner wall due to plastic deformation in the first pass. Hu (2018) also carried out a rigorous FEM modeling process and observed a 30–35% reduction in the average and peak mandrel driving forces in a two-pass swage autofrettage process compared to those obtained from a single pass. The initial first pass used an interference that was 0.707 times the maximum interference. The reduction in the mandrel driving force in a two-pass process is due to a lesser interference used per pass. In a single-pass operation, the maximum interference causes a greater degree of deformation, leading to a requirement for a greater degree of driving force. Moreover, a two-pass process causes a more uniform

Barbáchano et al. (2011) analyzed two types of double-pass swage autofrettage process: one with two consecutive passes in the same direction and another with two consecutive passes in opposite directions. A greater magnitude (as well as a more uniform distribution of the compressive hoop-residual stress) was obtained with the second arrangement with two opposite mandrel passes. It also reduced the edge effect in the cylinder. Edge effect is a phenomenon where the maximum hoop-residual stresses at the inner surface of the cylinder are lesser at the edges compared to those at the inner regions. It was also suggested that the two-pass swage autofrettage process with two consecutive opposite passes could be useful to autofrettage pipes or tubes with one closed end, e.g., in the case of pipes with bends and joints.

4.4 Closed-Form Model of Swage Autofrettage

Closed-form modeling of swage autofrettage is difficult because of the transient nature of the process. The deformation is localized and dynamic along the length of the tube. A few researchers have developed closed-form models (Chen, 1988b; Rees, 2011) based on simplified assumptions. The closed-form model by Chen (1988b) for the plane–strain analysis of a swage autofrettaged tube is presented in this chapter. The tube yields according to the Tresca yield criterion and subsequently flows without any strain hardening. The mandrel is made of tungsten carbide and is assumed to remain elastic throughout the process. The expressions for the distribution of stress after loading and residual stress after unloading are identical to those derived in Sections 3.4 and 3.5 for a hydraulically autofrettaged cylinder, except that the pressure term appearing in these equations corresponds to the contact pressure at the tube–mandrel interface in the case of a swage autofrettage process.

The elastic analysis of the tube is presented in Section 4.4.1. The contact pressure, as a function of the desired interference between the tube and the mandrel, is derived by using the strain–displacement equation and constitutive stress–strain relations in both the tube and the mandrel. Subsequent elasto-plastic analysis is presented in Section 4.4.2. Although the solution for the stresses has already been presented in Chapter 3, they are repeated in these sections for the sake of completeness and fair emphasis on the swage autofrettage model.

4.4.1 Elastic Analysis

The Lamé's stresses generated in the tube due to the elastic deformation as a result of contact pressure exerted by the mandrel as already expressed in Equations (3.3) and (3.4) are as follows:

$$\sigma_r^t = -p\left(\frac{\frac{b^2}{r_t^2}-1}{\frac{b^2}{a^2}-1}\right), \tag{4.2}$$

$$\sigma_\theta^t = p\left(\frac{\frac{b^2}{r_t^2}+1}{\frac{b^2}{a^2}-1}\right), \tag{4.3}$$

where σ_r^t is the radial stress component, σ_θ^t is the hoop-stress component, r_t is a particular radial position in the tube and p is the contact pressure at the tube–mandrel interface.

The strain–displacement equation relating the circumferential strain ε_θ and the radial displacement u as a function of the radial position r in axi-symmetric coordinates is expressed as (Chakrabarty, 2006):

$$\varepsilon_\theta = \frac{u}{r}. \tag{4.4}$$

From a generalized Hooke's law:

$$\varepsilon_\theta = -\nu\varepsilon_z + \frac{1+\nu}{E}\left\{(1-\nu)\sigma_r - \nu\sigma_r\right\}, \tag{4.5}$$

where E is Young's modulus and ν is Poisson's ratio of the material.

Substituting Equation (4.5) in (4.4) provides:

$$\frac{u}{r} = -\nu\varepsilon_z + \frac{1+\nu}{E}\left\{(1-\nu)\sigma_r - \nu\sigma_r\right\}, \tag{4.6}$$

For the tube, using $\varepsilon_z=0$ for the plane–strain condition and substituting Equations (4.2) and (4.3) in Equation (4.6), the radial displacement u^t for elastic deformation is (Chen, 1988b):

$$\frac{u_t}{r_t} = \frac{p}{E_t\left(1-\frac{a^2}{b^2}\right)}\left\{(1+\nu_t)(1-2\nu_t)\frac{a^2}{b^2} + (1+\nu_t)\frac{a^2}{r^2}\right\}, \tag{4.7}$$

where E_t is Young's modulus, ν_t is Poisson's ratio of the tube, and r_t is a radial position in the tube.

The stress distribution in the mandrel tube due to the contact pressure p is:

$$\sigma_r^m = \sigma_\theta^m = -p. \tag{4.8}$$

where σ_r^m is the radial stress component and σ_θ^m is the hoop-stress component in the mandrel.

Substituting Equation (4.8) in Equation (4.6) and using the plane–strain condition ($\varepsilon_z=0$) provides:

$$\frac{u_m}{r_m} = -\frac{p}{E_m}(1+\nu_m)(1-2\nu_m). \tag{4.9}$$

where E_m is Young's modulus, ν_m is Poisson's ratio of the mandrel, and u_m is the radial displacement in the mandrel as a function of the radial position r_m in the parallel section mandrel.

At the tube–mandrel interface i.e., $r=r_1=a$, compatibility requires that the radial displacement in the tube be equal to the sum of the displacement in the mandrel and the interference between the two. Therefore,

$$I = u^t\Big|_{r=a} - u^m\Big|_{r=a}, \tag{4.10}$$

where I is the interference between the tube and the mandrel. Substituting Equations (4.7) and (4.9) in (4.10) provides:

$$I=\frac{pa}{E\left(1-\frac{a^2}{b^2}\right)}\left[\left\{(1+\nu)(1-2\nu)\frac{a^2}{b^2}+(1+\nu)\right\}+\frac{E}{E_1}\left(1-\frac{a^2}{b^2}\right)(1+\nu_1)(1-2\nu_1)\right]. \tag{4.11}$$

Rearranging Equation (4.11), the contact pressure as a function of the interference can be expressed as:

$$p=\frac{\frac{EI}{a}\left(1-\frac{a^2}{b^2}\right)}{\left[\left\{(1+\nu)(1-2\nu)\frac{a^2}{b^2}+(1+\nu)\right\}+\frac{E}{E_1}\left(1-\frac{a^2}{b^2}\right)(1+\nu_1)(1-2\nu_1)\right]}. \tag{4.12}$$

4.4.2 Elasto-Plastic Analysis

Based on the Tresca yield criterion, the minimum contact pressure p^* for the initiation of yielding in the tube (analogous to Equation (3.13) for hydraulic pressure) is:

$$p^*=\frac{\sigma_Y}{2}\left(1-\frac{a^2}{b^2}\right). \tag{4.13}$$

where σ_Y is the yield strength of the tube. Substituting Equation (4.13) in Equation (4.11), the minimum interference I^* for the initiation of yielding in the tube is:

$$I^*=\frac{\sigma_Y a}{2E}\left[\left\{(1+\nu)(1-2\nu)\frac{a^2}{b^2}+(1+\nu)\right\}+\frac{E}{E_1}\left(1-\frac{a^2}{b^2}\right)(1+\nu_1)(1-2\nu_1)\right]. \tag{4.14}$$

In the elastic zone $c\le r\le b$, the distribution of the radial and hoop stresses in the tube are given by:

$$\sigma_r^t=-\frac{\sigma_Y c^2}{2b^2}\left(\frac{b^2}{r^2}-1\right), \tag{4.15}$$

$$\sigma_\theta^t=\frac{\sigma_Y c^2}{2b^2}\left(\frac{b^2}{r^2}+1\right). \tag{4.16}$$

In the plastic zone $a\le r\le c$, the distribution of the radial and hoop stresses as already derived in Section 3.5.1:

$$\sigma_r^t=-\frac{\sigma_Y}{2}\left(1-\frac{c^2}{b^2}+\ln\frac{c^2}{r^2}\right), \tag{4.17}$$

$$\sigma_\theta^t = \frac{\sigma_Y}{2}\left(1 + \frac{c^2}{b^2} - \ln\frac{c^2}{r^2}\right). \tag{4.18}$$

The radial displacement in the plastic zone is given by (Chen, 1988b):

$$\frac{u^t}{r} = \frac{1}{E}\left\{\sigma_Y(1-\nu^2)\frac{c^2}{r^2} + (1+\nu)(1-2\nu)\sigma_r^t\right\}. \tag{4.19}$$

Since the mandrel remains elastic throughout the process, the interference between the tube and the mandrel is obtained by applying Equation (4.10) with Equations (4.7) and (4.19) at $r = r_1 = a$, which provides:

$$I = \frac{a}{E}\left\{\sigma_Y\left(1-\nu^2\right)\frac{c^2}{a^2}\right\} - p\left\{(1+\nu)(1-2\nu) - \frac{E}{E_1}(1+\nu_1)(1-2\nu_1)\right\}. \tag{4.20}$$

The contact pressure (analogous to Equation (3.13) in case of hydraulic pressure) required to achieve the desired level of autofrettage with an inner plastic zone up to the radius c is given by:

$$p = \frac{\sigma_Y}{2}\left(1 - \frac{c^2}{b^2} + \ln\frac{c^2}{a^2}\right). \tag{4.21}$$

Substituting Equation (4.21) in Equation (4.20), the interference as a function of the elasto-plastic interface radius c becomes:

$$\frac{EI}{a} = \sigma_Y\left[\left\{\left(1-\nu^2\right)\frac{c^2}{a^2}\right\} - \frac{1}{2}\left(1 - \frac{c^2}{b^2} + \ln\frac{c^2}{a^2}\right)\left\{(1+\nu)(1-2\nu) - \frac{E}{E_1}(1+\nu_1)(1-2\nu_1)\right\}\right]. \tag{4.22}$$

Equation (4.22) can be solved numerically for the radius c for a prescribed interference I. Substituting c in Equation (4.21) provides the corresponding contact pressure, which can again be subsequently used to obtain the stress distributions in the elastic and plastic zones using Equations (4.2 and 4.3).

After the unloading of the tube, residual stresses are induced. The expressions for the distribution of the residual stresses can be readily modified from those already derived in Chapter 3, Section 3.5.2, and are rewritten as follows:

$$\left.\begin{aligned}\sigma_r^t &= -\frac{\sigma_Y}{2}\left(\frac{c^2}{a^2} - \frac{p}{p_Y}\right)\left(\frac{a^2}{r^2} - \frac{a^2}{b^2}\right),\\ \sigma_\theta^t &= \frac{\sigma_Y}{2}\left(\frac{c^2}{a^2} - \frac{p}{p_Y}\right)\left(\frac{a^2}{r^2} + \frac{a^2}{b^2}\right).\end{aligned}\right\}\text{(Elastic zone)} \tag{4.23}$$

$$\left.\begin{aligned} \sigma_r^t &= -\frac{\sigma_Y}{2}\left\{\frac{p}{p_Y}\left(1-\frac{a^2}{r^2}\right)-\ln\frac{r^2}{a^2}\right\}, \\ \sigma_\theta^t &= -\frac{\sigma_Y}{2}\left\{\frac{p}{p_Y}\left(1+\frac{a^2}{r^2}\right)-\left(2+\ln\frac{r^2}{a^2}\right)\right\}. \end{aligned}\right\}\text{(Plastic zone)} \tag{4.24}$$

Similar to the condition already expressed by Equation (3.28) for the case of hydraulic autofrettage, reverse yielding will occur in the swage autofrettaged tube when the contact pressure is at least twice the minimum contact pressure (Equation 4.13) for initial yielding in the tube. For a known contact pressure whose magnitude is twice the contact pressure for initial yielding, i.e., $p = 2p^*$, the elasto-plastic interface radius c can be determined by solving the following modified form of Equation (4.21):

$$2p^* = \frac{\sigma_Y}{2}\left(1-\frac{c^2}{b^2}+\ln\frac{c^2}{a^2}\right). \tag{4.25}$$

Subsequently, this value of c can be used to compute the minimum interference for causing reverse yielding by solving the Equation (4.21). The mandrel remains in the elastic state and the stress distribution is still by Equation (4.8).

4.5 Typical Results in Swage Autofrettage

This section presents the typical results of swage autofrettage in a tube. A tube with an inner radius of 10 mm and an outer radius of 15 mm is considered. The material of the tube is strainless Steel SS304; it has a Young's modulus of 193 GPa, a Poisson's ratio of 0.3, and a yield strength of 205 MPa (Shufen and Dixit, 2017). The mandrel is made of Tungsten Carbide with a Young's modulus of 610 GPa and a Poisson's ratio of 0.258 (Chen, 1988b).

The yield contact pressure of the cylinder calculated using Equation (4.13) is 86.1 MPa. The corresponding minimum interference for the initiation of yielding according to Equation (4.10) is 0.0082 mm. The swage autofrettage is carried out using an iterference of 0.0205 mm. The distribution of stresses after loading and residual stresses after unloading are shown in Figure 4.2a,b, respectively. Figure 4.2a also shows the distribution of equivalent stress denoted by σ_{eq} in the legend. The elasto-plastic interface radius obtained by solving Equation (4.18) is 15.46 mm. The maximum compressive hoop-residual stress is 158.37 MPa. The minimum interference for the commencement of reverse yielding in the tube obtained by sequentially solving

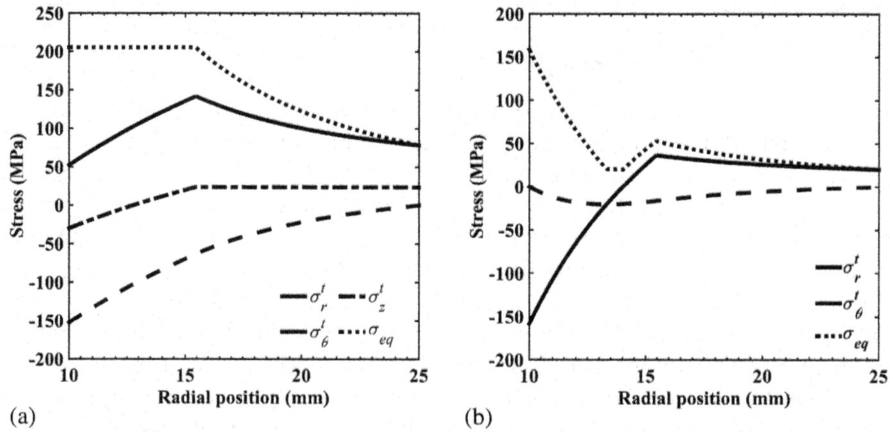

FIGURE 4.2
Distribution of (a) stresses after loading and (b) residual stresses after unloading in a swage autofrettaged tube for an interference of 0.0205 mm.

Equations (4.21) and (4.18) is 0.03 mm. The elasto-plastic interface radius is 18.46 mm. The distribution of stresses after loading and residual stresses after unloading in a swage autofrettaged tube on the verge of commencing reverse yielding are shown in Figure 4.3a,b, respectively. Figure 4.3a also shows that the distribution of the equivalent stress its value is equal to is the yield strength (205 MPa in this case) at the inner wall. Quite obviously, the maximum compressive hoop-residual stress at the inner wall is also 205 MPa.

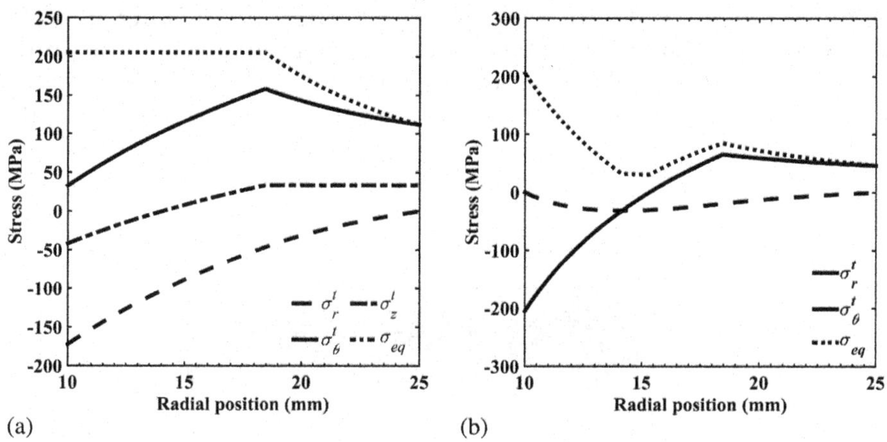

FIGURE 4.3
Distribution of (a) stresses after loading and (b) residual stresses after unloading in a swage autofrettaged tube on the verge of reverse yielding.

4.6 Explosive Autofrettage

A schematic of a typical setup of explosive autofrettage is shown in Figure 4.4. The earliest analytical model for the feasibility analysis of explosive autofrettage was reported by Mote et al. (1971) based on a strain rate formulation of plasticity using the von Mises yield criterion and its associated flow rule; a material behavior without any strain or strain rate hardening was assumed. A rigorous experimental analysis was also carried out along with the theoretical formulations. Explosive autofrettage was carried out on a cannon barrel with a bore diameter of 152 mm made up of steel alloy 4340. A preliminary experiment was first carried out on 40% of the breech section and the entire barrel. The breech section is the portion of the canon where the canon shell is loaded. The scaled models gave acceptable results and the experiment was extended to a full model gun barrel with a bore diameter of 105 mm. A simple apparatus of the experimental setup consisted of an array of explosive charges lined up along the axis of the barrel that was held in place by end plugs. The residual stresses were measured by using the Sachs boring technique. The Sachs boring technique is a method for the quantitative determination of residual strains induced in deformed specimens like that in autofrettage. It involves the layer-wise removal of the inner surface of the autofrettaged cylinder through a boring operation that causes surface strain fluctuation, which is measured by strain gauges attached to the outer surface. The corresponding stresses are calculated from the basic equations of elasticity (Lambert, 1954).

Zhan et al. (2005) studied the effect of explosive autofrettage on the fatigue life of different autofrettaged components using two different explosion media—sticking oil and dynamic liquid medium. The former was a medium of higher viscosity compared to the latter. The sticking oil explosion was similar to the explosion hardening technique, where the initial shock wave from the explosion was used to deform the cylinder, whereas the dynamic liquid used the gradual expansion of the medium after detonation. The residual stress was measured using an X-ray diffraction

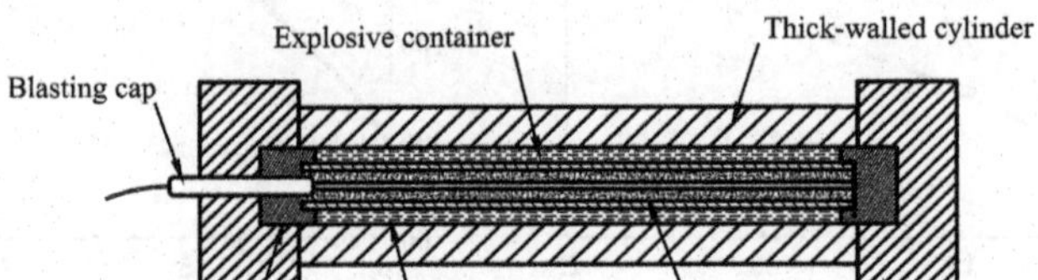

FIGURE 4.4
Schematic of a typical setup of explosive autofrettage process. Modified figure from Kamal (2016).

technique, and it was shown that the components autofrettaged by the sticking oil explosion gave better fatigue life improvement. Zhan et al. (1999) performed explosive autofrettage on a thick-walled cylinder made of material 43CrNi2MoVA steel and used the Sachs boring technique to measure residual stresses.

4.7 Conclusion

This chapter presented the modeling of the swage autofrettage process and a review of the explosive autofrettage process. The reader was briefly introduced to the swage autofrettage process, followed by a detailed explanation of its various practical aspects—mandrel geometry, mandrel material behavior, friction between the mandrel and the autofrettage specimen, and the number of passes of the mandrel. A closed-form solution of swage autofrettage is difficult; hence, a simple analytical model of swage autofrettage was presented. The model is analogous to the model for a hydraulically autofrettaged cylinder developed on the basis of the Tresca yield criterion, except that the pressure term signifies the contact pressure exerted by the mandrel on the tube. The contact pressure is computed as a function of the mechanical interference between the tube and the mandrel. Typical results based on this model were presented. Due to the scarcity of available research, the explosive autofrettage process was not explained in detail. From the little literature available, the construction and working of the process was described with a brief review.

References

Barbáchano, H., Alegre, J. and Cuesta, I., (2011), FEM simulation of the swage tube forming (STF) in cylinders subjected to internal pressure, *Anales de Mecanica de la Fractura*, 28, 481–486.

Bihamta, R., Movahhedy, M.R. and Mashreghi, A.R., (2007), A numerical study of swage autofrettage of thick-walled tubes, *Materials & Design*, **28**, 804–815. https://doi.org/10.1016/j.matdes.2005.11.012

Chaboche, J.L., (2008), A review of some plasticity and viscoplasticity constitutive theories, *International Journal of Plasticity, Special Issue in Honor of Jean-Louis Chaboche*, **24**, 1642–1693. https://doi.org/10.1016/j.ijplas.2008.03.009

Chakrabarty, J., (2006), *Theory of Plasticity*, Butterworth-Heinemann, Amsterdam, the Netherlands.

Chang, L., Pan, Y. and Ma, X., (2013), Residual stress calculation of swage autofrettage gun barrel, *International Journal of Computer Science Issues*, **10**, 52–59.

Chen, P.C., (1988a), Finite element analysis of the swage autofrettage process (No. ARCCB-TR-88037), Army Armament Research Development and Engineering Center, Benet Weapons Lab, Watervliet, NY.

Chen, P.C., (1988b), A simple analysis of the swage autofrettage process (No. ARCCB-TR-88037), Army Armament Research Development and Engineering Center, Benet Weapons Lab, Watervliet, NY.

Davidson, T.E., Barton, C.S., Reiner, A.N. and Kendall, D.P., (1962), New approach to the autofrettage of high-strength cylinders, *Experimental Mechanics*, **2**, 33–40. https://doi.org/10.1007/BF02325691

Gibson, M., Hameed, A. and Hetherington, J.G., (2012), Investigation of driving force variation during swage autofrettage, using finite element analysis, *Journal of Pressure Vessel Technology*, **134**, 051203–051203-7. https://doi.org/10.1115/1.4006922

Gibson, M.C., Hameed, A. and Hetherington, J.G., (2014), Investigation of residual stress development during swage autofrettage, using finite element analysis, *Journal of Pressure Vessel Technology*, **136**, 021206–021206-7. https://doi.org/10.1115/1.4025968

Hu, Z., (2018), Design of two-pass swage autofrettage processes of thick-walled cylinders by computer modeling, *Proceedings of the Institution of Mechanical Engineers, Part C: Journal of Mechanical Engineering Science*, **233**, 1312–1333. https://doi.org/10.1177/0954406218770221

Hua, Z. and Penumarthy, C., (2014), Computer modeling and optimization of swage autofrettage process of a thick-walled cylinder incorporating bauschinger effect, *American Transactions on Engineering & Applied Sciences*, **3**, 31–63.

Iremonger, M.J. and Kalsi, G.S., (2003), A numerical study of swage autofrettage, *Journal of Pressure Vessel Technology*, **125**, 347–351. https://doi.org/10.1115/1.1593073

Kamal, S.M., (2016), A theoretical and experimental study of thermal autofrettage process, Ph.D. Thesis, IIT Guwahati, India.

Lambert, J., (1954), A method of deriving residual stress equations, *Proceedings of the Society for Experimental Stress Analysis*, **12**, 91–96.

Mote, J.D., Ching, L.K., Knight, R.E., Fay, R.J. and Kaplan, M.A., (1971), Explosive autofrettage of cannon barrels (No. AMMRC CR 70-25), Army Materials and Research Center, Watertown, MA.

O'Hara, G.P., (1992), Analysis of the Swage Autofrettage Process (No. ARCCB-TR-92016), Army Armament Research Development and Engineering Center, Benet Weapons Lab, Watervliet, NY.

Perry, J. and Perl, M., (2008), A 3-D model for evaluating the residual stress field due to swage autofrettage, *Journal of Pressure Vessel Technology*, **130**, 041211–041211-6. https://doi.org/10.1115/1.2967741

Rees, D.W.A., (2011), A theory for swaging of discs and lugs, *Meccanica*, **46**, 1213–1237. https://doi.org/10.1007/s11012-010-9377-x

Shufen, R. and Dixit, U.S., (2017), A finite element method study of combined hydraulic and thermal autofrettage process, *Journal of Pressure Vessel Technology*, **139**, 041204–041204-9. https://doi.org/10.1115/1.4036143

Till, E.T. and Rammerstorfer, F.G., (1983), Nonlinear finite element analysis of an autofrettage process, *Computers & Structures*, **17**, 857–864. https://doi.org/10.1016/0045-7949(83)90099-8

Zhan, R., Tao, C., Han, L., Huang, Y. and Han, D., (2005), The residual stress and its influence on the fatigue strength induced by explosive autofrettage, *Explosion and Shock Waves*, **25**, 239–243.
Zhan, R., Tao, C. and Zhao, G., (1999), Elasto-plastical dynamic analysis of explosive autofrettage, *Journal-Southwest Petroleum Institute*, **21**, 82–85.

Appendix A

Derivation of Sachs Equations (Lambert, 1954)

The Sachs equations can be derived by using the equations from the theory of elasticity. A cylindrical specimen with inner radius a and outer radius b is considered. When the cylinder is progressively bored out from the inside, it is assumed that it possesses a radial position r. Hoop stress (σ_θ) at the outer surface of a thick-walled cylinder due a uniform internal pressure p_i is given by:

$$\sigma_\theta = \frac{2r^2 p_i}{b^2 - r^2}, \tag{A.1}$$

The generalized Hooke's law provides the axial (σ_a) and hoop (σ_θ) stress components as:

$$\sigma_z = \frac{E}{1-\nu^2}\left(\varepsilon_z + \nu\varepsilon_\theta\right) = \frac{E}{1-\nu^2}\Lambda, \tag{A.2}$$

$$\sigma_\theta = \frac{E}{1-\nu^2}\left(\varepsilon_\theta + \nu\varepsilon_a\right) = \frac{E}{1-\nu^2}\theta. \tag{A.3}$$

The above relations are to be applied on the surface of the cylinder where the radial stress $\sigma_r = 0$. The expression $\varepsilon_\theta + \nu\varepsilon_a$ is defined as θ and expression $\varepsilon_z + \nu\varepsilon_\theta$ is defined as Λ.

Now consider that the cylinder has been bored out to an arbitrary radius r. A uniform pressure p_i has been removed from the inside boundary due to the removal of material containing polar symmetric residual stresses. At radius r, the pressure p_i removed is equal to the original radial stress at that point before boring started. Thus,

$$\sigma_r = p_i \tag{A.4}$$

Substituting p_i from Equation (A.1) in Equation (A.4) and then using Equation (A.3), the expression for residual radial stress is obtained as:

$$\sigma_r = \frac{b^2 - r^2}{2r^2}\sigma_\theta' = \frac{b^2 - r^2}{2r^2}\frac{E}{1-\nu^2}\theta. \tag{A.5}$$

Multiplying the numerator and denominator by π, the above equation can be expressed in the form as given by Sachs as:

$$\sigma_r = \frac{E}{1-\nu^2}\left\{\frac{(F_b - F_r)}{2F_r}\theta\right\}, \tag{A.6}$$

where:

$$F_b = \pi b^2, \quad F_r = \pi r^2. \tag{A.7}$$

Equation (A.6) provides the radial residual stress as a function of r and the strains measured at various stages of boring.

To derive the expression for residual hoop stress, consider a thin cylindrical layer of material of radius r, which has been bored out. This layer was also subjected to a uniform pressure p_i. The relationship between the hoop stress and the external pressure p_i on the outer periphery of the removed layer can be obtained from a free body diagram of the half of the cylindrical layer. If the layer is split by a plane through its axis and any diameter, the equilibrium of the forces acting provides:

$$rp_i = \int_0^r \sigma_\theta(\rho)\,\mathrm{d}\rho \tag{A.8}$$

Substituting p_i from Equation (A.1) and σ_θ from Equation (A.3):

$$\int_0^r \sigma_\theta(\rho)\,\mathrm{d}\rho = \frac{E}{1-\nu^2}\frac{b^2 - r^2}{2r}\theta. \tag{A.9}$$

Differentiating both sides of Equation (A.9) with respect to r, the following equation is obtained:

$$\sigma_\theta(r) = \frac{E}{1-\nu^2}\left\{\left(-\frac{b^2}{2r^2} - \frac{1}{2}\right)\theta + \frac{b^2 - r^2}{2r}\frac{\mathrm{d}\theta}{\mathrm{d}r}\right\},$$

$$\sigma_\theta(r) = \frac{E}{1-\nu^2}\left(\frac{-b^2 - r^2}{2r^2}\theta + \frac{b^2 - r^2}{2r}\frac{\mathrm{d}\theta}{\mathrm{d}r}\right). \tag{A.10}$$

Using chain rule, the term $\frac{\mathrm{d}\theta}{\mathrm{d}r}$ can be expressed as:

$$\frac{\mathrm{d}\theta}{\mathrm{d}r} = \frac{\mathrm{d}\theta}{\mathrm{d}F_r}\frac{\mathrm{d}F_r}{\mathrm{d}r} = 2\pi r\frac{\mathrm{d}\theta}{\mathrm{d}F_r}. \tag{A.11}$$

Thus, Equation (A.10) becomes:

$$\sigma_\theta = \frac{E}{1-\nu^2}\left\{\frac{-b^2-r^2}{2r^2}\theta + \left(\pi b^2 - \pi r^2\right)\frac{d\theta}{dF_r}\right\}. \tag{A.12}$$

Multiplying the numerator and denominator of the term, $\frac{-b^2-r^2}{2r^2}$, by π, Equation (A.12) can be expressed as:

$$\sigma_\theta = \frac{E}{1-\nu^2}\left\{(F_b - F_r)\frac{d\theta}{dF_r} - \frac{(F_b + F_r)}{2F_r}\theta\right\}. \tag{A.13}$$

Equation (A.13) provides the residual hoop stress as given by Sachs.

For the axial residual stress, it is assumed that the specimen is long enough so that the plane–strain conditions are valid. When the specimen is bored out to a radius r, a longitudinal boundary force F is removed from the inside of the remaining hollow cylinder.

Considering the removed cylindrical layer of radius r, F is given by:

$$F = \int_0^r \sigma_z(\rho)\,dF_\rho = 2\pi\int_0^r \sigma_z(\rho)\rho\,d\rho. \tag{A.14}$$

The axial stress due to F is given by:

$$\sigma_z = \frac{F}{\pi(b^2 - r^2)}. \tag{A.15}$$

From Equations (A.2) and (A.15),

$$\frac{F}{\pi(b^2-r^2)} = \frac{E}{1-\nu^2}\Lambda. \tag{A.16}$$

Substituting Equation (A.14) in Equation (A.16):

$$2\int_0^r \sigma_z(\rho)\rho\,d\rho = \frac{E}{1-\nu^2}(b^2-r^2)\Lambda. \tag{A.17}$$

Differentiating Equation (A.17) on both sides, the following equation is obtained:

$$2r\sigma_z(r) = \frac{E}{1-\nu^2}\left\{-2r\Lambda + (b^2-r^2)\frac{d\Lambda}{dr}\right\},$$

$$\sigma_z = \frac{E}{1-\nu^2}\left(-\Lambda + \frac{(b^2-r^2)}{2r}\frac{d\Lambda}{dr}\right). \tag{A.18}$$

The term $\frac{d\Lambda}{dr}$ can be expressed, in a similar way as in Equation (A.11), as:

$$\frac{d\Lambda}{dr} = 2\pi r \frac{d\Lambda}{dF_r}. \tag{A.19}$$

Using Equation (A.19) in Equation (A.18), the expression of Sachs residual axial stress is obtained as:

$$\sigma_z = \frac{E}{1-\nu^2}\left\{(F_b - F_r)\frac{d\Lambda}{dr} - \Lambda\right\}. \tag{A.20}$$

5

Thermal Autofrettage

5.1 Introduction

Thermal autofrettage is an autofrettage process in which the autofrettage load is a thermally induced load. In a typical autofrettage process, a thick-walled vessel is first subjected to a radial temperature gradient along the thickness direction of the wall during loading. During unloading, the cylinder is cooled down to room temperature. This causes elastic unloading, leading to the inducement of beneficial compressive residual stresses in the vicinity of the inner wall. The pioneering concept of the thermal autofrettage process dates back to the 19th century. General Thomas Rodman of the United States Army used a differential cooling technique in the 1850s to fabricate a gun barrel casting by simultaneously heating the outer surface and cooling the bore with cold water (Bastable, 1992). However, the method could not capture attention back then as gun barrels were usually fabricated by cold working rather than casting. Much later, in 2002, Barbero and Wen (2002) first suggested the term thermal autofrettage process for the autofrettage of composite metal-lined cryogenic pressure vessels based on the same principle. However, no detailed analyses of the process were reported either theoretically or experimentally. The rigorous analysis of utilizing thermal load as a potential process of achieving autofrettage in a thick-walled cylinder or disk was first carried out by Kamal and Dixit (2015a, 2015b). The authors developed two closed-form models—one based on plane–stress assumption applicable in the analysis of thin disks, and another based on generalized plane–strain assumption applicable in the analysis of long cylinders. The studies demonstrated the potential of employing thermal loads for the autofrettage of thick-walled cylinders in inducing beneficial compressive residual stresses. The theoretical results were supported by experimental studies in a later article by Kamal et al. (2016). The article demonstrated the practical evidence of residual stresses in the thermal autofrettage of a thick-walled cylindrical specimen. These studies suggested thermal autofrettage as a simpler and economical alternative to the costlier hydraulic autofrettage process. In this chapter, a detailed study of the thermal autofrettage of thick-walled disks and cylinders is presented. The chapter is organized as follows. Section 5.2 describes the concept of thermal autofrettage. Section 5.3 presents the plane–stress model of thermal autofrettage. Section 5.4 presents the generalized plane–strain model of thermal autofrettage. Section 5.5 presents validation of the closed-form models with FEM models. Section 5.6 presents the closed-form model of thermal autofrettage of a thick-walled sphere. Section 5.7 presents an experimental study on thermal autofrettage. Section 5.8 presents a comparison of the thermal and hydraulic autofrettage processes. Section 5.8 concludes the chapter.

5.2 Concept and Practical Aspect of Thermal Autofrettage

A typical schematic of the thermal autofrettage setup has been depicted in Figure 1.6. The outer surface of the cylinder is subjected to a temperature T_b and the inner surface to a temperature T_a ($<T_b$). On reaching a certain threshold temperature difference called the yield temperature difference, the inner wall starts yielding. As the temperature is further increased, a plastic zone spreads outwards from the inner wall up to an intermediate radius beyond which the cylinder remains elastic. When the temperature difference again reaches a second threshold, the outer wall starts yielding. In that situation the cylinder contains three deformed zones—an inner plastic zone, an intermediate elastic zone, and an outer plastic zone. When the cylinder is cooled to room temperature, compressive residual stresses are induced in the vicinity of the inner wall of the cylinder, whereas tensile residual stresses are induced in the vicinity of the outer wall.

The practical implementation of thermal autofrettage was demonstrated by Kamal et al. (2016). The researchers heated the outer wall of the thick-walled cylinder using an electrical heating coil while cold water was circulated through the bore. The simultaneous heating of the outer wall and the convection heat dissipation from the inner wall created the required temperature gradient between the inner and outer wall.

5.3 Plane–Stress Modeling of Thermal Autofrettage

This section presents a closed-form model of thermal autofrettage considering plane–stress assumption. The model is based on the work carried out by Kamal and Dixit (2015b) and Kamal et al. (2017). The plane–stress model is applicable for the analysis of thick-walled hollow circular disks or very short cylinders such as fastener holes and the outer wall of a butterfly valve used in high-pressure pipelines. Like in the case of hydraulic autofrettage, the solution for the elastic analysis of a thick-walled hollow cylinder subjected to a radial temperature is presented first. Based on this solution, the mathematical model for the elastic–plastic analysis is derived. It is assumed that the material follows the Tresca yield criterion and its associated flow rule. The effect of strain hardening is incorporated using Ludwik's hardening law.

5.3.1 Elastic Analysis of a Thick-Walled Disk Subjected to a Radial Temperature Gradient

The solution for the thermo-elastic analysis of a thick-walled disk subjected to a radial temperature gradient is already available in several books of thermal stresses, e.g., (Noda et al., 2003). Figure 5.1 shows the cross-section of a hollow cylinder having an inner radius a and outer radius b. Let the temperatures of the inner and outer wall be T_a and T_b, respectively. The steady state temperature distribution is given by (Noda et al., 2003)

$$T = T_b + (T_a - T_b)\frac{\ln\left(\frac{b}{r}\right)}{\ln\left(\frac{b}{a}\right)}. \tag{5.1}$$

For a sufficiently low temperature difference (T_b–T_a), the entire wall thickness of the disk remains in the elastic state under the influence of thermal stresses. At this stage, the stresses and strains in the disk are governed by the generalized Hooke's law. With zero temperature as a reference, the generalized Hooke's law provides (Noda et al., 2003):

$$\varepsilon_r = \frac{1}{E}\{\sigma_r - \nu(\sigma_\theta + \sigma_z)\} + \alpha T, \tag{5.2}$$

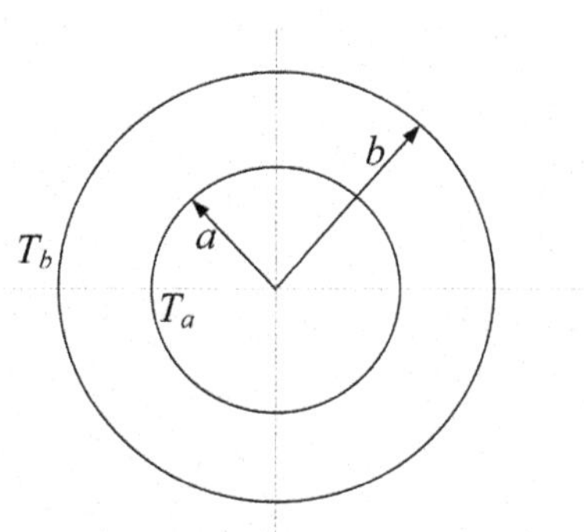

FIGURE 5.1
Cross-section of a thick-walled hollow cylinder subjected to a radial temperature gradient (Kamal, 2016).

$$\varepsilon_\theta = \frac{1}{E}\{\sigma_\theta - \nu(\sigma_r + \sigma_z)\} + \alpha T, \tag{5.3}$$

$$\varepsilon_z = \frac{1}{E}\{\sigma_z - \nu(\sigma_r + \sigma_\theta)\} + \alpha T, \tag{5.4}$$

where σ and ε refers to stress and strain, respectively, ν is the Poisson's ratio, E is the Young's modulus of elasticity, and α is the coefficient of thermal expansion. The subscripts r, θ, and z to σ and ε, represent their radial, hoop, and axial components, respectively. In the plane–stress case, the axial component of stress σ_z is zero.

For an axisymmetric case, the strain–displacement relation in the cylindrical coordinates is (Chakrabarty, 2006):

$$\varepsilon_r = \frac{du}{dr}, \quad \varepsilon_\theta = \frac{u}{r}, \tag{5.5}$$

where u is the radial displacement. Substituting u from the second relation to the first relation in Equation (5.5) provides the following expression for strain compatibility:

$$\varepsilon_r = \frac{d(r\varepsilon_\theta)}{dr}, \quad \text{or,} \quad \varepsilon_r - \varepsilon_\theta = r\frac{d\varepsilon_\theta}{dr}. \tag{5.6}$$

Substituting the expressions of ε_r and ε_θ in Equation (5.6) and using it in the equilibrium equation (Equation 3.16) provides the following differential equation:

$$\frac{d}{dr}\left\{\frac{1}{r}\frac{d}{dr}\left(r^2\sigma_r\right)\right\} + E\alpha\frac{dT}{dr} = 0, \tag{5.7}$$

where $2A$ is the constant of integration. Solving it after substituting the expression for T from Equation (5.1), the radial stress is obtained as:

$$\sigma_r = A - \frac{B}{r^2} + \frac{E\alpha(T_b - T_a)}{2\ln\left(\frac{b}{a}\right)}\left\{\frac{1}{2} - \ln\left(\frac{r}{a}\right)\right\}, \tag{5.8}$$

where A and B are the constants of integration. Substituting Equation (5.8) in the equilibrium equation yields:

$$\sigma_\theta = A + \frac{B}{r^2} - \frac{E\alpha(T_b - T_a)}{2\ln\left(\frac{b}{a}\right)}\left\{\frac{1}{2} + \ln\left(\frac{r}{a}\right)\right\}. \tag{5.9}$$

The constants A and B are obtained by using the boundary condition of zero radial stress condition at both inner and outer walls, i.e., at $r=a$ and b, $\sigma_r=0$. Thus, the elastic radial and hoop stresses in the disk under sufficiently low temperature difference are obtained as (Noda et al., 2003):

$$\sigma_r = \frac{E\alpha}{2}(T_b - T_a)\left\{-\frac{\ln\left(\frac{r}{a}\right)}{\ln\left(\frac{b}{a}\right)} + \left(1 - \frac{a^2}{r^2}\right)\frac{b^2}{b^2 - a^2}\right\}, \tag{5.10}$$

$$\sigma_\theta = \frac{E\alpha}{2}(T_b - T_a)\left\{-\frac{1+\ln\left(\frac{r}{a}\right)}{\ln\left(\frac{b}{a}\right)} + \left(1 + \frac{a^2}{r^2}\right)\frac{b^2}{b^2 - a^2}\right\}. \tag{5.11}$$

A typical distribution of stresses due to thermo-elastic deformation in an aluminum disk of $a=10$ mm, $b=20$ mm subjected to a temperature gradient $(T_b - T_a)=30$ °C is shown in Figure 5.2. Due to plane–stress assumption, the axial stress is zero throughout the cylinder. The hoop stress is tensile at the inner wall; it gradually reduces and becomes compressive after some radial distance. This behavior can be easily understood. Due to the higher temperature of the outer wall, it has a tendency to expand more than the inner wall. Thus, the outer portion of the cylinder tries to pull the inner portion, whilst the inner portion tries to suppress the expansion of the outer portion.

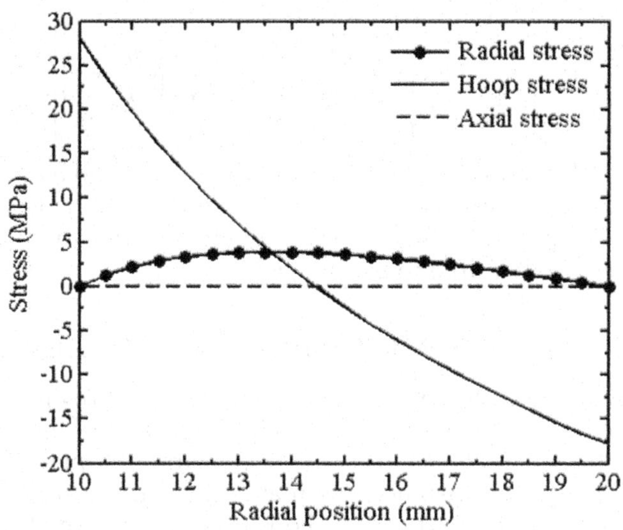

FIGURE 5.2
Distribution of stresses due to thermo-elastic deformation in a typical aluminum disk subjected to a radial temperature gradient of 30 °C ($E=69$ GPa, $\sigma_Y=50.3$ MPa and $\alpha=22.2\times10^{-6}$/°C).

Hence, the inner portion faces tensile stress and the outer portion faces compressive stress. At the same time, due to increasing temperature and consequently increasing thermal expansion from the inner wall to the outer wall, the radial stress is always tensile.

5.3.2 Onset of Yielding in the Disk

The stress distribution depicted in Figure 5.2 suggests that near the inner surface of the disk, the thermo-elastic stresses follow the inequality $\sigma_\theta > \sigma_r > \sigma_z$. The stresses at and around the outer surface follow the inequality $\sigma_r > \sigma_z > \sigma_\theta$. The difference $|\sigma_\theta - \sigma_z|$ in the vicinity of the inner wall is always greater than the difference $|\sigma_r - \sigma_\theta|$ in the vicinity of the outer radius. Hence, for the increasing temperature difference, yielding initiates first at the inner radius of the disk as per the Tresca yield criterion. Thus, for the onset of yielding:

$$(\sigma_\theta - \sigma_z)\Big|_{r=a} = \operatorname{sgn}(T_b - T_a)\sigma_Y, \tag{5.12}$$

where σ_Y is the yield stress and $\operatorname{sgn}(T_b - T_a)$ is a sign function that can be +1 or −1 depending on whether $T_b > T_a$ or $T_b < T_a$. The case $T_b = T_a$ is irrelevant as it will not cause yielding. For the sake of brevity, $\operatorname{sgn}(T_b - T_a)$ is denoted as k_1 in the subsequent equation. As $\sigma_z = 0$, evaluating σ_θ at $r = a$ from Equation (5.11) and substituting it into Equation (5.12), the temperature difference to initiate yielding is obtained as:

$$(T_b - T_a)_{Y_i} = \frac{2k_1\sigma_Y}{E\alpha\left\{\dfrac{\ln\left(\dfrac{b}{a}\right) - 1}{\ln\left(\dfrac{b}{a}\right)} + \left(\dfrac{a^2}{b^2 - a^2}\right)\left(\dfrac{b^2}{a^2} + 1\right)\right\}}. \tag{5.13}$$

Equation (5.13) indicates that the larger the values of E and α are, the smaller the temperature difference required to cause initial yielding is. On the other hand, obviously, the larger the yield stress of the material, the larger the temperature difference required to initiate yielding is.

5.3.3 Thermo-Elastic–Plastic Deformation in the Disk after Loading

As the temperature difference $(T_b - T_a)$ exceeds the temperature difference given by Equation (5.13), the material at the inner side of the disk starts to plastically deform. As $(T_b - T_a)$ keeps on increasing, the plastic zone keeps spreading. At a typical instance, the material is plastic starting from the inner wall up to an intermediate radial position ($r = c$). The material beyond c up to the outer radius b is subjected to elastic deformation. Hence, the wall of the disk consists of an inner plastic zone $a \le r \le c$ and an outer elastic zone $c \le r \le b$ as shown in Figure 5.3a. This is the first stage of elastic–plastic

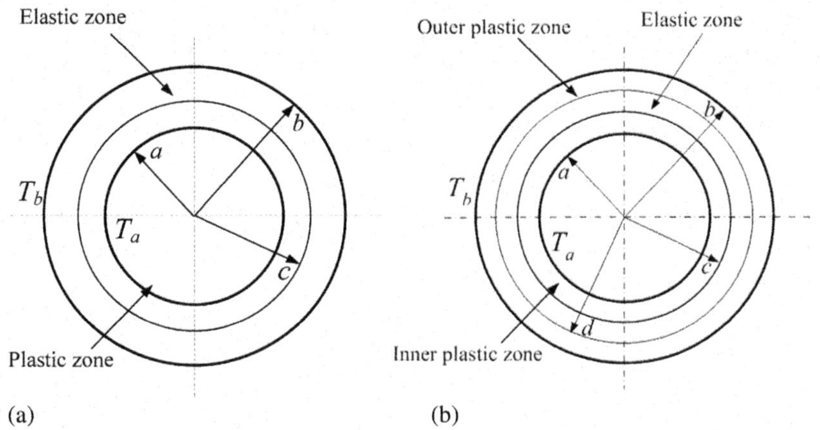

FIGURE 5.3
The elastic and plastic zones in the disk after the (a) first stage and (b) second stage of elastic–plastic deformation.

deformation. Upon further increase of temperature difference, yielding commences at the outer surface of the disk due to compressive hoop stresses crossing a threshold. When the temperature difference exceeds this threshold, both the inner and outer wall of the disk are subjected to simultaneous plastic deformation. After that, one more plastic zone propagating inward from the outer surface, up to a radial position d, emerges in the disk. Thus, the wall of the disk gets divided into an inner plastic zone $a \le r \le c$ and an outer plastic zone $d \le r \le b$, separated by an intermediate elastic zone $c \le r \le d$ as shown in Figure 5.3b. This stage of deformation is the second stage of elastic–plastic deformation. In the following subsections, the analysis of different zones during the first and second stage of elastic–plastic deformation is carried out.

5.3.4 Analysis of the First Stage of Elastic–Plastic Deformation

During the first stage of elastic–plastic deformation, the stresses in the outer elastic zone $c \le r \le b$ (refer Figure 5.3a) are given by Equations (5.8) and (5.9). Using boundary condition of vanishing radial stress at the outer radius and $(\sigma_\theta)_{r=c} = k_1\sigma_Y$ the expressions for radial and hoop stresses in the elastic zone are given by:

$$\sigma_r = \frac{E\alpha(T_b - T_a)}{2\ln\left(\dfrac{b}{a}\right)}\left[\ln\left(\frac{b}{r}\right) + \left(\frac{c^2}{b^2+c^2}\right)\left(1-\frac{b^2}{r^2}\right)\left\{1-\ln\left(\frac{b}{c}\right)\right\}\right] + \frac{c^2 k_1\sigma_Y}{b^2+c^2}\left(1-\frac{b^2}{r^2}\right), \tag{5.14}$$

$$\sigma_\theta = \frac{E\alpha(T_b - T_a)}{2\ln\left(\frac{b}{a}\right)}\left[\ln\left(\frac{b}{r}\right) - 1 + \left(\frac{c^2}{b^2+c^2}\right)\left(1+\frac{b^2}{r^2}\right)\left\{1-\ln\left(\frac{b}{c}\right)\right\}\right] + \frac{c^2 k_1 \sigma_Y}{b^2+c^2}\left(1+\frac{b^2}{r^2}\right). \tag{5.15}$$

With the help of the generalized Hooke's law (Equations 5.2–5.4), the radial displacement u in the elastic zone can be obtained from the strain–displacement relation (Equation 5.5). This provides:

$$u = \frac{\alpha(T_b - T_a)}{2\ln\left(\frac{b}{a}\right)}\left[\begin{array}{l}(1-\nu)r\ln\left(\frac{b}{r}\right) - r + (1-\nu)r\left(\frac{c^2}{b^2+c^2}\right)\left\{1-\ln\left(\frac{b}{c}\right)\right\} \\ +\left(\frac{c^2}{b^2+c^2}\right)\left\{1-\ln\left(\frac{b}{c}\right)\right\}\left\{(1-\nu)r + (1+\nu)\frac{b^2}{r}\right\}\end{array}\right] \tag{5.16}$$

$$+\frac{c^2 k_1 \sigma_Y}{E\left(b^2+c^2\right)}\left\{(1-\nu)r + (1+\nu)\frac{b^2}{r}\right\} + r\alpha T.$$

For stress analysis in the plastic zone ($a \le r \le c$), the effect of strain hardening is taken into account. When the material strain hardens, the Tresca yield criterion in the inner plastic zone becomes:

$$\sigma_\theta = k_1 \sigma_{\text{eq}}, \tag{5.17}$$

where σ_{eq} ($>\sigma_Y$) is the equivalent stress in uniaxial tension or compression. The Bauschinger effect is not considered in this model. It is assumed that the hardening of the disk material is governed by Ludwik's hardening law (Dixit and Dixit, 2008), given by:

$$\sigma_{\text{eq}} = \sigma_Y + K\left(\varepsilon_{\text{eq}}^p\right)^n, \tag{5.18}$$

where K is the strain hardening coefficient and n is the strain hardening exponent. Using Equation (5.17), the stress equilibrium equation (Equation 3.16) can be expressed as:

$$\frac{\mathrm{d}}{\mathrm{d}r}(r\sigma_r) = k_1\left\{\sigma_Y + K\left(\varepsilon_{\text{eq}}^p\right)^n\right\}. \tag{5.19}$$

Integrating Equation (5.19) in the plastic zone from r to $r=c$ and using the value of σ_r at $r=c$ from elastic zone (Equation 5.14), the solution for radial stress in the plastic zone is obtained as:

$$\sigma_r = k_1\sigma_Y\left\{1-\left(\frac{2b^2}{b^2+c^2}\right)\left(\frac{c}{r}\right)\right\}$$
$$+\left(\frac{c}{r}\right)\frac{E\alpha(T_b-T_a)}{2\ln\left(\frac{b}{a}\right)}\left[\ln\left(\frac{b}{c}\right)+\left(\frac{c^2}{b^2+c^2}\right)\left(1-\frac{b^2}{c^2}\right)\left\{1-\ln\left(\frac{b}{c}\right)\right\}\right] \tag{5.20}$$
$$-k_1\frac{K}{r}\int_r^c\left(\varepsilon_{\mathrm{eq}}^p\right)^n \mathrm{d}r_1,$$

where r_1 is a variable radius. The thermal hoop stress in the plastic zone is given by Equation (5.17).

In the plastic zone, the total strain is composed of elastic and plastic strains. Hence,

$$\varepsilon_r = \frac{\mathrm{d}u}{\mathrm{d}r} = \varepsilon_r^e + \varepsilon_r^p, \quad \varepsilon_\theta = \frac{u}{r} = \varepsilon_\theta^e + \varepsilon_\theta^p, \tag{5.21}$$

where the elastic components ε_θ^e and ε_r^e are given by the generalized Hooke's law. In view of the associated flow rule based on the Tresca yield criterion (Chakrabarty, 2006), $\mathrm{d}\varepsilon_r^p = 0, \quad \mathrm{d}\varepsilon_\theta^p = -\mathrm{d}\varepsilon_z^p$. Thus, ε_r is entirely elastic. Using Equation (5.21) in the generalized Hooke's law and substituting them in the equilibrium equation (Equation 3.16), the following differential equation is obtained:

$$\frac{\mathrm{d}}{\mathrm{d}r}\left\{\frac{1}{r}\frac{\mathrm{d}}{\mathrm{d}r}(ur)\right\} = \nu\frac{\mathrm{d}\varepsilon_\theta^p}{\mathrm{d}r} + \alpha(1+\nu)\frac{\mathrm{d}T}{\mathrm{d}r} - \left(\frac{1-\nu}{r}\right)\varepsilon_\theta^p. \tag{5.22}$$

For the plastic zone, the strain compatibility relation (Equation 5.6) may be written as:

$$r\frac{\mathrm{d}}{\mathrm{d}r}\left(\varepsilon_\theta^e + \varepsilon_\theta^p\right) = \left(\varepsilon_r^e - \varepsilon_\theta^e\right) - \varepsilon_\theta^p. \tag{5.23}$$

Using the generalized Hooke's law in Equation (5.23), the following differential equation is obtained:

$$\frac{1}{E}\frac{\mathrm{d}\sigma_\theta}{\mathrm{d}r} + \frac{1}{E}\frac{\mathrm{d}\sigma_r}{\mathrm{d}r} + \alpha\frac{\mathrm{d}T}{\mathrm{d}r} + \frac{\mathrm{d}\varepsilon_\theta^p}{\mathrm{d}r} = -\frac{\varepsilon_\theta^p}{r}. \tag{5.24}$$

Substituting ε_θ^p / r from Equation (5.24) in Equation (5.22) and then integrating one obtains:

$$\frac{1}{r}\frac{\mathrm{d}}{\mathrm{d}r}(ur)=\frac{1-\nu}{E}(\sigma_\theta+\sigma_r)+2\alpha\left\{T_a+(T_b-T_a)\frac{\ln\left(\frac{r}{a}\right)}{\ln\left(\frac{b}{a}\right)}\right\}+\varepsilon_\theta^p+C, \tag{5.25}$$

where C is a constant of integration. To determine the constant C in Equation (5.25), the right- and left-hand side expressions are evaluated at $r=c$ by taking the corresponding values from the elastic zone. Also, the value of ε_θ^p at the elastic–plastic interface, $r=c$ is zero. Insertion of all expressions in Equation (5.25) provides $C=0$. Thus, integrating Equation (5.25) in the plastic domain from r to $r=c$ provides the total hoop strain component as:

$$\frac{u}{r}=\varepsilon_\theta=\frac{\alpha(T_b-T_a)}{2\ln\left(\frac{b}{a}\right)}\left[\left\{\frac{2c^2b^2}{r^2\left(b^2+c^2\right)}\right\}\left\{1-\ln\left(\frac{b}{c}\right)\right\}+2\ln\left(\frac{r}{a}\right)-1\right]$$
$$+\frac{2}{E}\frac{k_1c^2b^2\sigma_Y}{r^2\left(b^2+c^2\right)}+\frac{1-\nu}{E}\sigma_r+\alpha T_a-\frac{1}{r^2}\int_r^c r_1\varepsilon_\theta^p\mathrm{d}r_1. \tag{5.26}$$

The plastic hoop strain is obtained by subtracting the elastic part from the total hoop strain component and is given by:

$$\varepsilon_\theta^p=\frac{\alpha(T_b-T_a)}{2\ln\left(\frac{b}{a}\right)}\left[\begin{array}{l}\left\{\frac{2c^2b^2}{r^2\left(b^2+c^2\right)}\right\}\left\{1-\ln\left(\frac{b}{c}\right)\right\}-1+\frac{c}{r}\ln\left(\frac{b}{c}\right)\\+\left\{\frac{c^3}{r\left(b^2+c^2\right)}\right\}\left(1-\frac{b^2}{c^2}\right)\left\{1-\ln\left(\frac{b}{c}\right)\right\}\end{array}\right]+\frac{2}{E}\frac{k_1c^2b^2\sigma_Y}{r^2\left(b^2+c^2\right)}$$
$$+\frac{k_1c^3\sigma_Y}{Er\left(b^2+c^2\right)}\left(1-\frac{b^2}{c^2}\right)-\frac{k_1\sigma_Y}{E}\left(\frac{c}{r}\right)$$
$$-k_1\frac{K}{Er}\int_r^c\left(\varepsilon_{\mathrm{eq}}^p\right)^n\mathrm{d}r_1-\frac{1}{r^2}\int_r^c r_1\varepsilon_\theta^p\mathrm{d}r_1-k_1\frac{K}{E}\left(\varepsilon_{\mathrm{eq}}^p\right)^n. \tag{5.27}$$

For this case, the equivalent plastic strain field $\varepsilon_{\mathrm{eq}}^p$ is given as:

$$\varepsilon_{\mathrm{eq}}^p=\frac{2}{\sqrt{3}}\varepsilon_\theta^p. \tag{5.28}$$

Thus, Equation (5.28) becomes:

$$\varepsilon_\theta^p = \frac{\alpha(T_b - T_a)}{2\ln\left(\frac{b}{a}\right)}\left[\begin{array}{l}\left\{\frac{2c^2b^2}{r^2\left(b^2+c^2\right)}\right\}\left\{1-\ln\left(\frac{b}{c}\right)\right\}-1+\frac{c}{r}\ln\left(\frac{b}{c}\right)\\ +\left\{\frac{c^3}{r\left(b^2+c^2\right)}\right\}\left(1-\frac{b^2}{c^2}\right)\left\{1-\ln\left(\frac{b}{c}\right)\right\}\end{array}\right]+\frac{2}{E}\frac{k_1c^2b^2\sigma_Y}{r^2\left(b^2+c^2\right)}$$

$$+\frac{k_1c^3\sigma_Y}{Er\left(b^2+c^2\right)}\left(1-\frac{b^2}{c^2}\right)-\frac{k_1\sigma_Y}{E}\left(\frac{c}{r}\right)$$

$$-k_1\left(\frac{2}{\sqrt{3}}\right)^n\frac{K}{Er}\int_r^c\left(\varepsilon_\theta^p\right)^n \mathrm{d}r_1-\frac{1}{r^2}\int_r^c r_1\varepsilon_\theta^p \mathrm{d}r_1-k_1\left(\frac{2}{\sqrt{3}}\right)^n\frac{K}{E}\left(\varepsilon_\theta^p\right)^n. \tag{5.29}$$

Equation (5.29) can be solved for determining plastic strains at any radial positions in the plastic zone. An iterative solution procedure may be employed for this. For example, one can use fixed iteration method to solve Equation (5.29).

5.3.5 Analysis of the Second Stage of Elastic–Plastic Deformation

During the second stage of elastic–plastic deformation, the wall of the disk consists of three zones as shown in Figure 5.3b. At this stage, the criterion for the plastic deformation at the inner zone is same as that in the first zone. However, at the outer zone, the difference between radial and hoop stresses is responsible for yielding as per the Tresca yield criterion. The stress distribution in the different elastic and plastic zones are obtained as follows:

Intermediate elastic zone: $c \le r \le d$:

Using the boundary conditions $(\sigma_\theta)_{r=c} = k_1\sigma_Y$ and $(\sigma_\theta - \sigma_r)_{r=d} = -k_1\sigma_Y$, the stresses in the elastic zone $c \le r \le d$ are obtained from Equations (5.8), and (5.9), and are given by:

$$\sigma_r = \frac{E\alpha(T_b - T_a)}{2\ln\left(\frac{b}{a}\right)}\left\{1+\ln\left(\frac{c}{r}\right)-\frac{d^2}{2c^2}-\frac{d^2}{2r^2}\right\}+k_1\sigma_Y\left(1+\frac{d^2}{2c^2}+\frac{d^2}{2r^2}\right), \tag{5.30}$$

$$\sigma_\theta = \frac{E\alpha(T_b - T_a)}{2\ln\left(\frac{b}{a}\right)}\left\{\ln\left(\frac{c}{r}\right)-\frac{d^2}{2c^2}+\frac{d^2}{2r^2}\right\}+k_1\sigma_Y\left(1+\frac{d^2}{2c^2}-\frac{d^2}{2r^2}\right). \tag{5.31}$$

The displacement component for this elastic zone is obtained in a similar manner as discussed in Subsection 5.3.4 and is given by:

$$u=\frac{a(T_b-T_a)}{2\ln\left(\frac{b}{a}\right)}\left\{(1+\nu)\frac{d^2}{2r}-(1-\nu)\frac{d^2r}{2c^2}+(1-\nu)r\ln\left(\frac{c}{r}\right)-\nu r\right\}$$

$$+\frac{k_1\sigma_Y}{E}\left\{(1-\nu)r+(1-\nu)\frac{d^2r}{2c^2}-(1+\nu)\frac{d^2}{2r}\right\}+ar\left\{T_a+(T_b-T_a)\frac{\ln\left(\frac{r}{a}\right)}{\ln\left(\frac{b}{a}\right)}\right\}. \tag{5.32}$$

Inner plastic zone: $a\le r\le c$:

Integrating Equation (5.19) and using the value of σ_r at $r=c$ from elastic zone (Equation 5.30), the expression for radial thermal stress is obtained as:

$$\sigma_r=\left(\frac{c}{r}\right)\frac{Ea(T_b-T_a)}{2\ln\left(\frac{b}{a}\right)}\left(1-\frac{d^2}{c^2}\right)+k_1\sigma_Y\left(1+\frac{d^2}{cr}\right)-k_1\frac{K}{r}\int_r^c\left(\varepsilon_{\text{eq}}^p\right)^n \mathrm{d}r_1. \tag{5.33}$$

Following the similar procedure as discussed in Subsection 5.3.4, the plastic part of the hoop strain component in the inner plastic zone is obtained as:

$$\varepsilon_\theta^p=\frac{a(T_b-T_a)}{2\ln\left(\frac{b}{a}\right)}\left\{\frac{d^2}{r^2}-1+\left(\frac{c}{r}\right)\left(1-\frac{d^2}{c^2}\right)\right\}+\frac{k_1\sigma_Y}{E}\left(\frac{d^2}{cr}-\frac{d^2}{r^2}\right)$$

$$-\frac{1}{r^2}\int_r^c r_1\varepsilon_\theta^p \mathrm{d}r_1-k_1\left(\frac{2}{\sqrt{3}}\right)^n\frac{K}{rE}\int_r^c\left(\varepsilon_\theta^p\right)^n \mathrm{d}r_1-k_1\left(\frac{2}{\sqrt{3}}\right)^n\frac{K}{E}\left(\varepsilon_\theta^p\right)^n. \tag{5.34}$$

The hoop stress component is given by Equation (5.17).

Outer plastic zone: $d\le r\le b$:

In the outer plastic zone ($d\le r\le b$), the Tresca yield criterion incorporating the effect of strain hardening is given by:

$$\sigma_\theta-\sigma_r=-k_1\sigma_{\text{eq}} \tag{5.35}$$

where σ_{eq} is given by Equation (5.18). Using Equation (5.35) in equilibrium equation (Equation 3.16) and then integrating the resulting expression in the plastic domain from $r=d$ to r, the solution for radial stress is obtained as:

$$\sigma_r = \frac{E\alpha(T_b - T_a)}{2\ln\left(\frac{b}{a}\right)}\left\{\frac{1}{2} + \ln\left(\frac{c}{d}\right) - \frac{d^2}{2c^2}\right\} + k_1\sigma_Y\left\{\frac{3}{2} + \frac{d^2}{2c^2} - \ln\left(\frac{r}{d}\right)\right\} + k_1K\int_r^d \frac{1}{r_1}\left(\varepsilon_{\text{eq}}^p\right)^n \mathrm{d}r_1. \tag{5.36}$$

Equation (5.35) provides the expression for hoop stress as:

$$\sigma_\theta = \frac{E\alpha(T_b - T_a)}{2\ln\left(\frac{b}{a}\right)}\left\{\frac{1}{2} + \ln\left(\frac{c}{d}\right) - \frac{d^2}{2c^2}\right\} + k_1\sigma_Y\left\{\frac{1}{2} + \frac{d^2}{2c^2} - \ln\left(\frac{r}{d}\right)\right\} + k_1K\int_r^d \frac{1}{r_1}\left(\varepsilon_{\text{eq}}^p\right)^n \mathrm{d}r_1 - k_1K\left(\varepsilon_{\text{eq}}^p\right)^n. \tag{5.37}$$

In the outer plastic zone, associated flow rule based on the Tresca yield criterion provides, $\mathrm{d}\varepsilon_z^p = 0, \quad \mathrm{d}\varepsilon_\theta^p = -\mathrm{d}\varepsilon_r^p$. Hence, due to plastic incompressibility,

$$\varepsilon_r^p + \varepsilon_\theta^p = 0 \tag{5.38}$$

Using the additive decomposition of elastic and plastic parts of strains in Equation (5.38) and with the help of generalized Hooke's law along with the strain–displacement relation (Equation 5.5), the following differential equation is obtained:

$$\frac{\mathrm{d}}{\mathrm{d}r}(ur) = 2\alpha\left\{T_a + (T_b - T_a)\frac{\ln\left(\frac{r}{a}\right)}{\ln\left(\frac{b}{a}\right)}\right\}r + \frac{1}{E}(1-\nu)\frac{\mathrm{d}}{\mathrm{d}r}\left(r^2\sigma_r\right). \tag{5.39}$$

Integrating Equation (5.39) one obtains,

$$ur = \alpha T_a r^2 + 2\alpha\frac{(T_b - T_a)}{\ln\left(\frac{b}{a}\right)}\left\{\ln\left(\frac{r}{a}\right)\frac{r^2}{2} - \frac{r^2}{4}\right\} + \frac{1}{E}(1-\nu)r^2\sigma_r + D, \tag{5.40}$$

where D is a constant of integration. To find out the constant D, the left- and right-hand side of Equation (5.40) are evaluated at $r = c$ taking the corresponding values of σ_r and u from the elastic zone ($c \le r \le d$). Thus, the value of D is obtained as:

$$D = \frac{\alpha(T_b - T_a)}{2\ln\left(\frac{b}{a}\right)}d^2 - \frac{k_1\sigma_Y}{E}d^2. \tag{5.41}$$

Substituting the value of D from Equation (5.41) in Equation (5.41), the total hoop strain component in the plastic zone $d \le r \le b$ is obtained as:

$$\frac{u}{r} = \varepsilon_\theta = \alpha T_a + \frac{\alpha(T_b - T_a)}{2\ln\left(\frac{b}{a}\right)}\left\{2\ln\left(\frac{r}{a}\right) - 1 + \frac{d^2}{r^2}\right\} + \frac{1}{E}(1-\nu)\sigma_r - \frac{k_1\sigma_Y}{E}\frac{d^2}{r^2}. \quad (5.42)$$

The plastic part of hoop strain is obtained by subtracting the elastic parts from Equation (5.42) as:

$$\varepsilon_\theta^p = \frac{\alpha(T_b - T_a)}{2\ln\left(\frac{b}{a}\right)}\left\{\frac{d^2}{r^2} - 1\right\} + \frac{k_1\sigma_Y}{E}\left(1 - \frac{d^2}{r^2}\right) + \frac{k_1 K}{E}\left(\frac{2}{\sqrt{3}}\right)^n \left(\varepsilon_\theta^p\right)^n. \quad (5.43)$$

5.3.6 Distribution of Residual Thermal Stresses after Unloading

When the disk is gradually cooled to room temperature, thermal unloading takes place and residual thermal stresses are induced in the disk. Assuming the unloading process to be completely elastic, linear, and devoid of the Bauschinger effect, the residual thermal stresses generated in different zones within the wall of the disk are obtained by subtracting elastic stresses (Equations 5.10 and 5.11) from the respective loading stresses. The residual stresses in the elastic and plastic zone during the first stage of elastic–plastic deformation are given as:

Elastic zone, $c \le r \le b$:

$$(\sigma_r)_{\text{res}} = \frac{E\alpha(T_b - T_a)}{2\ln\left(\frac{b}{a}\right)}\left[\left(\frac{c^2}{b^2+c^2}\right)\left(1 - \frac{b^2}{r^2}\right)\left\{1 - \ln\left(\frac{b}{c}\right)\right\} + \left(\frac{a^2}{b^2-a^2}\right)\ln\left(\frac{b}{a}\right)\left(\frac{b^2}{r^2} - 1\right)\right] + \frac{c^2 k_1\sigma_Y}{b^2+c^2}\left(1 - \frac{b^2}{r^2}\right), \quad (5.44)$$

$$(\sigma_\theta)_{\text{res}} = \frac{E\alpha(T_b - T_a)}{2\ln\left(\frac{b}{a}\right)}\left[\left(\frac{c^2}{b^2+c^2}\right)\left(1 + \frac{b^2}{r^2}\right)\left\{1 - \ln\left(\frac{b}{c}\right)\right\} - \left(\frac{a^2}{b^2-a^2}\right)\ln\left(\frac{b}{a}\right)\left(1 + \frac{b^2}{r^2}\right)\right] + \frac{c^2 k_1\sigma_Y}{b^2+c^2}\left(1 + \frac{b^2}{r^2}\right). \quad (5.45)$$

Plastic zone, $a \le r \le c$:

$$(\sigma_r)_{\text{res}} = k_1\sigma_Y\left\{1-\left(\frac{2b^2}{b^2+c^2}\right)\left(\frac{c}{r}\right)\right\} - k_1\frac{K}{r}\int_r^c\left(\varepsilon_{\text{eq}}^p\right)^n \mathrm{d}r$$

$$+\frac{E\alpha(T_b-T_a)}{2\ln\left(\dfrac{b}{a}\right)}\left[\begin{aligned}&\left(\frac{c}{r}\right)\ln\left(\frac{b}{c}\right)+\left(\frac{c^2}{b^2+c^2}\right)\left(\frac{c}{r}\right)\left(1-\frac{b^2}{r^2}\right)\left\{1-\ln\left(\frac{b}{c}\right)\right\}\\&-\ln\left(\frac{b}{r}\right)-\left(\frac{a^2}{b^2-a^2}\right)\ln\left(\frac{b}{a}\right)\left(\frac{b^2}{r^2}-1\right)\end{aligned}\right], \tag{5.46}$$

$$(\sigma_\theta)_{\text{res}} = k_1\left\{\sigma_Y + K\left(\varepsilon_{\text{eq}}^p\right)^n\right\} + \frac{E\alpha(T_b-T_a)}{2\ln\left(\dfrac{b}{a}\right)}\left\{1+\ln\left(\frac{r}{a}\right)-\left(1+\frac{a^2}{r^2}\right)\ln\left(\frac{b}{a}\right)\left(\frac{b^2}{b^2-a^2}\right)\right\}. \tag{5.47}$$

The residual stresses in the different elastic and plastic zones during the second stage of elastic–plastic deformation are obtained as:

Elastic zone, $c \le r \le d$:

$$(\sigma_r)_{\text{res}} = \frac{E\alpha(T_b-T_a)}{2\ln\left(\dfrac{b}{a}\right)}\left\{1+\ln\left(\frac{c}{a}\right)-\frac{d^2}{2c^2}-\frac{d^2}{2r^2}-\left(1-\frac{a^2}{r^2}\right)\ln\left(\frac{b}{a}\right)\left(\frac{b^2}{b^2-a^2}\right)\right\}$$

$$+k_1\sigma_Y\left\{1+\frac{d^2}{2c^2}+\frac{d^2}{2r^2}\right\}, \tag{5.48}$$

$$(\sigma_\theta)_{\text{res}} = \frac{E\alpha(T_b-T_a)}{2\ln\left(\dfrac{b}{a}\right)}\left\{1+\ln\left(\frac{c}{a}\right)-\frac{d^2}{2c^2}+\frac{d^2}{2r^2}-\left(1+\frac{a^2}{r^2}\right)\ln\left(\frac{b}{a}\right)\left(\frac{b^2}{b^2-a^2}\right)\right\}$$

$$+k_1\sigma_Y\left\{1+\frac{d^2}{2c^2}-\frac{d^2}{2r^2}\right\}. \tag{5.49}$$

Inner plastic zone, $a \le r \le c$:

$$(\sigma_r)_{\text{res}} = \left(\frac{c}{r}\right)\frac{E\alpha(T_b - T_a)}{2\ln\left(\dfrac{b}{a}\right)}\left\{\frac{c}{r} - \frac{d^2}{cr} + \ln\left(\frac{r}{a}\right) - \left(1 - \frac{a^2}{r^2}\right)\ln\left(\frac{b}{a}\right)\left(\frac{b^2}{b^2 - a^2}\right)\right\}$$
$$+ k_1\sigma_Y\left(1 + \frac{d^2}{cr}\right) - k_1\frac{K}{r}\int_r^c \left(\varepsilon_{\text{eq}}^p\right)^n \mathrm{d}r, \tag{5.50}$$

The residual hoop stress in the inner plastic zone is given by Equation (5.47).

Outer plastic zone, $d \le r \le b$:

$$(\sigma_r)_{\text{res}} = \frac{E\alpha(T_b - T_a)}{2\ln\left(\dfrac{b}{a}\right)}\left\{\frac{1}{2} + \ln\left(\frac{c}{d}\right) - \frac{d^2}{2c^2} + \ln\left(\frac{r}{a}\right) - \left(1 - \frac{a^2}{r^2}\right)\ln\left(\frac{b}{a}\right)\left(\frac{b^2}{b^2 - a^2}\right)\right\}$$
$$+ k_1\sigma_Y\left\{\frac{3}{2} + \frac{d^2}{2c^2} - \ln\left(\frac{r}{d}\right)\right\} + k_1K\int_r^d \frac{1}{r_1}\left(\varepsilon_{\text{eq}}^p\right)^n \mathrm{d}r_1, \tag{5.51}$$

$$(\sigma_\theta)_{\text{res}} = \frac{E\alpha(T_b - T_a)}{2\ln\left(\dfrac{b}{a}\right)}\left\{\frac{3}{2} + \ln\left(\frac{c}{d}\right) - \frac{d^2}{2c^2} + \ln\left(\frac{r}{a}\right) - \left(1 + \frac{a^2}{r^2}\right)\ln\left(\frac{b}{a}\right)\left(\frac{b^2}{b^2 - a^2}\right)\right\}$$
$$+ k_1\sigma_Y\left\{\frac{1}{2} + \frac{d^2}{2c^2} - \ln\left(\frac{r}{d}\right)\right\} + k_1K\int_r^d \frac{1}{r_1}\left(\varepsilon_{\text{eq}}^p\right)^n \mathrm{d}r_1 - k_1K\left(\varepsilon_{\text{eq}}^p\right)^n. \tag{5.52}$$

5.3.7 Evaluation of Unknown Elastic–Plastic Interface Radii

After the first stage of elastic–plastic deformation, the radius of elastic–plastic interface c is unknown. The elastic–plastic interface radii c and d are the unknowns during the second stage of elastic–plastic deformation. The boundary condition of vanishing radial stress at the inner radius during the first stage of elastic–plastic deformation provides:

$$k_1\sigma_Y\left\{1 - \left(\frac{2b^2}{b^2 + c^2}\right)\left(\frac{c}{a}\right)\right\} + \left(\frac{c}{a}\right)\frac{E\alpha(T_b - T_a)}{2\ln\left(\dfrac{b}{a}\right)}\left[\ln\left(\frac{b}{c}\right) + \left(\frac{c^2}{b^2 + c^2}\right)\left(1 - \frac{b^2}{c^2}\right)\left\{1 - \ln\left(\frac{b}{c}\right)\right\}\right]$$
$$- k_1\frac{K}{a}\int_a^c \left(\varepsilon_{\text{eq}}^p\right)^n \mathrm{d}r_1 = 0. \tag{5.53}$$

Solution of Equation (5.53) will provide the value of the unknown radius c. The equation can be solved by a suitable numerical method, e.g., the bisection method. In the bisection method, the first two values of c are assumed. Substitution of one value of c in the left-hand side of Equation (5.53) provides a positive value, while the substitution of other value provides a negative value. The arithmetic mean of these values is calculated. If the function value at the arithmetic mean is positive, the guess value corresponding to the positive function value is replaced by the mean. If the function value at the arithmetic mean is negative, the guess value corresponding to the negative function value is replaced by the mean. The procedure is repeated till the difference between the guess values is sufficiently small. The mean of the final guess values can be taken as the proper value of c. The boundary conditions of vanishing radial stress both at the inner and the outer radius during the second stage of elastic–plastic deformation gives:

$$\left(\frac{c}{a}\right)\frac{E\alpha(T_b - T_a)}{2\ln\left(\frac{b}{a}\right)}\left(1-\frac{d^2}{c^2}\right)+k_1\sigma_Y\left(1+\frac{d^2}{ca}\right)-k_1\frac{K}{a}\int_a^c\left(\varepsilon_{eq}^p\right)^n dr_1 = 0, \tag{5.54}$$

$$\frac{E\alpha(T_b - T_a)}{2\ln\left(\frac{b}{a}\right)}\left\{\frac{1}{2}+\ln\left(\frac{c}{d}\right)-\frac{d^2}{2c^2}\right\}+k_1\sigma_Y\left\{\frac{3}{2}+\frac{d^2}{2c^2}-\ln\left(\frac{b}{d}\right)\right\}+k_1K\int_b^d\frac{1}{r_1}\left(\varepsilon_{eq}^p\right)^n dr_1 = 0. \tag{5.55}$$

The unknown elastic–plastic interface radii c and d can be obtained by solving the two non- linear equations numerically. Newton's method can be used for this purpose. Alternatively, one can use an optimization method for reducing the error in both of the equations by treating c and d as design variables.

5.3.8 Procedure for Calculating the Unknown Elastic–Plastic Interface Radii

The unknown elastic–plastic interface radii at various stages of plastic deformation must be calculated to obtain the distribution of stresses. A step by step numerical solution methodology to estimate the value of c or c and d is as follows:

Step 1: For the first stage of elastic–plastic deformation, the initial estimate of c is obtained by making $K=0$ in Equation (5.53) and solving it by a numerical method, e.g., the bisection method. During the second stage of elastic–plastic deformation, the initial estimates for c and d are obtained by making $K=0$ in Equations (5.54) and (5.55) and solving them for c and d by using a numerical technique such as the least square method of solving simultaneous nonlinear equations. The initial guess value of ε_θ^p is taken as zero everywhere.

Step 2: The updated plastic strain field ε_θ^p at different radial positions is found out from Equation (5.29) during the first stage of elastic–plastic deformation and from Equations (5.34) and (5.43) during the second stage of elastic–plastic deformation. For the fixed c in Equation (5.29) and for the fixed values of c and d in Equations (5.34) and (5.43), values of ε_θ^p are updated further and this process is repeated till the convergence in ε_θ^p is achieved. This basically amounts to solving for ε_θ^p using the fixed-point iteration method (Gerald and Wheatley, 1994). The integrations involved in Equations (5.29), (5.34), and (5.55) can be evaluated numerically. Some numerical integration methods are the trapezoidal rule, Simpson's rule, and Gauss quadrature. Using Equation (5.28), the updated estimate of $\varepsilon_{\text{eq}}^p$ is obtained.

Step 3: Now, using the values of ε_θ^p at different radial positions in the corresponding plastic zones, the integral terms in Equations (5.53), (5.54), and (5.55) are evaluated and the expressions are solved for new estimated values of c or c and d. If the new estimated values of c or c and d are same as the previously estimated values, go to step 4. Otherwise go to Step 2 and repeat the process till the convergence for c or c and d is achieved.

Step 4: Using the latest updated values of ε_θ^p, c or c and d, the radial and hoop stresses in the different plastic zones of the first and second stage of elastic–plastic deformation during loading are calculated. The residual thermal stresses are estimated using the relevant equations obtained in Subsection 5.3.6.

5.3.9 Typical Results of Plane–Stress Model of Thermal Autofrettage

This section presents the typical result of the plane–stress model of thermal autofrettage. A disk with inner radius 10 mm and outer radius 30 mm made of mild steel having $E=200$ GPa, $\sigma_Y=324$ MPa and $\alpha=13\times10^{-6}/°\text{C}$ is taken as the autofrettage specimen. The hardening coefficient K and the strain hardening exponent n for the disk are taken as 226.98 MPa and 0.43, respectively. Using Equation (5.13), the yield onset temperature difference of the disk is obtained as 186°, which is much lower than the recrystallization temperature of mild steel. The disk is thermally autofrettaged with a temperature difference of 270 °C. At a temperature difference of 270 °C, the disk undergoes the first phase of elastic–plastic deformation. Following the methodology discussed in Subsection 5.3.8, the radius of elastic–plastic interface c is estimated as 11.709 mm. The distribution of stresses after thermal loading obtained using the equations discussed in Subsection 5.3.4 and are shown in Figure 5.4a. The distribution of

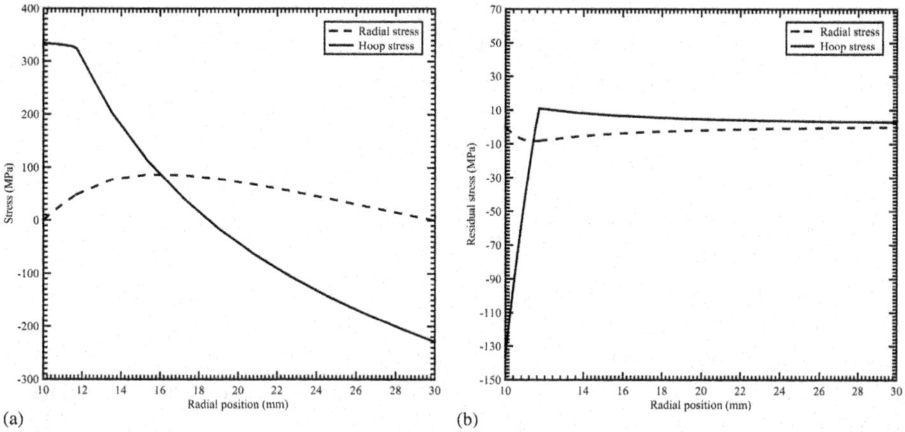

FIGURE 5.4
Distribution of (a) stresses after loading and (b) residual stresses after unloading in mild steel disk (Kamal, 2016).

residual stress after thermal unloading using the relevant equations presented in Subsection 5.3.6 are shown in Figure 5.4b.

5.4 Generalized Plane–Strain Modeling of Thermal Autofrettage

This section presents the closed-form model of thermal autofrettage considering generalized plane–strain assumption and is applicable for the analysis of long cylinders. The yield criterion is based on Tresca yield criterion and its associated flow rule. Like in the case of the plane–stress model, the elastic analysis of a thick-walled cylinder subjected to temperature gradient is presented first. This is followed by the solution of the elastic–plastic analysis of a thick-walled cylinder that has undergone thermal autofrettage.

5.4.1 Elastic Analysis of a Thick-Walled Cylinder Subjected to Radial Temperature Gradient

The distribution of stresses due to thermo-elastic deformation of a thick-walled cylinder under radial temperature gradient is (Noda et al., 2003):

$$\sigma_r = \frac{E\alpha}{2(1-\nu)}(T_b - T_a)\left\{-\frac{\ln\left(\frac{r}{a}\right)}{\ln\left(\frac{b}{a}\right)} + \left(1-\frac{a^2}{r^2}\right)\frac{b^2}{b^2-a^2}\right\}, \tag{5.56}$$

$$\sigma_\theta = \frac{Ea}{2(1-\nu)}(T_b - T_a)\left\{-\frac{1+\ln\left(\frac{r}{a}\right)}{\ln\left(\frac{b}{a}\right)} + \left(1+\frac{a^2}{r^2}\right)\frac{b^2}{b^2-a^2}\right\}, \tag{5.57}$$

$$\sigma_z = -\alpha E T_a + \frac{Ea}{2(1-\nu)}(T_b - T_a)\left\{-\frac{\nu+2\ln\left(\frac{r}{a}\right)}{\ln\left(\frac{b}{a}\right)} + \frac{2\nu b^2}{b^2-a^2}\right\} + E\varepsilon_0. \tag{5.58}$$

In Equation (5.58), ε_0 is the constant axial strain in the z-direction. The constant axial strain ε_0 can be determined from the free end condition (zero total axial force) given by:

$$\int_a^b \sigma_z 2\pi r dr = 0. \tag{5.59}$$

After obtaining the expression for ε_0 and then substituting it in Equation (5.58), the resulting expression for axial stress is given by:

$$\sigma_z = \frac{Ea}{2(1-\nu)}(T_b - T_a)\left\{-\frac{1+2\ln\left(\frac{r}{a}\right)}{\ln\left(\frac{b}{a}\right)} + \frac{2b^2}{b^2-a^2}\right\}. \tag{5.60}$$

5.4.2 Initiation of Yielding

A typical thermo-elastic stress distribution in the same aluminum cylinder as considered in Subsection 5.3.1 is shown in Figure 5.5 using Equations (5.56), (5.57), and (5.60). The distribution suggests that the initiation of yielding takes place first at the inner radius of the cylinder. At the inner radius, the stress distribution follows the inequality: $\sigma_\theta = \sigma_z > \sigma_r$. Thus, for the initiation of yielding at the inner radius of the cylinder, the Tresca yield criterion provides:

$$\left.(\sigma_\theta - \sigma_r)\right|_{r=a} = k_1\sigma_Y \quad \text{or} \quad \left.(\sigma_z - \sigma_r)\right|_{r=a} = k_1\sigma_Y, \tag{5.61}$$

where σ_Y is the yield stress of the material and k_1 is a sign factor as discussed in Section 5.2. By evaluating the corresponding stresses from Equations (5.56), (5.57), and (5.60) in Equation (5.61) at $r = a$, the temperature

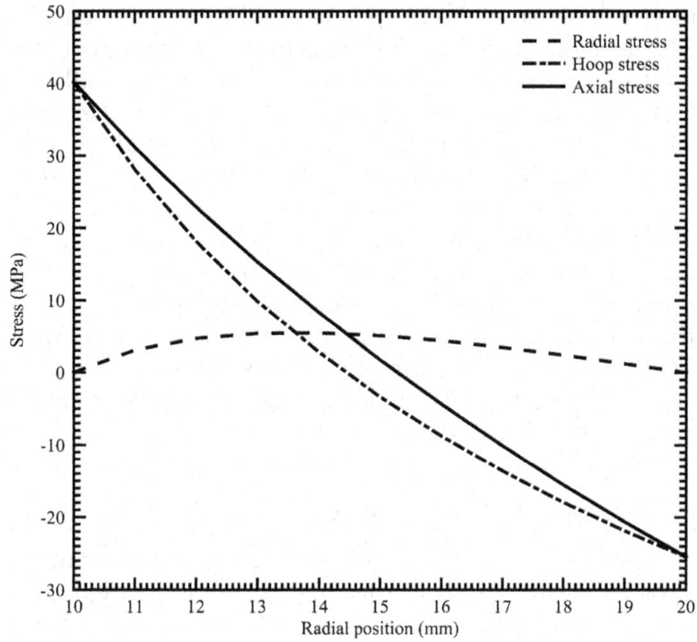

FIGURE 5.5
Typical thermo-elastic stress distributions in an aluminum cylinder (Kamal, 2016).

difference required for the initiation of yielding at the inner radius is obtained as:

$$(T_b - T_a)_{Y_i} = \frac{2(1-\nu)k_1\sigma_Y}{Ea\left\{-\frac{1}{\ln\left(\frac{b}{a}\right)} + \frac{2b^2}{b^2 - a^2}\right\}}. \tag{5.62}$$

When the temperature difference exceeds the yield onset temperature difference given by Equation (5.62), the cylinder undergoes the first stage of elastic–plastic deformation.

Due to the induced temperature difference in the cylinder, there can be three different states of deformation. When the temperature difference crosses the first threshold, the inner wall yields and undergoes the first stage of elastic–plastic deformation. Here, the cylinder wall consists of three deformed zones (Figure 5.5a), namely plastic zone I: $a \le r \le c$, plastic zone II: $c \le r \le d$, and the outer elastic zone: $d \le r \le b$. These two plastic zones correspond to different sides of the Tresca yield locus. In the case of the von Mises yield criterion, the yield locus is smooth (without any discontinuity) and it would not be necessary to consider two different plastic zones.

Upon further increase of the temperature difference, the outer wall starts yielding. This is the second stage of elastic–plastic deformation. At this stage, an additional two consecutive plastic zones at the outer wall (Figure 5.5b), namely plastic zone III: $f \le r \le b$ and plastic zone IV: $e \le r \le f$ emerge. In between this there is an intermediate elastic zone: $d \le r \le e$. The analytical modeling of the thermal stresses is based on the assumption that the maximum allowable temperature T_b in the cylinder is below the recrystallization temperature (homologous temperature ≈0.4) of the material to avoid phase transformation. For most of the materials, the material properties do not vary much at such a low temperature and the percentage change is below 10%. It is also to be noted that the temperature rapidly drops from the outer to the inner surface of the cylinder and that limiting the outer surface up to recrystallization temperature does not alter overall mechanical and metallurgical properties. The present modeling is based on the small strain plasticity.

5.4.3 The First Stage of Elastic–Plastic Deformation

During the first stage of elastic–plastic deformation, the cylinder wall has two consecutive plastic zones at the inner side of the cylinder as shown in Figure 5.6a. The first plastic zone (plastic zone I) is formed according to the stress inequality $\sigma_\theta = \sigma_z > \sigma_r$ existing at the inner radius of the cylinder. Beyond the inner radius, the stresses follow the inequality $\sigma_z > \sigma_\theta > \sigma_r$ and a second plastic zone (plastic zone II) is formed. The expressions for different elastic–plastic zones are as follows.

Plastic zone I, $a \le r \le c$:

Considering material strain hardening with the Tresca criterion, the plastic zone I, $a \le r \le c$ propagates needs to follow either (or both) of the following relations:

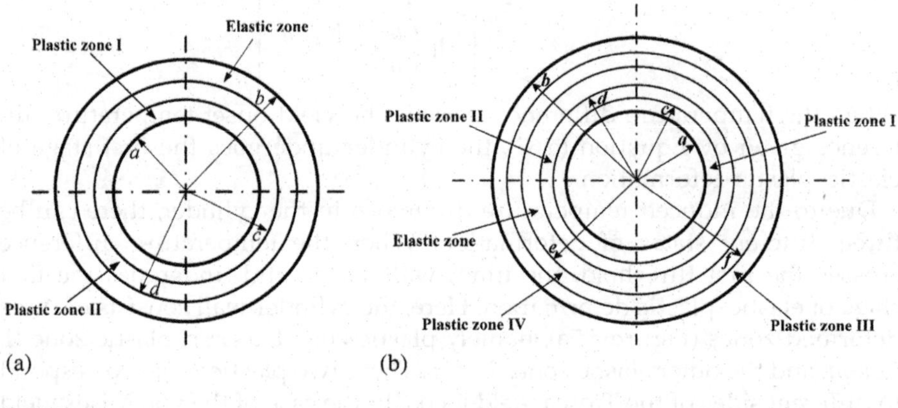

(a) (b)

FIGURE 5.6
The elastic and plastic zones in the cylinder during (a) first and (b) second stage of elastic–plastic deformation.

$$\sigma_\theta - \sigma_r = k_1\sigma_{\text{eq}}, \quad \sigma_z - \sigma_r = k_1\sigma_{\text{eq}}, \tag{5.63}$$

where σ_{eq} is the equivalent yield stress in uniaxial tension or compression given by Ludwik's hardening law (Equation 5.18). Using the first expression of Equation (5.63) in the equilibrium equation (Equation 3.16) and integrating the radial stress is obtained as:

$$\sigma_r = k_1\sigma_Y \ln r + k_1 K \int_a^r \frac{\left(\varepsilon_{\text{eq}}^p\right)^n}{r_1} \mathrm{d}r_1 + C_1, \tag{5.64}$$

where $\varepsilon_{\text{eq}}^p$ is a function of radius, r_1 is a dummy variable and C_1 is a constant of integration. From Equation (5.63),

$$\sigma_\theta = \sigma_z = k_1\sigma_Y\left(1+\ln r\right) + k_1 K\left(\varepsilon_{\text{eq}}^p\right)^n + k_1 K \int_a^r \frac{\left(\varepsilon_{\text{eq}}^p\right)^n}{r_1} \mathrm{d}r_1 + C_1. \tag{5.65}$$

Using plastic incompressibility, the additive decomposition of strains into elastic and plastic parts and generalized Hooke's law, one obtains:

$$E\left(\varepsilon_\theta + \varepsilon_r + \varepsilon_0\right) = \left(1-2\nu\right)\left(\sigma_r + \sigma_\theta + \sigma_z\right) + 3E\alpha T. \tag{5.66}$$

Using the strain–displacement relations (Equation 5.5) in Equation (5.66), the following differential equation is obtained:

$$\frac{\mathrm{d}u}{\mathrm{d}r} + \frac{u}{r} = \frac{1-2\nu}{E}\left(\sigma_r + \sigma_\theta + \sigma_z\right) + 3\alpha T - \varepsilon_0. \tag{5.67}$$

The solution of differential Equation (5.67) provides the value of radial displacement u. Once the value of u is known, one can obtain the value of the total hoop and radial strain components using strain–displacement relations. The plastic parts of the strains are obtained by subtracting the elastic parts (given by generalized Hooke's law) from the total strain components. They are given by:

$$\begin{aligned}\varepsilon_\theta^p = \frac{1-2\nu}{E}&\left[\frac{1}{2}\left(k_1\sigma_Y \ln r + C_1\right) + \frac{2k_1K}{r^2}\int_a^r r_1\left(\varepsilon_{\text{eq}}^p\right)^n \mathrm{d}r_1 - k_1K\int_a^r \frac{\left(\varepsilon_{\text{eq}}^p\right)^n}{r_1}\mathrm{d}r_1 + \frac{3k_1K}{r^2}\int_a^r\left\{r_2\int_a^{r_2}\frac{\left(\varepsilon_{\text{eq}}^p\right)^n}{r_1}\mathrm{d}r_1\right\}\mathrm{d}r_2\right] \\ &- \frac{3-2\nu}{4E}k_1\sigma_Y - \frac{1-\nu}{E}k_1K\left(\varepsilon_{\text{eq}}^p\right)^n + \frac{1}{2}\alpha T_a + \frac{\alpha\left(T_b - T_a\right)}{2\ln\left(\dfrac{b}{a}\right)}\left\{\ln\left(\frac{r}{a}\right) - \frac{3}{2}\right\} - \frac{1}{2}\varepsilon_0 + \frac{C_2}{r^2},\end{aligned} \tag{5.68}$$

$$\varepsilon_r^p = \frac{1-2\nu}{E}\left[\frac{1}{2}(k_1\sigma_Y \ln r + C_3) + 2k_1K\int_a^r \frac{\left(\varepsilon_{\text{eq}}^p\right)^n}{r_1}\text{d}r_1 - \frac{2k_1K}{r^2}\int_a^r r_1\left(\varepsilon_{\text{eq}}^p\right)^n \text{d}r_1 - \frac{3k_1K}{r^2}\int_a^r\left\{r_2\int_a^{r_2}\frac{\left(\varepsilon_{\text{eq}}^p\right)^n}{r_1}\text{d}r_1\right\}\text{d}r_2\right]$$

$$+\frac{7-6\nu}{4E}k_1\sigma_Y + \frac{2(1-\nu)}{E}k_1K\left(\varepsilon_{\text{eq}}^p\right)^n + \frac{1}{2}\alpha T_a + \frac{\alpha(T_b - T_a)}{2\ln\left(\frac{b}{a}\right)}\left\{\ln\left(\frac{r}{a}\right)+\frac{3}{2}\right\} - \frac{1}{2}\varepsilon_0 - \frac{C_2}{r^2},$$

(5.69)

$$\varepsilon_\theta^p = \varepsilon_0 - \frac{1-2\nu}{E}\left\{k_1\sigma_Y \ln r + C_1 + k_1K\int_a^r \frac{\left(\varepsilon_{\text{eq}}^p\right)^n}{r_1}\text{d}r_1\right\} - \frac{1-\nu}{E}k_1\sigma_Y - \frac{1-\nu}{E}k_1K\left(\varepsilon_{\text{eq}}^p\right)^n$$

$$-\alpha T_a - \alpha(T_b - T_a)\frac{\ln\left(\frac{r}{a}\right)}{\ln\left(\frac{b}{a}\right)},$$

(5.70)

where r_1 and r_2 are variable radii. The equivalent plastic strain in the plastic zone I is given by:

$$\varepsilon_{\text{eq}}^p = \sqrt{\frac{2}{3}\varepsilon_{ij}^p\varepsilon_{ij}^p}, \tag{5.71}$$

where ε_{ij}^p is the plastic parts of the strain tensor.

Plastic zone II, $c \le r \le d$:

The plastic zone II, $c \le r \le d$ is formed according to the Tresca yield criterion:

$$\sigma_z - \sigma_r = k_1\sigma_{\text{eq}} \tag{5.72}$$

The associated flow rule based on the Tresca yield criterion provides $\varepsilon_r^p = -\varepsilon_z^p,\ \ \varepsilon_\theta^p = 0$. Hence, in the plastic zone II, the hoop strain component is wholly elastic. Using the additive decomposition of strains, the strain compatibility condition for the plastic zone II can be written as:

$$\varepsilon_r^e + \varepsilon_z^e - \varepsilon_0 = \frac{\text{d}}{\text{d}r}\left(\varepsilon_\theta^e r\right). \tag{5.73}$$

Using the generalized Hooke's law, the Tresca yield criterion (Equation 5.72) in Equation (5.73), and the equilibrium equation (Equation 3.16), the following differential equation is obtained:

$$\frac{r^2}{E}\frac{d^2\sigma_r}{dr^2}+\frac{3r}{E}\frac{d\sigma_r}{dr}-\frac{1-2\nu}{E}\sigma_r=\frac{1}{E}k_1\sigma_{eq}+k_1\frac{\nu}{E}r\frac{d\sigma_{eq}}{dr}$$

$$+\alpha T_a+\alpha(T_b-T_a)\frac{\ln\left(\frac{r}{a}\right)}{\ln\left(\frac{b}{a}\right)}-\alpha\frac{(T_b-T_a)}{\ln\left(\frac{b}{a}\right)}-\varepsilon_0. \tag{5.74}$$

The solution of Equation (5.74) provides:

$$\sigma_r=C_3r^{-1+\sqrt{2(1-\nu)}}+C_4r^{-1-\sqrt{2(1-\nu)}}+\frac{k_1\sigma_Y}{(2\nu-1)}+\frac{k_1Kr^{-1+\sqrt{2(1-\nu)}}\left\{1-\nu+\nu\sqrt{2(1-\nu)}\right\}}{2\sqrt{2(1-\nu)}}\int_c^r r_1^{-\sqrt{2(1-\nu)}}\left(\varepsilon_{eq}^p\right)^n dr_1$$

$$-\frac{k_1Kr^{-1-\sqrt{2(1-\nu)}}\left\{1-\nu-\nu\sqrt{2(1-\nu)}\right\}}{2\sqrt{2(1-\nu)}}\int_c^r r_1^{\sqrt{2(1-\nu)}}\left(\varepsilon_{eq}^p\right)^n dr_1+\frac{E\alpha T_a}{(2\nu-1)}+\frac{E\alpha(T_b-T_a)}{(2\nu-1)}\frac{\ln\left(\frac{r}{a}\right)}{\ln\left(\frac{b}{a}\right)}$$

$$-\frac{2E\alpha(T_b-T_a)}{(2\nu-1)^2\ln\left(\frac{b}{a}\right)}-\frac{E\alpha(T_b-T_a)}{(2\nu-1)\ln\left(\frac{b}{a}\right)}-\frac{E\varepsilon_0}{(2\nu-1)}, \tag{5.75}$$

Using Equation (5.75), in equilibrium equation (Equation 3.16), the hoop stress is obtained as:

$$\sigma_\theta=C_3\sqrt{2(1-\nu)}r^{\sqrt{2(1-\nu)}-1}-C_4\sqrt{2(1-\nu)}r^{-\sqrt{2(1-\nu)}-1}+\frac{k_1\sigma_Y}{(2\nu-1)}+\frac{E\alpha T_a}{(2\nu-1)}+\frac{E\alpha(T_b-T_a)}{(2\nu-1)}\frac{\ln\left(\frac{r}{a}\right)}{\ln\left(\frac{b}{a}\right)}$$

$$-\frac{2E\alpha(T_b-T_a)}{(2\nu-1)^2\ln\left(\frac{b}{a}\right)}-\frac{E\varepsilon_0}{(2\nu-1)}+\frac{k_1K\sqrt{2(1-\nu)}r^{\sqrt{2(1-\nu)}-1}\left\{1-\nu+\nu\sqrt{2(1-\nu)}\right\}}{2\sqrt{2(1-\nu)}}\int_c^r r_1^{-\sqrt{2(1-\nu)}}\left(\varepsilon_{eq}^p\right)^n dr_1$$

$$+\frac{k_1K\sqrt{2(1-\nu)}r^{-\sqrt{2(1-\nu)}-1}\left\{1-\nu-\nu\sqrt{2(1-\nu)}\right\}}{2\sqrt{2(1-\nu)}}\int_c^r r_1^{\sqrt{2(1-\nu)}}\left(\varepsilon_{eq}^p\right)^n dr_1+\nu k_1K\left(\varepsilon_{eq}^p\right)^n. \tag{5.76}$$

Knowing the expression for σ_r from Equation (5.75), the expression for axial stress σ_z can be obtained from Equation (5.72).

The radial plastic strain field is given by:

$$\varepsilon_r^p = -\varepsilon_z^p = \varepsilon_z^e - \varepsilon_0 \quad , (5.77)$$

Using ε_z^e from the generalized Hooke's law in Equation (5.77) and then substituting the corresponding expressions of stresses from plastic zone II, the plastic part of radial strain is obtained as:

$$\begin{aligned}\varepsilon_r^p = -\varepsilon_0^p = {} & \frac{C_3}{E} r^{-1+\sqrt{2(1-\nu)}}\left\{1-\nu-\nu\sqrt{2(1-\nu)}\right\}+\frac{C_4}{E} r^{-1-\sqrt{2(1-\nu)}}\left\{1-\nu+\nu\sqrt{2(1-\nu)}\right\} \\ & +\frac{k_1 K r^{-1+\sqrt{2(1-\nu)}}\left\{1-\nu-\nu\sqrt{2(1-\nu)}\right\}\left\{1-\nu+\nu\sqrt{2(1-\nu)}\right\}}{2\sqrt{2(1-\nu)}E}\int_c^r r_1^{-\sqrt{2(1-\nu)}}\left(\varepsilon_{eq}^p\right)^n dr_1 \\ & -\frac{k_1 K r^{-1-\sqrt{2(1-\nu)}}\left\{1-\nu+\nu\sqrt{2(1-\nu)}\right\}\left\{1-\nu-\nu\sqrt{2(1-\nu)}\right\}}{2\sqrt{2(1-\nu)}E}\int_c^r r_1^{\sqrt{2(1-\nu)}}\left(\varepsilon_{eq}^p\right)^n dr_1 \\ & +\left(1-\nu^2\right)\frac{k_1 K}{E}\left(\varepsilon_{eq}^p\right)^n+\frac{\alpha\left(T_b-T_a\right)}{\ln\left(\frac{b}{a}\right)}\frac{(1+\nu)}{(2\nu-1)},\end{aligned} \quad (5.78)$$

where ε_{eq}^p is given by:

$$\varepsilon_{eq}^p = \frac{2}{\sqrt{3}}\varepsilon_r^p . \quad (5.79)$$

Elastic zone, $d \le r \le b$:

Using the generalized Hooke's law, strain compatibility (Equation 5.6), stress equilibrium (Equation 3.16) along with the boundary conditions $\left(\sigma_z-\sigma_r\right)\big|_{r=d} = k_1\sigma_Y$ and vanishing of radial stress at the outer radius, the stresses in the elastic zone are given by:

$$\begin{aligned}\sigma_r = {} & \frac{E\alpha}{2(1-\nu)}\frac{\left(T_b-T_a\right)}{\ln\left(\frac{b}{a}\right)}\left[\ln\left(\frac{b}{r}\right)-\left\{\frac{d^2}{b^2+d^2(2\nu-1)}\right\}\left\{\ln\left(\frac{b}{a}\right)(2\nu-1)-\nu-\ln\left(\frac{d}{a}\right)\right\}\left(1-\frac{b^2}{r^2}\right)\right] \\ & +\left\{\frac{d^2}{b^2+d^2(2\nu-1)}\right\}k_1\sigma_Y\left(1-\frac{b^2}{r^2}\right)+\left\{\frac{d^2}{b^2+d^2(2\nu-1)}\right\}E\alpha T_a\left(1-\frac{b^2}{r^2}\right) \\ & -\left\{\frac{d^2}{b^2+d^2(2\nu-1)}\right\}E\varepsilon_0\left(1-\frac{b^2}{r^2}\right),\end{aligned} \quad (5.80)$$

$$\sigma_\theta = \frac{E\alpha}{2(1-\nu)}\frac{(T_b - T_a)}{\ln\left(\frac{b}{a}\right)}\left[\ln\left(\frac{b}{r}\right) - 1 - \left\{\frac{d^2}{b^2 + d^2(2\nu-1)}\right\}\left\{\ln\left(\frac{b}{a}\right)(2\nu-1) - \nu - \ln\left(\frac{d}{a}\right)\right\}\left(1+\frac{b^2}{r^2}\right)\right]$$

$$+\left\{\frac{d^2}{b^2+d^2(2\nu-1)}\right\}k_1\sigma_Y\left(1+\frac{b^2}{r^2}\right)+\left\{\frac{d^2}{b^2+d^2(2\nu-1)}\right\}E\alpha T_a\left(1+\frac{b^2}{r^2}\right)$$

$$-\left\{\frac{d^2}{b^2+d^2(2\nu-1)}\right\}E\varepsilon_0\left(1+\frac{b^2}{r^2}\right), \tag{5.81}$$

$$\sigma_z = \frac{E\alpha\nu}{2(1-\nu)}\frac{(T_b - T_a)}{\ln\left(\frac{b}{a}\right)}\left[2\ln\left(\frac{b}{r}\right) - 1 - \left\{\frac{2d^2}{b^2 + d^2(2\nu-1)}\right\}\left\{\ln\left(\frac{b}{a}\right)(2\nu-1) - \nu - \ln\left(\frac{d}{a}\right)\right\}\right]$$

$$+\left\{\frac{2\nu d^2}{b^2+d^2(2\nu-1)}\right\}k_1\sigma_Y+\left\{\frac{d^2-b^2}{b^2+d^2(2\nu-1)}\right\}E\alpha T_a-\left\{\frac{d^2-b^2}{b^2+d^2(2\nu-1)}\right\}E\varepsilon_0 \tag{5.82}$$

$$-E\alpha(T_b-T_a)\frac{\ln\left(\frac{r}{a}\right)}{\ln\left(\frac{b}{a}\right)}.$$

5.4.4 The Second Stage of Elastic–Plastic Deformation

During the second stage of elastic–plastic deformation, the cylinder wall consists of two more plastic zones as shown in Figure 5.5b. The outermost plastic zone—plastic zone III: $f \le r \le b$ evolves according to the Tresca yield criterion given by:

$$\sigma_\theta - \sigma_r = -k_1\sigma_{eq} \quad \text{or} \quad \sigma_z - \sigma_r = -k_1\sigma_{eq} \tag{5.83}$$

Another plastic zone—plastic zone IV: $e \le r \le f$ emerges simultaneously along with plastic zone III, as per the Tresca yield criterion:

$$\sigma_\theta - \sigma_r = -k_1\sigma_{eq} \tag{5.84}$$

The stress and strain expressions for plastic zones I and II during the second stage of elastic–plastic deformation are same as in the case of first stage of elastic–plastic deformation. However, the expressions constants—C_1, C_2, C_3, and C_4 will change due to the change of boundary conditions. The stresses in the elastic zone and the stresses and plastic strains in plastic zone III and plastic zone IV are obtained as follows:

Elastic zone, $d \le r \le e$:

Using the generalized Hooke's law, strain compatibility (Equation 5.6), stress equilibrium (Equation 3.16), and the boundary conditions: $(\sigma_z - \sigma_r)\big|_{r=d} = k_1\sigma_Y$, and $(\sigma_\theta - \sigma_r)\big|_{r=e} = -k_1\sigma_Y$, the stresses in the elastic zone are given by:

$$\sigma_r = \frac{E\alpha}{(2\nu-1)}T_a + \frac{E\alpha(T_b - T_a)}{2(1-\nu)\ln\left(\frac{b}{a}\right)}\left[\frac{1}{(2\nu-1)}\left\{\ln\left(\frac{d}{a}\right) + \frac{1}{2} - \frac{e^2}{2d^2}\right\} - \ln\left(\frac{r}{a}\right) + \frac{1}{2} - \frac{e^2}{2r^2}\right]$$
$$+ \frac{k_1\sigma_Y}{2\nu-1}\left(1 + \frac{e^2}{2d^2}\right) + \frac{e^2}{2r^2}k_1\sigma_Y + \frac{E\varepsilon_0}{(1-2\nu)}, \quad (5.85)$$

$$\sigma_\theta = \frac{E\alpha}{(2\nu-1)}T_a + \frac{E\alpha(T_b - T_a)}{2(1-\nu)\ln\left(\frac{b}{a}\right)}\left[\frac{1}{(2\nu-1)}\left\{\ln\left(\frac{d}{a}\right) + \frac{1}{2} - \frac{e^2}{2d^2}\right\} - \ln\left(\frac{r}{a}\right) - \frac{1}{2} + \frac{e^2}{2r^2}\right]$$
$$+ \frac{k_1\sigma_Y}{2\nu-1}\left(1 + \frac{e^2}{2d^2}\right) - \frac{e^2}{2r^2}k_1\sigma_Y + \frac{E\varepsilon_0}{(1-2\nu)}, \quad (5.86)$$

$$\sigma_z = \frac{E\alpha}{(2\nu-1)}T_a + \frac{E\alpha(T_b - T_a)}{2(1-\nu)\ln\left(\frac{b}{a}\right)}\left[\frac{2\nu}{(2\nu-1)}\left\{\ln\left(\frac{d}{a}\right) + \frac{1}{2} - \frac{e^2}{2d^2}\right\} - 2\ln\left(\frac{r}{a}\right)\right]$$
$$+ \frac{2\nu}{2\nu-1}k_1\sigma_Y\left(1 + \frac{e^2}{2d^2}\right) + \frac{E\varepsilon_0}{(1-2\nu)}. \quad (5.87)$$

Plastic zone III, $f \le r \le b$:

Following a similar procedure as discussed for plastic zone I in Subsection 5.4.3, the stresses and the plastic strain fields in plastic zone III are obtained. The stress components are given by:

$$\sigma_r = -k_1\sigma_Y \ln r - k_1 K \int_f^r \frac{\left(\varepsilon_{eq}^p\right)^n}{r_1} dr_1 + C_5, \quad (5.88)$$

$$\sigma_\theta = \sigma_z = -k_1\sigma_Y(1+\ln r) - k_1K\left(\varepsilon_{eq}^p\right)^n - k_1K\int_f^r \frac{\left(\varepsilon_{eq}^p\right)^n}{r_1}dr_1 + C_5, \tag{5.89}$$

where C_5 is an integration constant. The expressions for plastic strain components are given as:

$$\varepsilon_\theta^p = \frac{1-2\nu}{E}\left[\frac{1}{2}(C_5 - k_1\sigma_Y \ln r) - \frac{2k_1K}{r^2}\int_f^r r_1\left(\varepsilon_{eq}^p\right)^n dr_1 + k_1K\int_f^r \frac{\left(\varepsilon_{eq}^p\right)^n}{r_1}dr_1 - \frac{3k_1K}{r^2}\int_f^r\left\{r_2\int_f^{r_2}\frac{\left(\varepsilon_{eq}^p\right)^n}{r_1}dr_1\right\}dr_2\right]$$
$$+\frac{3-2\nu}{4E}k_1\sigma_Y + \frac{1-\nu}{E}k_1K\left(\varepsilon_{eq}^p\right)^n + \frac{1}{2}\alpha T_a + \frac{\alpha(T_b - T_a)}{2\ln\left(\frac{b}{a}\right)}\left\{\ln\left(\frac{r}{a}\right) - \frac{3}{2}\right\} - \frac{1}{2}\varepsilon_0 + \frac{C_6}{r^2}, \tag{5.90}$$

$$\varepsilon_r^p = \frac{1-2\nu}{E}\left[\frac{1}{2}(C_5 - k_1\sigma_Y \ln r) - 2k_1K\int_f^r \frac{\left(\varepsilon_{eq}^p\right)^n}{r_1}dr_1 + \frac{2k_1K}{r^2}\int_f^r r_1\left(\varepsilon_{eq}^p\right)^n dr_1 + \frac{3k_1K}{r^2}\int_f^r\left\{r_2\int_f^{r_2}\frac{\left(\varepsilon_{eq}^p\right)^n}{r_1}dr_1\right\}dr_2\right]$$
$$-\frac{7-6\nu}{4E}k_1\sigma_Y - \frac{2(1-\nu)}{E}k_1K\left(\varepsilon_{eq}^p\right)^n + \frac{1}{2}\alpha T_a + \frac{\alpha(T_b - T_a)}{2\ln\left(\frac{b}{a}\right)}\left\{\ln\left(\frac{r}{a}\right) + \frac{3}{2}\right\} - \frac{1}{2}\varepsilon_0 - \frac{C_6}{r^2}, \tag{5.91}$$

$$\varepsilon_z^p = \varepsilon_0 - \frac{1-2\nu}{E}\left\{-k_1\sigma_Y \ln r + C_5 - k_1K\int_f^r \frac{\left(\varepsilon_{eq}^p\right)^n}{r_1}dr_1\right\} + \frac{1-\nu}{E}k_1\sigma_Y + \frac{1-\nu}{E}k_1K\left(\varepsilon_{eq}^p\right)^n$$
$$-\alpha T_a - \alpha(T_b - T_a)\frac{\ln\left(\frac{r}{a}\right)}{\ln\left(\frac{b}{a}\right)}, \tag{5.92}$$

where C_6 is integration constant.

Plastic zone IV, $e \le r \le f$:

In plastic zone IV, the Tresca yield criterion takes the following form:

$$\sigma_\theta - \sigma_r = -k_1\sigma_{eq}. \tag{5.93}$$

Proceeding in a similar manner as described in Subsection 5.4.3 plastic zone I, the radial stress in the plastic zone IV is obtained as:

$$\sigma_r = \frac{Ea}{(2\nu-1)}T_a + \frac{Ea(T_b - T_a)}{2(1-\nu)\ln\left(\frac{b}{a}\right)}\left[\frac{1}{(2\nu-1)}\left\{\ln\left(\frac{d}{a}\right)+\frac{1}{2}-\frac{e^2}{2d^2}\right\}-\ln\left(\frac{e}{a}\right)\right]$$

$$+\left\{\frac{1}{2\nu-1}\left(1+\frac{e^2}{2d^2}\right)+\frac{1}{2}-\ln\left(\frac{r}{e}\right)\right\}k_1\sigma_Y + \frac{E\varepsilon_0}{(1-2\nu)} - k_1K\int_e^r \frac{\left(\varepsilon_{\mathrm{eq}}^p\right)^n}{r_1}dr_1. \tag{5.94}$$

Knowing the expression for σ_r, the expression for σ_θ can be obtained from Equation (5.93). In view of the associated flow rule based on the Tresca yield criterion, $d\varepsilon_\theta^p = -d\varepsilon_r^p$, $d\varepsilon_z^p = 0$. This shows that the axial strain is composed of only elastic parts. Thus, Equation (5.4) is also valid in plastic zone IV. Using the expressions of σ_r and σ_θ obtained for plastic zone IV in Equation (5.4), provides the solution for axial stress as:

$$\sigma_z = \frac{Ea}{(2\nu-1)}T_a + \frac{Ea(T_b - T_a)}{2(1-\nu)\ln\left(\frac{b}{a}\right)}\left[\frac{2\nu}{(2\nu-1)}\left\{\ln\left(\frac{d}{a}\right)+\frac{1}{2}-\frac{e^2}{2d^2}\right\}+2\nu\ln\left(\frac{r}{e}\right)-2\ln\left(\frac{r}{a}\right)\right]$$

$$+\left\{\frac{2\nu}{2\nu-1}\left(1+\frac{e^2}{2d^2}\right)-2\nu\ln\left(\frac{r}{e}\right)\right\}k_1\sigma_Y + \frac{E\varepsilon_0}{(1-2\nu)} - 2\nu k_1K\int_e^r \frac{\left(\varepsilon_{\mathrm{eq}}^p\right)^n}{r_1}dr_1 - \nu k_1K\left(\varepsilon_{\mathrm{eq}}^p\right)^n. \tag{5.95}$$

Using plastic incompressibility and strain–displacement relations along with the generalized Hooke's law, the following differential equation is obtained:

$$\frac{d}{dr}(ur) = \frac{1}{E}(1-\nu)\frac{d}{dr}\left(r^2\sigma_r\right) - \frac{2\nu}{E}r\sigma_z + 2aT_ar + 2a(T_b - T_a)r\frac{\ln\left(\frac{r}{a}\right)}{\ln\left(\frac{b}{a}\right)}. \tag{5.96}$$

Solution of Equation (5.96) provides the value of radial displacement, u. Once u is known, the total hoop or radial strain component can be obtained by using the strain–displacement relations. The plastic part of the hoop or radial strain is obtained by subtracting its elastic part from the total strain. This gives:

$$\varepsilon_\theta^p = -\varepsilon_r^p = \frac{a(T_b - T_a)}{2(1-\nu)\ln\left(\frac{b}{a}\right)}\left(\nu^2-1\right)+\left(1-\nu^2\right)\frac{k_1\sigma_Y}{E} + \frac{4\nu^2}{Er^2}k_1K\int_e^r\left\{r_2\int_e^{r_2}\frac{\left(\varepsilon_{\mathrm{eq}}^p\right)^n}{r_1}dr_1\right\}dr_2$$

$$+\frac{2\nu^2}{Er^2}k_1K\int_e^r r_1\left(\varepsilon_{\mathrm{eq}}^p\right)^n dr_1 + \left(1-\nu^2\right)\frac{k_1K}{E}\left(\varepsilon_{\mathrm{eq}}^p\right)^n - \frac{2\nu^2k_1K}{E}\int_e^r\frac{\left(\varepsilon_{\mathrm{eq}}^p\right)^n}{r_1}dr_1 + \frac{C_7}{r^2}, \tag{5.97}$$

where C_7 is an integration constant.

5.4.5 Evaluation of Different Integration Constants and Constant Axial Strain, ε_0

The different constants—C_1, C_2, C_3, C_4, C_5, C_6, C_7 involved in the stress and strain equations of different zones in the first as well as in the second stage of elastic–plastic deformation can be obtained by invoking different boundary conditions as provided in Appendix A. The constant axial strain ε_0 in the first and second stage of elastic–plastic deformation can be obtained by using the free end condition. The detailed evaluation of ε_0 for both the first and the second stage of elastic–plastic deformation is also provided in Appendix A.

5.4.6 Procedure for Calculating the Unknown Elastic–Plastic Interface Radii

The boundary radii *c* and *d* in the first stage of elastic–plastic deformation are unknown and need to be evaluated numerically. Also, in the second stage of elastic–plastic deformation, two more unknown boundary radii *e* and *f* need to be evaluated. The numerical estimation of the unknown boundary radii involves an iterative approach.

The initial estimates for *c* and *d* can be obtained from the non-hardening case ($K=0$) by using the boundary conditions of vanishing radial stress at the inner radius and $\sigma_\theta^{(\text{plastic zone II})} = \sigma_z^{(\text{plastic zone II})}$ at $r=c$ and solving them numerically. One can use Newton's method or any optimization method for this purpose. The initial guess value of ε_{eq}^p is taken as zero everywhere in the plastic zones I and II. With these values of *c* and *d*, the values of ε_θ^p, ε_r^p and ε_0^p are updated from Equations (5.68–5.70) for plastic zone I. Equation (5.71) provides the updated values of ε_{eq}^p at any radial position in plastic zone I. Similarly, using Equation (5.78), the values of ε_r^p at different radial positions in plastic zone II are updated. The updated values of ε_{eq}^p at different radial positions in plastic zone II are obtained from Equation (5.79). The values of ε_{eq}^p are updated further for fixed *c* and *d* in both of the plastic zones and the procedure is repeated till the convergence in ε_{eq}^p is achieved. The integral terms involved in the expressions can be evaluated numerically by using Gauss quadrature. Now, using the converged values of ε_{eq}^p, the boundary conditions of vanishing radial stress at the inner radius and $\sigma_\theta^{(\text{plastic zone II})} = \sigma_z^{(\text{plastic zone II})}$ at $r=c$ are solved again to get the new estimates of *c* and *d*. If these new estimated values of *c* and *d* are same as the previously estimated values of *c* and *d*, the procedure is stopped, otherwise the whole procedure is repeated till the convergence for *c* and *d* is achieved. A similar numerical procedure is followed in order to estimate the unknown boundary radii *c*, *d*, *e*, and *f* appearing in the second stage of elastic–plastic deformation.

5.4.7 Typical Results of Generalized Plane–Strain Model of Thermal Autofrettage

A typical SS304 cylinder with radial dimensions $a=10$ mm and $b=30$ mm is subjected to thermal autofrettage. The material properties of SS304 are taken as follows: Young's modulus of elasticity $E=193$ GPa, yield stress $\sigma_Y=205$ MPa, Poisson's ratio $\nu=0.3$, and coefficient of thermal expansion $\alpha=17.2\times10^{-6}$/°C. The strain hardening parameters K and n are taken as 1425 MPa and 0.70, respectively. A temperature difference of 130 °C is considered for achieving thermal autofrettage in the cylinder. The inner wall of the cylinder is maintained at 25 °C. At 130 °C, the cylinder undergoes the first stage of elastic–plastic deformation. The plastic zone I propagates outward to a radial position $c=12.1523$ mm, and a consecutive plastic zone II propagates up to a radial position $d=13.2690$ mm. The distribution of elastic–plastic stresses under the action of 130 °C radial temperature difference and the residual stresses distribution after vanishing the temperature difference by cooling along the thickness direction are shown in Figure 5.7a,b, respectively.

It is observed from Figure 5.7a that the magnitudes of hoop and axial stresses are the same in plastic zone I. However, in a significant portion of the wall, the axial stress is more than the hoop stress. Along the positive radial direction the hoop and axial stresses change from tensile to compressive. The maximum tensile stress in the cylinder during loading exists at the radius of interface between the plastic zone II and elastic zone. The magnitudes of radial stresses throughout the wall of the cylinder are smaller compared to the magnitudes of hoop and axial stresses. The radial stresses

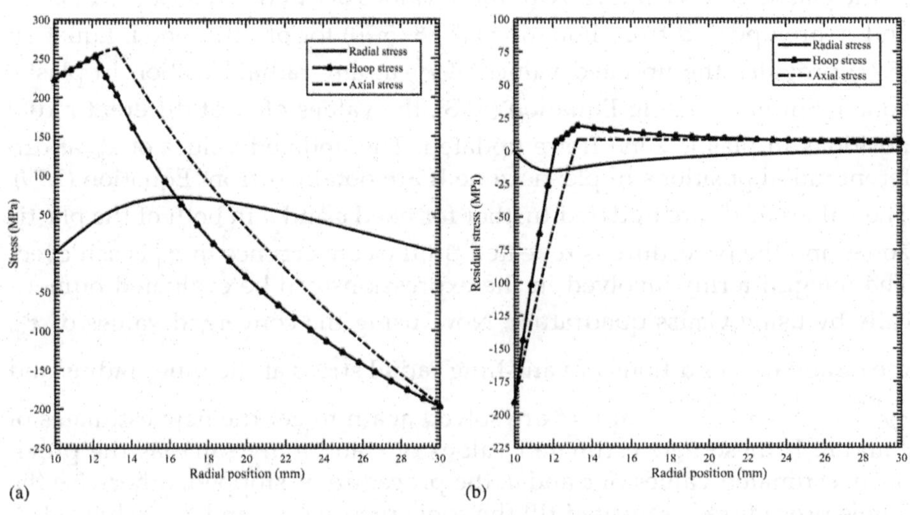

FIGURE 5.7
Distribution of (a) stresses after loading and (b) residual stresses after unloading in thermal autofrettage of SS304 cylinder with a temperature gradient $(T_b-T_a)=130$ °C.

are always tensile in nature. The plastic strains in the plastic zones I and II as a function of radial position are evaluated using the relevant equations of Subsection 5.4.3, and it is observed that the maximum plastic strain is of the order of 10^{-4}. This justifies the assumption of small strains. On the removal of the temperature difference, the trend of the residual stresses generated in the wall of the cylinder gets reversed to the elastic–plastic stresses as shown in Figure 5.7b. The maximum magnitude of residual hoop and axial stress generated at the inner surface of the cylinder is −190.9 MPa ($-0.93\sigma_Y$), which is significantly large.

5.5 An Assessment on the Validity of the Analytical Models through a Numerical Comparison with Three-Dimensional (3D) Finite Element Method (FEM) Analysis

The plane–stress and the generalized plane–strain analytical model of thermal autofrettage were presented in Sections 5.3 and 5.4, respectively. It is stated that the plane–stress model is applicable for very short cylinders or thin disks and the generalized plane–strain model is applicable for long cylinders in general. However, from the analysis it is not clear which model is valid for a particular length of the cylinder. Kamal et al. (2017) carried out a FEM-based study to assess the applicability of the analytical models to different length to wall thickness ratios of a cylinder. A 3D FEM analysis was carried out for a thick-walled aluminum cylinder of radial dimensions $a=10$ mm and $b=20$ mm. The analyses are carried out for different lengths of the cylinder for a fixed radial temperature difference of 75 °C. The commercial package ABAQUS® was used. The material properties of aluminum are as provided in Section 5.3.

5.5.1 Boundary Conditions and Mesh Generation

This section presents the thermal and mechanical boundary conditions used in the FEM model. The inner wall is assigned a temperature $T_a=25$ °C and that of the outer surface is assigned a temperature $T_b=100$ °C. A mechanical constraint of zero circumferential displacement is imposed. This ensures that the cylinder displaces uniformly and is free to expand axially. A schematic of the cylinder with the thermal and displacement boundary conditions is shown in Figure 5.8. Kamal et al. 2017 used an eight-node continuum C3D8T thermally coupled brick, trilinear displacement and temperature element available in ABAQUS®. The optimum mesh was decided on the basis of a mesh sensitivity analysis; the cylinder was divided into 38400 elements comprising 24 divisions in the radial direction, 32 divisions in the circumferential direction and 50 divisions in the axial direction. It needs to be

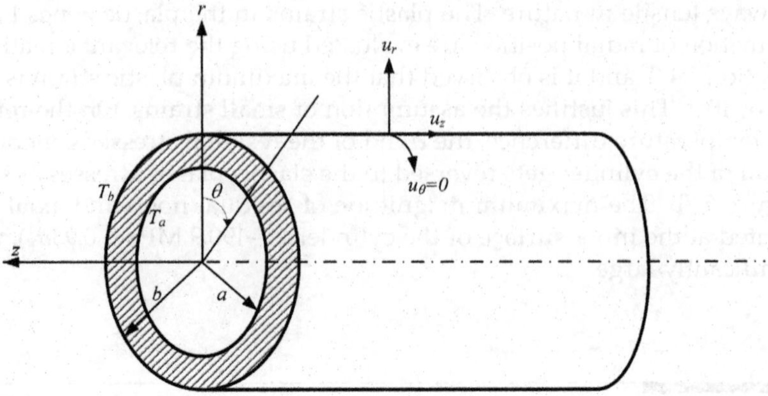

FIGURE 5.8
Schematic of the cylinder with the prescribed boundary conditions (Kamal et al., 2017).

mentioned that the analytical modes used the Tresca yield criterion, while the FEM model used the von Mises yield criterion.

5.5.2 Comparison of 3D Finite Element Elastic–Plastic Thermal Stresses with the Plane–Stress and Generalized Plane–Strain Models for Different Length to Wall Thickness Ratios of the Cylinder

Kamal et al. (2017) carried out 3D FEM simulations for different length to wall thickness ratios of the cylinder and calculated the variations from the results obtained with analytical plane–stress and generalized plane–strain model on the basis of the L_2 norm displayed in Table 5.1. As observed, the L_2 norm of error in stresses between 3D FEM and generalized plane–strain are reasonably small when the length to wall thickness ratio of the cylinder is greater than or equal to 6. Thus, the generalized plane–strain model provides a realistic solution for $L/(b-a) \geq 6$. Further when $L/(b-a) \leq 1$, the L_2 norm of error between 3D FEM and plane–stress becomes smaller. This shows that the plane–stress analytical model is valid for a length to wall thickness ratio of less than one, i.e., for $L/(b-a) \leq 1$. The comparison of 3D finite element solutions with the generalized plane–strain (GPS) model for $L/(b-a) = 10$ is shown in Figure 5.9a. Figure 5.9b shows the comparison of 3D finite element results with the plane–stress model for $L/(b-a) = 0.5$ in the aluminum cylinder. It is observed that the 3D FEM stresses are smoother at the interface radii and slightly deviate from the analytical solutions in the plastic zones. This may be due to the incorporation of different yield criteria in FEM modeling and analytical modeling. As mentioned, the plane–stress and plane–strain models used the Tresca criterion, whilst the FEM model used the von Mises criterion.

TABLE 5.1

Comparison of Stresses between 3D FEM and Analytical Models (Kamal et al., 2017)

	L_2 norm of error in stresses between FEM and plane–stress (MPa)			L_2 norm of error in stresses between FEM and generalized plane–strain (MPa)		
$L/(b-a)$	σ_r	σ_θ	σ_z	σ_r	σ_θ	σ_z
10	5.5849	26.7094	134.5956	**2.4418**	**4.3992**	**2.5369**
8	5.5148	27.3775	136.0487	**2.5578**	**4.7597**	**2.7440**
6	5.9687	29.4939	137.3141	**2.4754**	**7.6960**	**4.1740**
5	7.4358	28.3009	133.9557	4.9296	8.7777	11.7016
4	7.8224	40.6167	112.6305	3.8787	31.2258	24.3158
2	4.6443	36.8936	30.4968	4.9293	43.5744	110.0771
1	**3.4278**	**5.1446**	**3.6140**	8.2861	33.3531	135.7536
0.5	**2.7080**	**3.7979**	**1.3881**	7.1190	33.2992	136.0458

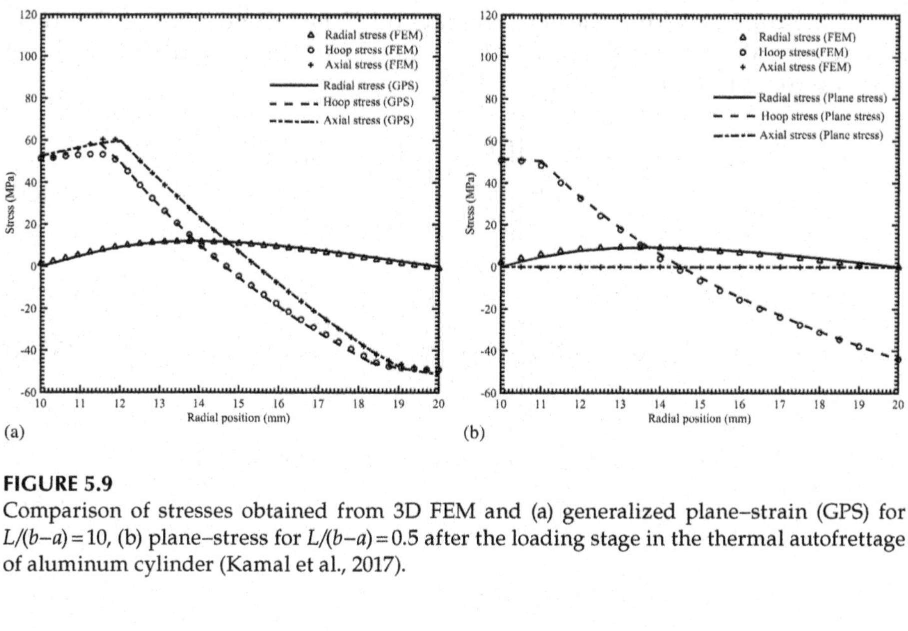

(a) (b)

FIGURE 5.9

Comparison of stresses obtained from 3D FEM and (a) generalized plane–strain (GPS) for $L/(b-a) = 10$, (b) plane–stress for $L/(b-a) = 0.5$ after the loading stage in the thermal autofrettage of aluminum cylinder (Kamal et al., 2017).

5.6 Thermal Autofrettage of a Thick-Walled Sphere

This section presents the closed-form model of the thermal autofrettage of a sphere. The stress analysis due to the elastic deformation of a thick-walled sphere subjected to a radial temperature gradient is first presented. This is followed by the stress analysis due to elastic–plastic deformation of the sphere due to thermal autofrettage.

5.6.1 Elastic Analysis of a Thick-Walled Sphere Subjected to a Radial Temperature Gradient

The solution for the distribution of stresses due to elastic deformation in a thick-walled sphere subjected to a radial temperature gradient are as follows (Chakrabarty, 2006):

$$\sigma_r = A + B\frac{b^3}{r^3} - \beta\frac{\frac{b}{r}}{\left(\frac{b}{a}-1\right)}, \tag{5.98}$$

$$\sigma_\theta = A - B\frac{b^3}{2r^3} - \beta\frac{\frac{b}{2r}}{\left(\frac{b}{a}-1\right)}. \tag{5.99}$$

where $\beta = \alpha(T_a - T_b)/(1-\nu)$, and σ_r and σ_θ are the radial and circumferential stress components, respectively, in the sphere. The constants A and B can be obtained by invoking the zero traction boundary conditions at the inner and outer wall surfaces, i.e., $(\sigma_r)_{r=a} = 0$ and $(\sigma_r)_{r=b} = 0$. Thus, the distributions of stresses are obtained as follows:

$$\sigma_r = -\beta E\left\{\left(\frac{\frac{b}{r}-1}{\frac{b}{a}-1}\right) - \left(\frac{\frac{b^3}{r^3}-1}{\frac{b^3}{a^3}-1}\right)\right\}, \tag{5.100}$$

$$\sigma_\theta = -\beta E\left\{\left(\frac{\frac{b}{2r}-1}{\frac{b}{a}-1}\right) + \left(\frac{\frac{b^3}{2r^3}+1}{\frac{b^3}{a^3}-1}\right)\right\}. \tag{5.101}$$

5.6.2 Elastic–Plastic Analysis Due to Thermal Autofrettage

For a sphere, both the von Mises and Tresca yield criteria reduce to an identical form as follows:

$$\sigma_\theta - \sigma_r = k_1\sigma_Y, \tag{5.102}$$

where σ_Y is the yield stress in uniaxial tension or compression, and k_1 is a constant whose value is 1 in case of $T_b > T_a$, and –1 case of $T_a > T_b$. Using Equations (5.100) and (5.101) provides:

$$\sigma_\theta - \sigma_r = \beta E\left\{\left(\frac{\frac{3b}{2r}}{\frac{b^3}{a^3}-1}\right) - \left(\frac{\frac{b}{2r}}{\frac{b}{a}-1}\right)\right\}. \tag{5.103}$$

which shows that the magnitude of the stress difference is at the maximum at the inner radius $r=a$. Thus, yielding initiates first at the inner radius. The minimum temperature difference required to cause the initial yielding at the inner radius is obtained by evaluating $(\sigma_\theta-\sigma_r)$ at $r=a$ using Equation (5.103) and then substituting it in the Tresca yield criterion (Equation 5.102). Hence,

$$T_b - T_a = -k_1\sigma_Y \frac{(1-\nu)}{Ea}\left(\frac{1+\dfrac{a}{b}+\dfrac{a^2}{b^2}}{1+\dfrac{a}{2b}}\right). \tag{5.104}$$

5.6.2.1 Stress Distribution after Loading

In the elastic zone $c \le r \le b$, the radial and hoop stresses are given by Lame's equations (Equations 3.8 and 3.9). Applying the boundary conditions $(\sigma_r)_{r=b}=0$, and $(\sigma_\theta-\sigma_r)_{r=c}=k_1\sigma_Y$, the radial and hoop stress distributions in the elastic zone are given by:

$$\sigma_r = \left\{\frac{2}{3}k_1\sigma_Y - \beta E\frac{\dfrac{b}{c}}{3\left(\dfrac{b}{a}-1\right)}\right\}\left(\frac{c^3}{b^3}-\frac{c^3}{r^3}\right)+\beta E\frac{\left(1-\dfrac{b}{r}\right)}{\left(\dfrac{b}{a}-1\right)}, \tag{5.105}$$

$$\sigma_\theta = \left\{\frac{2}{3}k_1\sigma_Y - \beta E\frac{\dfrac{b}{c}}{3\left(\dfrac{b}{a}-1\right)}\right\}\left(\frac{c^3}{b^3}+\frac{c^3}{r^3}\right)+\beta E\frac{\left(1-\dfrac{b}{2r}\right)}{\left(\dfrac{b}{a}-1\right)}. \tag{5.106}$$

In the plastic zone, the solution for the stresses is derived from the stress equilibrium equation that remains satisfied through all stages of deformation. The equation of equilibrium in spherical coordinates is given by

$$\frac{2}{r}(\sigma_\theta-\sigma_r)=\frac{d\sigma_r}{dr}. \tag{5.107}$$

It is assumed that the material of the sphere does not strain harden. Considering σ_z to be the intermediate principal stress throughout the plastic zone, $a \le r \le c$ and applying Tresca yield criterion in the equilibrium equation (Eq. 5.107) and integrating provides the following:

$$\sigma_r = 2k_1\sigma_Y \ln r + C, \tag{5.108}$$

where C is a constant of integration and can be evaluated using the continuity of radial stress at $r=c$. Evaluating the radial stress component for elastic zone from Equation (5.105) and equating it with the plastic radial stress component from Equation (5.108) at $r=c$ leads to:

$$C=\left\{\frac{2}{3}k_1\sigma_Y-\frac{\beta Eb}{3c(\rho-1)}\right\}\left(\frac{c^3}{b^3}-1\right)+\frac{\beta E}{\rho-1}\left(1-\frac{b}{c}\right)-2k_1\sigma_Y\ln c, \tag{5.109}$$

Substituting C in Equation (5.108), the radial stress distribution in the plastic zone, $a\le r\le c$, is given by:

$$\sigma_r=\left\{\frac{2}{3}k_1\sigma_Y-\frac{\beta Eb}{3c(\rho-1)}\right\}\left(\frac{c^3}{b^3}-1\right)+\frac{\beta E}{\rho-1}\left(1-\frac{b}{c}\right)+k_1\sigma_Y\ln\frac{r^2}{c^2}, \tag{5.110}$$

From the Tresca yield criterion, the hoop stress distribution is given by:

$$\sigma_\theta=\left\{\frac{2}{3}k_1\sigma_Y-\frac{\beta Eb}{3c(\rho-1)}\right\}\left(\frac{c^3}{b^3}-1\right)+\frac{\beta E}{\rho-1}\left(1-\frac{b}{c}\right)+k_1\sigma_Y\left(1+\ln\frac{r^2}{c^2}\right). \tag{5.111}$$

The unknown radius of elastic–plastic interface, c can be obtained by using the boundary condition of zero traction condition at the inner radius, *i.e.*, $\sigma_r=-p$ at $r=a$. Thus, from Equation (5.110):

$$0=\left\{\frac{2}{3}k_1\sigma_Y-\frac{\beta Eb}{3c(\rho-1)}\right\}\left(\frac{c^3}{b^3}-1\right)+\frac{\beta E}{\rho-1}\left(1-\frac{b}{c}\right)+k_1\sigma_Y\ln\frac{a^2}{c^2}, \tag{5.112}$$

which can be numerically solved for the unknown radius, c.

5.6.2.2 Residual Stress Distribution after Unloading

The residual stress distribution after unloading is obtained by subtracting Equations (5.100) and (5.101) from (5.105) and (5.106), respectively, for the elastic zone. Similarly, the residual stress distribution in the plastic zone is obtained by subtracting Equations (5.100) and (5.101) from Equations (5.110) and (5.111), respectively.

5.6.3 Typical Result of the Thermal Autofrettage of a Sphere

This section presents the typical results of the thermal autofrettage of a sphere. The material is taken as being stainless steel SS304, having a Young's modulus of 193 GPa, a Poisson's ratio of 0.3, a yield strength of 205 MPa, and a coefficient of thermal expansion 17.2×10^{-6}. A thick-walled sphere with an inner radius of 10 mm and an outer radius of 35 mm is taken as the autofrettage specimen and is thermally autofrettaged with a radial temperature gradient of 80 °C across the thickness. The sphere yields at a temperature difference of 51.72 °C that is to be calculated using Equation (5.104). The stress distribution due to elastic–plastic deformation after loading is shown

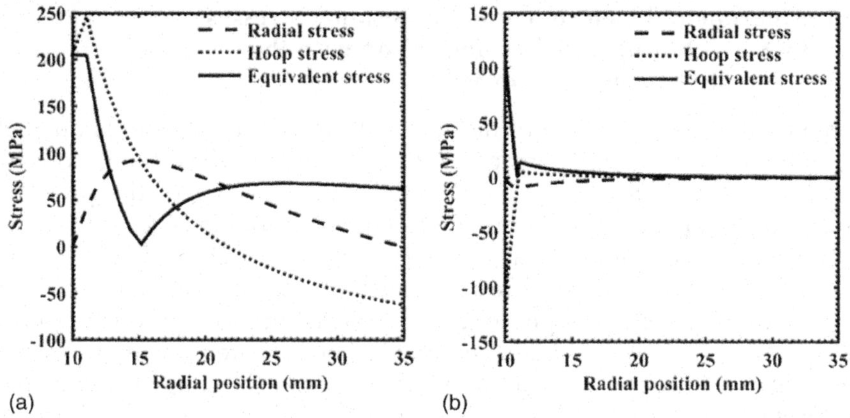

FIGURE 5.10
Distribution of (a) stresses after loading and (b) residual stresses after unloading in a thermally autofrettaged sphere.

in Figure 5.10a and residual stress distribution after unloading is shown in Figure 5.10b. The figures also show the equivalent stresses. The elastic–plastic interface radius calculated using Equation (5.112) is 11.06 mm.

5.7 Experimental Studies on Thermal Autofrettage

This section presents an experimental investigation of the thermal autofrettage process. The experimental determination of the residual stresses is necessary in order to get confidence in the analytical and numerical model of the process. Kamal et al. (2016) carried out rigorous experimental studies on the thermal autofrettage of thick-walled cylinders. An experimental setup has been developed by Kamal et al. (2016). They measured the residual stress field in the thermally autofrettaged cylinders experimentally using the Sachs boring technique. The measured residual stresses were compared with the generalized plane–strain analytical model presented in Section 5.4. The authors also established the generation of residual stresses in the thermally autofrettaged cylinders on the basis of the Vickers microhardness test. Moreover, the presence of residual stresses was also confirmed by measuring the opening angle after cutting a thermally autofrettaged cylinder along a radius and comparing it with the theoretical opening angle based on the residual hoop stress distribution predicted by the generalized plane–strain model. Before discussing the different experimental verification of the residual stresses, the experimental setup for achieving thermal autofrettage is briefly described in Subsection 5.7.1.

5.7.1 Experimental Determination of Residual Stresses by Sachs Boring and Its Comparison with the Generalized Plane–Strain Model

Kamal et al. (2016) carried out thermal autofrettage experiments on SS304 and mild steel cylinders considering different geometrical dimensions and temperature differences. The construction and working of the process is explained as follows. The setup consisted of a ceramic jacketed heater, a cold-water storage tank, a water pump, a polyvinyl chloride (PVC) pipe, ceramic tube sections, a variable autotransformer, and ammeter and digital temperature controllers. A photograph of the assembled setup is presented in Figure 5.11. A well-insulated ceramic jacketed heater was employed for the heating of the outer wall of the cylinder to be autofrettaged. *K*-type thermocouples were attached on the outer and inner surfaces and connected to temperature controllers (and the display) in order to read the corresponding temperatures. Simultaneously, the inner surface was cooled by the continuous flow of cold water through the bore. The cold water was stored in a tank and was pumped through a PVC pipeline. At the inlet and outlet, the cylinder to be heated was connected to the PVC pipeline using ceramic tube. The ceramic tube is a good insulator of heat and electricity and prevents the heating of PVC pipe. The joint between the cylinder and the ceramic tube section was sealed by using a high temperature sealant. After achieving the desired temperature difference, the heater was switched off, but the water pump continued pumping the cold water through the bore till the cylinder was brought down to room temperature.

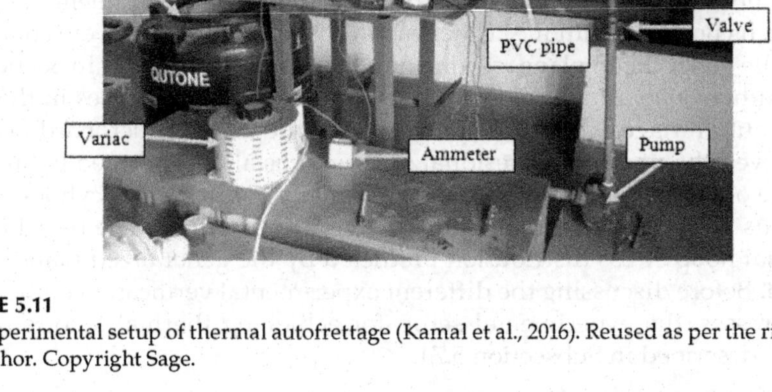

FIGURE 5.11
The experimental setup of thermal autofrettage (Kamal et al., 2016). Reused as per the rights of the author. Copyright Sage.

The material properties of SS304 and mild steel are listed in Table 5.2. The geometrical dimensions of the cylinders along with the autofrettage temperature difference are provided in Table 5.3. During the operation of the thermal autofrettage process, the temperature variation was observed to be less than 3 °C along the length of the cylinder. The residual thermal stresses generated in the cylinder after the whole cylinder attained room temperature were evaluated experimentally using the standard Sachs boring procedure. The experimental results are compared with the generalized plane–strain model of thermal autofrettage.

A schematic of Sachs boring along with the positions of strain gauges pasted on the outer periphery of the cylinder is shown in Figure 5.12a. One strain gauge was pasted aligning in the circumferential direction and the other was pasted aligning in the longitudinal direction. The strain gages used were of type 10/120LG11 (make HBM) with 120 Ω gauge resistance with a gauge factor of 2. The lead wires connected to the strain gages were a four-wired system. The four-wired system helps in eliminating the temperature effect in the lead wires as well as in eliminating any measurement error due to gauge factor correction and contact resistance.

The removal of the layers from the inside diameter of the cylinders was performed in a lathe machine. The arrangement for the Sachs boring procedure in lathe is shown in Figure 5.11b. The boring was carried out at a low spindle speed using a sharp boring tool. The depth of cut varied from 0.1 mm to 0.5 mm during the machining at a feed rate of 0.04 mm/rev. The sharp tool and low RPM during boring minimizes vibration, and reduces the possibility of generating residual stresses in the component due to machining (Smith et al., 1998). A liquid coolant was employed continuously on the cylindrical specimen during the boring operation to minimize the generation of heat.

TABLE 5.2

Material Properties of Specimens (Kamal, 2016)

Material	Yield stress, σ_Y (MPa)	Modulus of elasticity, E (GPa)	Poisson's ratio, ν	Coefficient of thermal expansion, α (/°C)
SS304	215	200	0.29	17.8×10^{-6}
Mild steel	352	219	0.30	13×10^{-6}

TABLE 5.3

Geometry and Autofrettage Temperature Difference of Specimens (Kamal, 2016)

Material	Inner radius a (mm)	Outer radius b (mm)	Length L (mm)	Temperature difference (T_b–T_a) (°C)
SS304	10	25	90	120
Mild steel	12.65	25.30	75	230

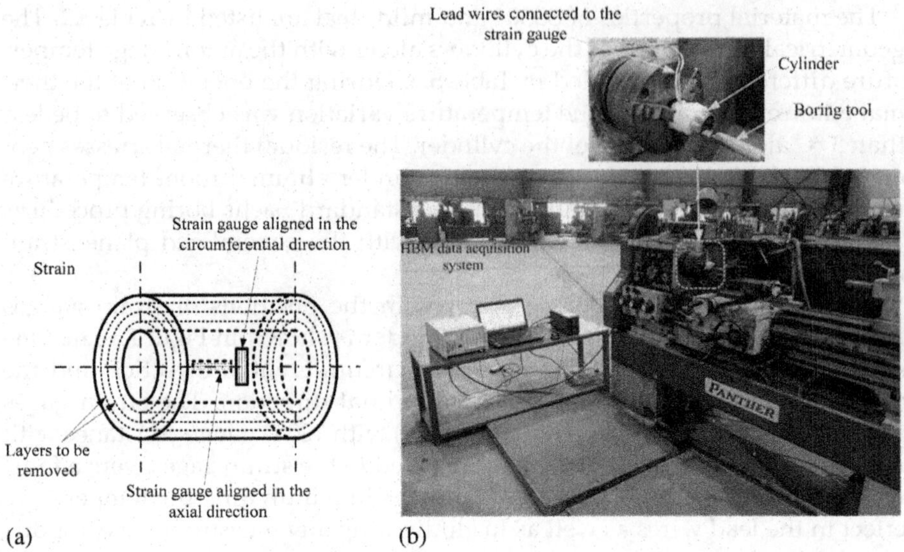

FIGURE 5.12
(a) Schematic of Sachs boring (b) Sachs boring experimental setup (Kamal, 2016).

The released hoop and axial strains, after the boring out of each layer of material, measured on the outer radius of the cylinder were recorded and stored. During machining, the lead wires from the strain gages were wound around the circumference of the chuck of the lathe machine and fixed on the circumference with the help of adhesive tape. This enabled the wires to rotate with the chuck during the material removal process. All the strains were set to zero before the boring was carried out. When one layer of material was removed completely from the cylinder, the machine was stopped and lead wires were taken out from the chuck by removing the adhesive tape. The lead wires were then plugged into the data acquisition system for recording strain data. After recording the strain data, the lead wires were disconnected and again fixed on the circumference of the chuck. Then the machine was started and the removal of the next layer of material was carried out. The strain readings were recorded fifteen minutes after boring. This was to allow the specimen to cool to room temperature. The procedure was repeated till about 80% of the thickness was removed from the cylinder. During measurement, it was observed that the strains increased from zero to a certain value in compression and then gradually became tensile. The experimental Sachs boring residual stresses in the SS304 and mild steel cylinders are shown in Figures 5.13 and 5.14, respectively, as a function of radial distance along with the corresponding analytical residual stresses obtained from a generalized plane–strain model.

It is observed from Figures 5.13 and 5.14 that the experimental Sachs boring residual stresses are in good agreement with the residual stresses

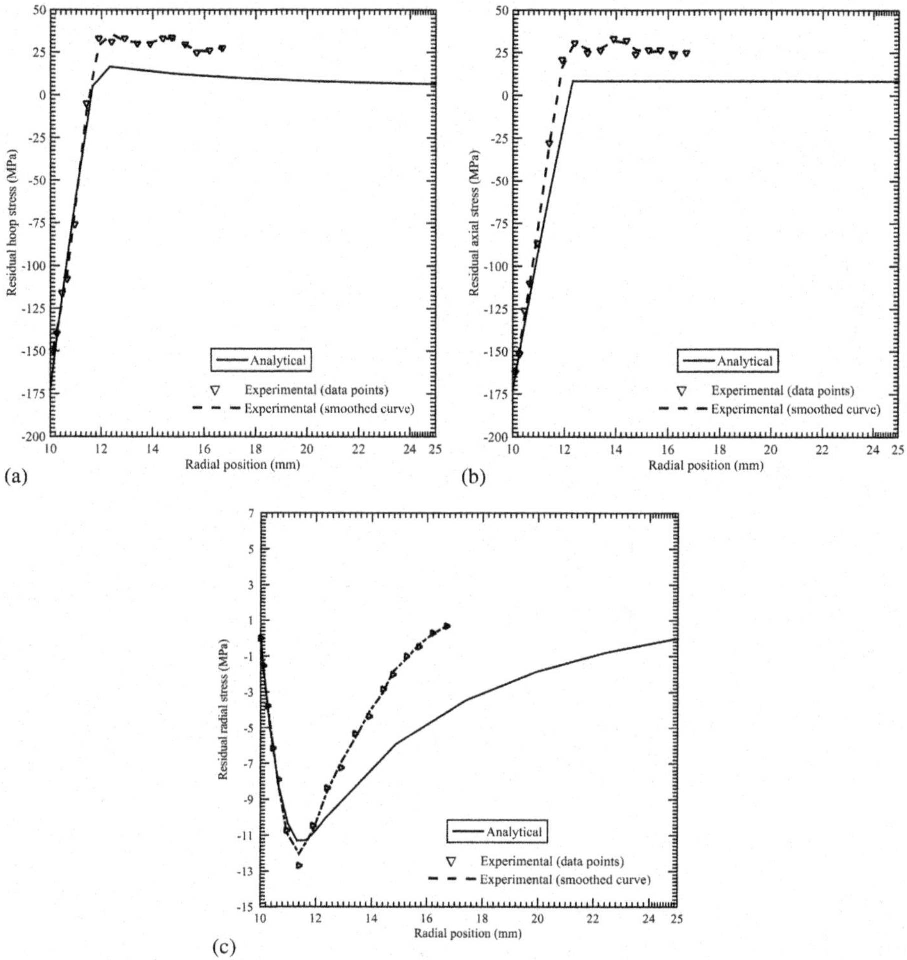

FIGURE 5.13
Residual (a) hoop, (b) axial and (c) radial stress distributions in SS304 cylinder ($b/a = 2.5$) (Kamal et al., 2016). Reused as per the rights of the author. Copyright Sage.

predicted by the generalized plane–strain model. In all of the cases, very near to the region of inner surface, the experimental compressive residual stresses closely matched with the analytical residual stresses. Some differences between the experimental and analytical stresses may be attributed to the experimental and numerical errors. In the case of the SS304 cylinder, the average errors based on L_2 norm (i.e., root mean squared errors) for hoop, axial, and radial stresses are observed to be 6.12 MPa, 8.28 MPa, and 1.15 MPa, respectively. In the case of the mild steel cylinder, these errors were 7.52 MPa, 8.55 MPa, and 2.27 MPa, for hoop, axial, and radial stresses, respectively.

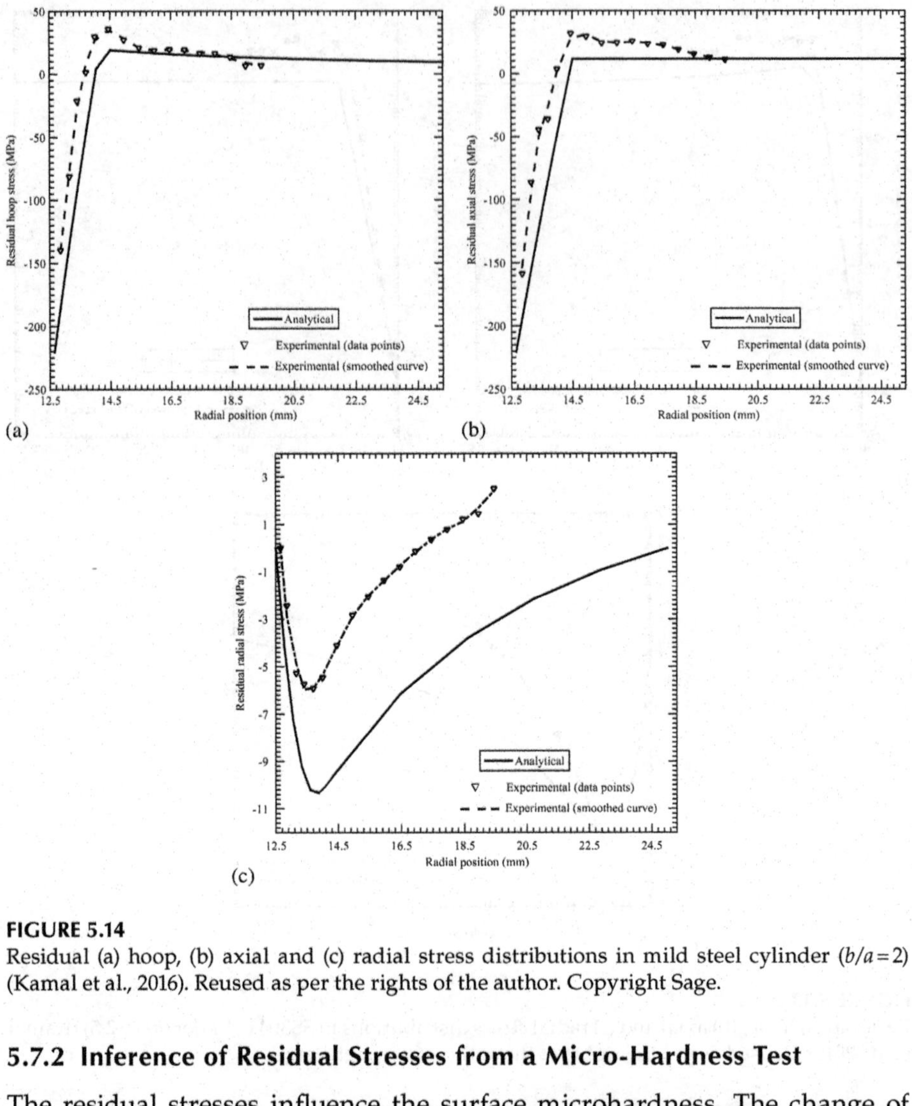

FIGURE 5.14
Residual (a) hoop, (b) axial and (c) radial stress distributions in mild steel cylinder ($b/a=2$) (Kamal et al., 2016). Reused as per the rights of the author. Copyright Sage.

5.7.2 Inference of Residual Stresses from a Micro-Hardness Test

The residual stresses influence the surface microhardness. The change of hardness of the surface after any metal working process may be interpreted as one of the qualifying tests for the presence of residual stresses. Researchers have been using the measurement of hardness as a means to detect the presence of surface residual stresses and showed that hardness decreases in the presence of tensile stresses whilst it increases in the presence of compressive stresses (Kokubo, 1932; Tosha, 2002; Simes et al., 1984).

The microhardness on the inner and outer surfaces of the thermally autofrettaged cylinders based on the Vickers diamond indentation test was carried out in order to envisage the generation of thermal residual stresses in the cylinders. To carry out the microhardness test, small test samples were extracted from the cylinders by cutting it radially. The Vickers indentation on the inner and outer surfaces of both the autofrettaged and non-autofrettaged samples was carried out at 500 gf. The measured microhardness values on the inner and outer surfaces of the samples showed that the hardness increases on the inner surface of the autofrettaged sample as compared to the non-autofrettaged sample due to the presence of compressive residual stresses. Similarly, due to the presence of tensile residual stresses on the outer surface, a decrease in the measured microhardness was observed. The microhardness tests were conducted on the inner and outer surfaces of the thermally autofrettaged SS304 cylinders ($b/a=2$, 2.5) and compared with the corresponding surfaces of non-autofrettaged cylinders. It was found that in the case of the SS304 cylinder with $b/a=2$, the average microhardness on the inner surface increases from 256.94 HV with a standard deviation of 3.03 to 279.36 HV with a standard deviation of 4.49. However, on the outer surface, the average microhardness decreases from 283.18 HV with a standard deviation of 3.21 to 224.79 HV with standard deviation of 3.50. In the case of the SS304 cylinder with $b/a=2.5$, the average microhardness on the inner surface increased to 290.95 HV with a standard deviation of 3.02, and that on the outer surface decreased to 236.22 HV with a standard deviation of 3.19. These results indicate that the inner surface of the cylinder is in compression and the outer surface is in tension.

5.7.3 Implication of Residual Stresses by the Measurement of the Opening Angle of a Radial Cut through the Wall Thickness

When the autofrettaged cylinders are cut radially through their wall thickness, the cylinders open by an angle β as shown in Figure 5.15. This opening angle β implies the presence of residual stresses in the cylinder (Parker et al., 1983). The measurement of opening angle β provides the pure bending

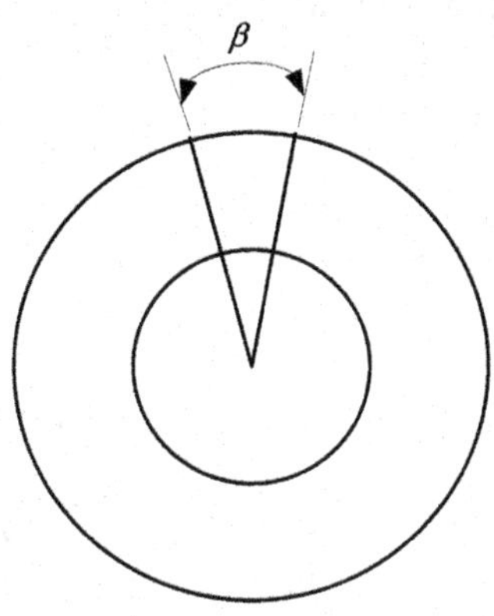

FIGURE 5.15
Opening angle in thick-walled autofrettaged cylinder with single radial cut (Kamal et al., 2016). Reused as per the rights of the author. Copyright Sage.

moment "locked in" to the cylinder during autofrettage. In an autofrettaged cylinder of unit length with an inner radius *a* and outer radius *b*, the total bending moment acting over any radial cut is given by:

$$M = \int_a^b \sigma_\theta r dr, \tag{5.113}$$

where σ_θ is the net residual hoop stress setup in the cylinder by autofrettage. In the thermally autofrettaged cylinders, the pure bending moment "locked in" to the cylinders is calculated based on the residual stress distribution predicted by the generalized plane–strain model.

The opening angle β due to the released bending moment M "locked in" to the cylinder can be evaluated by using the following equation (Parker and Farrow, 1980):

$$\beta = -\frac{4\pi M}{E}\frac{4\left(b^2 - a^2\right)}{N}, \tag{5.114}$$

where E is the Young's modulus of elasticity and N is given by:

$$N = \left(b^2 - a^2\right)^2 - 4a^2b^2\left\{\ln\left(\frac{b}{a}\right)\right\}^2. \tag{5.115}$$

The opening angles in the thermally autofrettaged SS304 cylinders of different wall thickness ratios (b/a) were measured in order to illustrate the existence of residual stresses in the cylinders. The theoretical angle of opening for the cylinders is calculated using Equation (5.114). The value of M required in Equation (5.114) is evaluated from Equation (5.113). For the experimental determination of the opening angle, a disk of 15 mm thickness was extracted from each of the autofrettaged cylinders. A radial cut was made in each disk and the opening angle was measured using a profile projector. The experimentally measured opening angles in SS304 cylinders are compared with the theoretical opening angles and are presented in Table 5.4

The comparison of theoretical and experimental measurement of opening angles in the cylinder shows that the experimentally determined opening

TABLE 5.4

Comparison of Experimental Opening Angle with the Theoretical Opening Angle (Kamal et al., 2016)

Sl. No.	Material	b/a	$(T_b - T_a)$ (°C)	Theoretical opening angle (β_{th})	Experimental opening angle (β_{exp})	% variation
1	SS304	2	130	1.69°	1.98°	14.64
2	SS304	2.5	120	0.91°	1.05°	13.33

angles are slightly greater than the theoretically predicted opening angles. The percentage variation between the experimental and theoretical opening angles is observed to be less than 15%. This implies the setting up of residual stresses in the thermally autofrettaged cylinders.

5.8 A Critical Comparison of Hydraulic and Thermal Autofrettage Processes

In this section, a critical comparison of thermal and hydraulic autofrettage is carried out for different materials. The comparison is based on the different influencing parameters of both the processes for high-pressure application. There are many situations where the thick-walled cylindrical pressure vessels are subjected to high temperature difference instead of pressure loading. In this case, the vessels may fail due to the generation of high thermal stresses in the wall of the vessel. Sometimes they may be subjected to both temperature difference and internal pressurization. The effect of autofrettage for such situations is also studied.

The different parameters such as wall thickness ratio (b/a), the yield strength (σ_Y), the product of Young's modulus (E) and the coefficient of thermal expansion (α) of the material plays a very significant role during autofrettage. A comparison of the thermal autofrettage with the conventional hydraulic autofrettage is carried out based on these influencing parameters in the following subsections.

5.8.1 Effect of Wall Thickness Ratio (*b/a*)

The wall thickness ratio (b/a) is an important parameter in the autofrettage of thick-walled cylinders. In both the hydraulic and thermal autofrettage process, the extent of the load requirement for achieving the desired elastic–plastic deformation within the wall of the cylinder depends on the wall thickness ratio. Using the Tresca yield criterion, the yield onset temperature difference for a cylinder is given by Equation (5.62). When the cylinder is subjected to internal pressure, the pressure required for initial yielding at the inner wall is given by Equation (3.13). Defining the non-dimensional yield onset temperature difference as $\theta_Y = E\alpha(T_b - T_a)_Y/(1-\nu)\sigma_Y$ from Equation (5.62) and the non-dimensional yield onset pressure as $\bar{p}_Y = p/p_Y$ from Equation (3.13), the variation of θ_Y versus b/a and $\bar{p}_Y$ versus b/a is shown in Figure 5.16. It is observed that the required non-dimensional temperature difference θ_Y decreases as b/a of the cylinder increases (Figure 5.16a). However, the reverse is the case when the cylinder is subjected to hydraulic pressure as observed in Figure 5.15b. This indicates that in the thermal autofrettage process, the maximum temperature difference required to cause the desired elastic–plastic

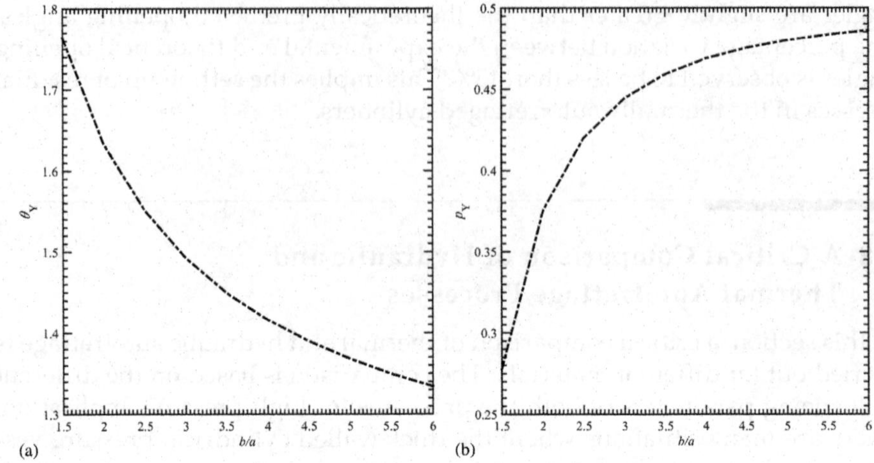

FIGURE 5.16
Variation of required non-dimensional yield onset loads with wall thickness ratio in (a) thermal autofrettage (b) hydraulic autofrettage (Kamal and Dixit, 2016a).

deformation in the wall of the cylinder decreases with b/a. On the other hand, the maximum pressure required for achieving hydraulic autofrettage increases with b/a. Hence, for high b/a cylinders, the thermal autofrettage process may be more convenient than the hydraulic autofrettage process.

As discussed in Subsection 3.5.2 of Chapter 3, the minimum pressure required for reyielding in a hydraulically autofrettaged non-hardening cylinder is $p=2p_Y$ for wall thickness ratios $b/a\geq 2.22$. This indicates that the maximum pressure-carrying capacity of a hydraulically autofrettaged non-hardening cylinder can be increased by 100% compared to the non-autofrettaged one for $b/a\geq 2.22$. In the case of thermal autofrettage, the achievable level of autofrettage is influenced by the temperature difference between the outer and inner wall of the cylinder. The increase of the maximum pressure-carrying capacity of the cylinder is more in the case of hydraulic autofrettage than the thermal autofrettage. Nevertheless, the thermal autofrettage also increases the maximum pressure-carrying capacity of the cylinder significantly. For example, a typical comparison of the maximum achievable increase in the pressure-carrying capacity of Al 6061-O and SS316 cylinders in both thermal and hydraulic autofrettage is shown in Table 5.5 for same wall thickness ratios. It is observed that the percentage increase in the maximum pressure-carrying capacity increases with the increase of the wall thickness ratio for the thermal autofrettage process.

5.8.2 Effect of $E\alpha$ and σ_Y

The thermal autofrettage process is influenced by $E\alpha$ and σ_Y of the material. Even in hydraulic autofrettage process, the load required is dependent on σ_Y.

TABLE 5.5

Simulation Results of Hydraulic and Thermal Autofrettage of Cylinders with Different Wall Thickness Ratio (b/a) (Kamal and Dixit, 2016a)

Cylinder materials	b/a	Maximum (T_b–T_a) required in thermal autofrettage (°C)	Maximum pressure required in hydraulic autofrettage (MPa)	% increase of maximum pressure-carrying capacity in autofrettage	
				Thermal	Hydraulic
Al 6061-O (E=68.9 GPa, σ_Y=55.2 MPa, α=25.2×10^{-6}/°C, ν=0.33)	2	69	38.25	21.84	84.76
	3	63	49.06	32.54	100
	4	60	51.74	40.37	100
	5	58	52.98	45.02	100
SS316 (E=193 GPa, σ_Y=290 MPa, α=16.2×10^{-6}/°C, ν=0.3)	2	212	200.93	23.41	84.76
	3	193.8	257.78	31.91	100
	4	183.9	271.88	39.41	100
	5	177.6	278.40	44.19	100

In the following subsections, the thermal and hydraulic autofrettage processes for the materials with different values of $E\alpha$ and σ_Y are discussed.

5.8.2.1 Autofrettage of Cylinders for Low $\sigma_Y/(E\alpha)$ Materials

Equation (5.62) suggests that for materials with low value of $\sigma_Y/(E\alpha)$, a small temperature difference can cause the yielding at the inner wall. Thus, cylinders made of materials with low value of $\sigma_Y/(E\alpha)$ are very much suitable for the thermal autofrettage process. The small temperature difference can be easily established, which avoids recrystallization and other microstructural changes in the materials. In hydraulic autofrettage, materials with low value of σ_Y require low pressure for initial yielding. In order to compare the autofrettage of materials with low $\sigma_Y/(E\alpha)$, numerical simulation of both the thermal and hydraulic autofrettage is carried out for some typical low $\sigma_Y/(E\alpha)$ materials. The results are presented in Table 5.6. For all simulations, cylinders with fixed $b/a=3$ are considered. It is observed that in hydraulic autofrettage, the required maximum pressure for achieving autofrettage increases with the increase of σ_Y. As the maximum pressure increases, the cost of the hydraulic power pack increases exponentially. On the other hand, the heater cost increases almost linearly with the power of the heater. Thus, the thermal autofrettage of cylinders is more economical for the materials with low $\sigma_Y/(E\alpha)$ materials than the hydraulic autofrettage. By employing the thermal autofrettage process, the maximum pressure-carrying capacity of the cylinders can be increased by about 32–34%, which is significant, although it is less than that achievable in the hydraulic autofrettage process.

TABLE 5.6

Maximum Load Required in Thermal and Hydraulic Autofrettage for Achieving Maximum Level of Autofrettage (Kamal and Dixit, 2016a)

	Maximum load required in		% increase in pressure in	
Cylinder materials	thermal autofrettage, $(T_b - T_a)$ (°C)	hydraulic autofrettage, pressure (p) (MPa)	thermal autofrettage	hydraulic autofrettage
ASTM 128 grade B2 ($E=200$ GPa, $\sigma_Y=145$ MPa, $\alpha=21.5\times10^{-6}$/°C, $\nu=0.3$)	184.70	128.88	32	100
Muntz metal ($E=110$ GPa, $\sigma_Y=345$ MPa, $\alpha=20.8\times10^{-6}$/°C, $\nu=0.34$)	124.80	306.67	33.24	100
Constantan ($E=163$ GPa, $\sigma_Y=74$ MPa, $\alpha=18.8\times10^{-6}$/°C, $\nu=0.33$)	47.58	65.77	33.54	100
Gun metal ($E=103$ GPa, $\sigma_Y=152$ MPa, $\alpha=19.8\times10^{-6}$/°C, $\nu=0.34$)	146.80	135.11	33.33	100

5.8.2.2 Autofrettage of Cylinders for High $\sigma_Y/(E\alpha)$ Materials

For high $\sigma_Y/(E\alpha)$ materials, the hydraulic autofrettage process requires an enormous amount of pressure, and the thermal autofrettage process requires a large radial temperature difference. For the thermal autofrettage process, either the inner wall of the cylinder can be subjected to a low temperature by cryogenic cooling, or the outer wall can be subjected to a high temperature to create the large temperature difference. There are practical limitations on the temperatures that the materials can sustain due to the effect on material properties. For example, Al6061-T6 alloy ($E=68.9$ GPa, $\sigma_Y=276$ MPa, $\alpha=25.2\times10^{-6}$/°C, $\nu=0.33$) and 17-4 PH stainless steel (condition H1150-M with $E=196$ GPa, $\sigma_Y=517$ MPa, $\alpha=7.1\times10^{-6}$/°C, $\nu=0.33$), possessing high $\sigma_Y/(E\alpha)$ are suitable for thermal autofrettage as they can sustain good material properties down to very low temperatures. In the thermal autofrettage of an Al6061-T6 cylinder, the maximum increase of pressure-carrying capacity of about 33% can be achieved by creating a temperature difference of 318 °C. This high temperature difference in the cylinder can be achieved by exposing the inner wall to cryogenic temperature using liquid hydrogen or liquid nitrogen and keeping the outer wall below its recrystallization temperature. In a 17-4 PH stainless steel cylinder, a maximum temperature difference of 512 °C may be achieved as it retains its good mechanical and thermal properties up to 316 °C and down to −196 °C (liquid nitrogen temperature). Thus, the thermal autofrettage of a 17-4 PH stainless steel cylinder provides the maximum increase of about 17% in the pressure-carrying capacity. Using hydraulic autofrettage, the pressure-carrying capacity can be increased by two times compared to the non-autofrettaged cylinders. The use of cryogenic fluids in the thermal autofrettage of cylinders makes the process expensive; albeit it is less expensive than the hydraulic autofrettage process. Moreover, cryogenic cooling of the inner wall of the cylinders for achieving the thermal autofrettage is a lot safer than applying a high pressure in the hydraulic autofrettage process. The most common cryogenic coolants such as liquid nitrogen and liquid hydrogen are not harmful to the environment. On the other hand, for creating hydraulic pressures, a hydraulic power pack utilizes a large amount of hydraulic oil that is harmful to the environment. Thus, the thermal autofrettage is a greener manufacturing process, although for some materials, e.g., 17-4 PH stainless steel, the maximum increase of pressure-carrying capacity is limited due to allowable temperature limits on the material.

5.8.3 Application of the Thermal Autofrettage Process to Cylinders Subjected to High Temperature Difference in Service and Its Comparison with the Hydraulic Autofrettage Process

In general, the thick-walled cylindrical vessels are autofrettaged for high-pressure applications. However, there are many applications in industries where thick-walled cylinders are subjected to high temperature differences, such as cryogenic storage tanks. For example, the Shuttle program requires very large cryogenic ground storage tanks to store liquid oxygen and hydrogen (Arens et al., 2010). There are other examples in the chemical and oil industries where thick-walled pipes are used to carry hot fluids. In such cases, the inner wall of the pipe is subjected to a higher temperature than that of the outer wall, which may create a large temperature difference across the wall of the pipe. Due to such large temperature differences in the wall of the thick-walled cylinders, they are subjected to large thermal stresses. The cylinder may fail when the generated thermal stresses exceed the yield strength of the material. Hence, the thermal load-carrying capacity should be enhanced. The autofrettage process can be employed to withstand high thermal stresses due to an induced temperature difference in service.

As a way to strengthen the cylinder material to withstand high temperature difference in service, both the thermal and hydraulic autofrettage processes are simulated for typical SS316 and Al6061-O cylinders with $b/a=3$. The thermal and hydraulic autofrettage of the cylinders is carried out by inducing the maximum load as stated in Table 5.5 for $b/a=3$. It is found that the thermally autofrettaged cylinders, when subjected to a radial temperature difference in the next stage of loading, can increase its temperature difference bearing capacity by 100% as compared to the corresponding non-autofrettaged cylinders. However, the hydraulically autofrettaged SS316 cylinder can increase its temperature difference bearing capacity by only 30%, and that the Al6061-O cylinder can increase its temperature difference bearing capacity by only 33% (as compared to the corresponding non-autofrettaged cylinders), when subjected to temperature difference during service (Kamal and Dixit, 2016). Thus, for the cylinders subjected to high temperature difference without pressure, the thermal autofrettage process is superior to the hydraulic autofrettage process.

5.8.4 Application of the Thermal Autofrettage Process to Cylinders Subjected to Temperature Difference with Pressure in Service and Its Comparison with the Hydraulic Autofrettage Process

There are situations where thick-walled cylindrical vessels are subjected to high pressure as well as temperature differences. Typical examples are thick-walled pipes carrying hot fluid under pressure in the oil or chemical industries. This needs the special attention of the vessel designers to increase the load-carrying capability of the cylinders subjected to the thermal gradient with internal pressure. In this section, both the hydraulically and thermally autofrettaged cylinders subjected to the temperature difference with internal pressure are studied and a comparison is made. For example, the thermal and hydraulic autofrettage of an SS316 cylinder ($b/a=3$) is considered. The combination of temperature difference and internal pressure loading in both the thermally and hydraulically autofrettaged cylinders for which the cylinder is safe is obtained by numerical simulation and is shown in Figure 5.17. It is observed from Figure 5.17a that, in the case of thermal autofrettage, the maximum equivalent Tresca stress occurs at the inner wall as the pressure is increased up to the temperature difference of 165 °C. After that, the safe pressure decreases with increasing temperature difference because the maximum equivalent Tresca stress exists at the outer wall. The hydraulically autofrettaged cylinders can withstand enhanced pressure in combination with the negative thermal gradient ($T_b < T_a$) as shown in Figure 5.17b. It is observed that, as the temperature difference is increased, the safe pressure decreases. For all combinations, the maximum equivalent Tresca stress occurs at the radius of elastic–plastic interface. In short, both the thermal and hydraulic autofrettage process can sustain temperature difference in combination with internal pressure. In the hydraulic autofrettage process, the pressure loading decreases with the temperature difference. On the other hand, in the thermal autofrettage process, the pressure loading increases with the temperature difference up to a certain temperature difference; thereafter it cannot sustain more pressure.

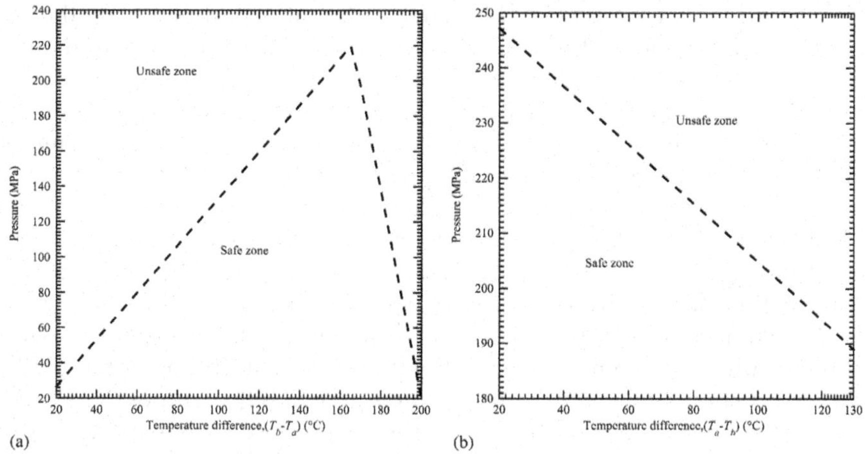

FIGURE 5.17
Safe and unsafe zones for different combinations of temperature difference and pressure loading for (a) thermally autofrettaged and (b) hydraulically autofrettaged SS316 cylinder (Kamal and Dixit, 2016a).

5.9 Conclusions

In this Chapter, a detailed analysis of thermal autofrettage was presented. The concept of thermal autofrettage was explained first, followed by the derivation of stress, strain, and a displacement solution in the thermal autofrettage of thin disks based on a plane–stress assumption. A generalized plane–strain model applicable for long thick hollow cylinders/pressure vessels was also presented. Both the plane–stress and generalized plane–strain models take into account the effect of strain hardening. Further, a 3D FEM analysis of the process was carried out and an assessment was made on the applicability of the analytical models by comparing the FEM results with the analytical solutions for different length to wall thickness ratios of the cylinders. It was found that the generalized plane–strain model is valid for the cylinders with length to wall thickness ratio, $L/(b-a) \geq 6$. The plane–stress model provides realistic solution for cylinders with $L/(b-a) \leq 1$. A detailed experimental study on thermal autofrettage of thick-walled cylinders was discussed. An experimental setup for creating desired temperature difference across the wall thickness of a cylinder in order to achieve thermal autofrettage was explained. The setup is simple and easy to handle. Moreover, it provides additional advantages over the hydraulic and explosive autofrettage processes by avoiding the use of explosives or ultra-high levels of hydraulic pressure. The experimental arrangement does not utilize any sliding parts as in the case of the swage autofrettage process, which avoids friction and retains a good surface finish at the bore surface. Further, the residual stresses after thermal autofrettage of different cylinders were assessed by three different experimental methods—the Sachs boring method, a microhardness test, and the measurement of opening angle due to a single radial cut through the wall thickness of the cylinders. The experimentally measured Sachs boring residual stresses closely match the theoretical results predicted by the generalized plane–strain model. The other two experimental methods provide a qualitative measure for the presence of residual stresses in the thermally autofrettaged cylinders. A critical comparison of the thermal autofrettage with the conventional hydraulic autofrettage process was presented considering different influencing parameters. The thermal autofrettage process was also studied for the thick-walled cylinders subjected to high temperature difference with or without pressure, and the results were compared with the hydraulic autofrettage process. The comparison shows that, for cylinders subjected to high temperature difference without pressure, the thermal autofrettage process is superior to the hydraulic autofrettage process.

Acknowledgments

This chapter is mainly based on the following articles:

(1) Kamal, S.M. and Dixit, U.S., (2015a), Feasibility study of thermal autofrettage of thick-walled cylinders, *ASME Journal of Pressure Vessel Technology*, **137**(6), 061207-1–061207-18.

(2) Kamal, S.M. and Dixit U.S., (2015b), Feasibility study of thermal autofrettage process, in *Advances in Material Forming and Joining, 5th International and 26th All India Manufacturing Technology, Design and Research Conference, AIMTDR 2014*, edited by R.G. Narayanan and U.S. Dixit, Springer, New Delhi, India.

(3) Kamal, S.M. and Dixit, U.S., (2016a), A comparative study of thermal and hydraulic autofrettage, *Journal of Mechanical Science and Technology*, **30**(6), 2483–2496.

(4) Kamal, S.M., Dixit, U.S., Roy, A., Liu, Q. and Silberschmidt, V.V., (2017), Comparison of plane-stress, generalized-plane-strain and 3D FEM elastic–plastic analyses of thick-walled cylinders subjected to radial thermal gradient, *International Journal of Mechanical Sciences*, **131–132**, 744–752.

(5) Kamal, S.M., Borsaikia, A.C. and Dixit, U.S., (2016), Experimental assessment of residual stresses induced by the thermal autofrettage of thick-walled cylinders, *The Journal of Strain Analysis for Engineering Design*, **51**(2), 144–160.

The authors are grateful to the publishers of these articles.

References

Arens, E., Youngquist, R., McFall, J., Simmons, S. and Cox, R., (2010), Developing NDE techniques for large cryogenic tanks, NASA Technical Report, Report/Patent Number: KSC-2010-251.

Barbero, E.J. and Wen E.W., (2002), Coefficient of thermal expansion compatibility through mechanical and thermal autofrettage in metal lined composite pipes, ASTM STP 1436, edited by C.E. Bakis, American Society for Testing and Materials, West Conshohocken, PA.

Bastable, M.J., (1992), From breechloaders to monster guns: Sir William Armstrong and the invention of modern artillery, 1854–1880. *Technology and Culture*, **33**, 213–247. https://doi.org/10.2307/3105857

Chakrabarty, J., (2006), *Theory of Plasticity*, 3rd edition, Butterworth-Heinemann, Burlington, MA.

Dixit, P.M. and Dixit, U.S., (2008), *Modeling of Metal Forming and Machining Processes: By Finite Element and Soft Computing Methods*, Springer, London, UK.

Gerald, C.F. and Wheatley, P.O., (1994), *Applied Numerical Analysis*, 5th edition, Addison-Wesley, Wokingham, England.

Kamal, S.M., (2016), A theoretical and experimental study of thermal autofrettage process, Ph.D. Thesis, IIT Guwahati.

Kamal, S.M., Borsaikia, A.C. and Dixit, U.S., (2016), Experimental assessment of residual stresses induced by the thermal autofrettage of thick-walled cylinders, *The Journal of Strain Analysis for Engineering Design*, **51**(2), 144–160.

Kamal, S.M. and Dixit, U.S., (2015a), Feasibility study of thermal autofrettage of thick-walled cylinders, *ASME Journal of Pressure Vessel Technology*, **137**(6), 061207-1–061207-18.

Kamal, S.M. and Dixit U.S., (2015b), Feasibility study of thermal autofrettage process, in *Advances in Material Forming and Joining, 5th International and 26th All India Manufacturing Technology, Design and Research Conference, AIMTDR 2014*, edited by R.G. Narayanan and U.S. Dixit, Springer, New Delhi, India.

Kamal, S.M. and Dixit, U.S., (2016), A comparative study of thermal and hydraulic autofrettage, *Journal of Mechanical Science and Technology*, **30**(6), 2483–2496.

Kamal, S.M., Dixit, U.S., Roy, A., Liu, Q. and Silberschmidt, V.V., (2017), Comparison of plane-stress, generalized-plane-strain and 3D FEM elastic–plastic analyses of thick-walled cylinders subjected to radial thermal gradient, *International Journal of Mechanical Sciences*, **131–132**, 744–752.

Kokubo, S., (1932), Changes in hardness of a plate caused by bending, *Science Reports of the Tohoku Imperial University, Japan, Series I*, **21**, 256–267.

Noda, N., Hetnarski, R.B. and Tanigawa, Y. (2003), *Thermal Stresses*, 2nd edition. Taylor and Francis, New York.

Parker, A.P. and Farrow, J.R., (1980), Technical note: on the equivalence of axisymmetric bending, thermal, and autofrettage residual stress fields, *Journal of Strain Analysis for Engineering Design*, **15**(1), 51–52.

Parker, A.P., Underwood, J.H., Throop, J.F. and Andrasic, C.P., (1983), Stress intensity and fatigue crack growth in a pressurized, autofrettaged thick cylinder, American Society for Testing and Materials 14th National Symposium on Fracture Mechanics, UCLA, ASTM STP 791, pp. 216–237.

Sachs, G., (1927), Der Nachweiss Immerer Spannungen in Stangen und Rohren, *Zeitschrift für Metallkunde*, **19**, 352–357.

Simes, T.R., Mellor, S.G. and Hills, D.A., (1984), Research note: a note on the influence of residual stress on measured hardness, *Journal of Strain Analysis for Engineering Design*, **19**(2), 135–137.

Smith, D.J., Pavier, M.J. and Poussard, C.P., (1998), An assessment of Sachs method for measuring residual stresses in cold worked fastener holes, *Journal of Strain Analysis for Engineering Design*, **33**, 263–274.

Tosha, K., (2002), Influence of residual stresses on the hardness number in the affected layer produced by shot peening, Proceedings of the Second Asia-Pacific Forum on Precision Surface Finishing and Deburring Technology, Seoul, Korea, July 22–24, pp. 48–54.

Appendix A

A.1 Evaluation of Different Integration Constants and Constant Axial Strain ε_0 in the First Stage of Elastic–Plastic Deformation

The constants— C_1, C_2, C_3, and C_4 can be evaluated by using the boundary conditions:

(i) At the radius of elastic–plastic interface, $r=d$

$$\sigma_r^{(\text{plastic zone II})} = \sigma_r^{(\text{elastic zone})}, \quad \varepsilon_r^{p(\text{plastic zone II})} = 0. \tag{A.1}$$

(ii) At $r=c$,

$$\sigma_r^{(\text{plastic zone I})} = \sigma_r^{(\text{plastic zone II})}, \quad \varepsilon_\theta^{p(\text{plastic zone I})} = \varepsilon_\theta^{p(\text{plastic zone II})}. \tag{A.2}$$

Boundary conditions of Equation (A.1) provides:

$$C_3 = Q + \frac{2\nu b^2}{(2\nu-1)\left\{b^2 + d^2(2\nu-1)\right\}} d^{1-\sqrt{2(1-\nu)}} \left\{\frac{1-\nu+\nu\sqrt{2(1-\nu)}}{2\nu\sqrt{2(1-\nu)}}\right\} E\varepsilon_0, \tag{A.3}$$

$$C_4 = P - \frac{2\nu b^2}{(2\nu-1)\left\{b^2 + d^2(2\nu-1)\right\}} \left\{\frac{1-\nu-\nu\sqrt{2(1-\nu)}}{d^{-1-\sqrt{2(1-\nu)}}\, 2\nu\sqrt{2(1-\nu)}}\right\} E\varepsilon_0, \tag{A.4}$$

where:

$$P=\frac{E\alpha\left(T_b-T_a\right)}{(2\nu-1)\ln\left(\frac{b}{a}\right)}\left\{\frac{1-\nu-\nu\sqrt{2(1-\nu)}}{d^{-1-\sqrt{2(1-\nu)}}2\nu\sqrt{2(1-\nu)}}\right\}\left\{\ln\left(\frac{d}{a}\right)-\frac{2\nu+1}{2\nu-1}-\frac{(1+\nu)}{1-\nu-\nu\sqrt{2(1-\nu)}}\right\}$$

$$+k_1K\left\{\frac{1-\nu-\nu\sqrt{2(1-\nu)}}{2\sqrt{2(1-\nu)}}\right\}\int_c^d r_1^{\sqrt{2(1-\nu)}}\left(\varepsilon_{\mathrm{eq}}^p\right)^n \mathrm{d}r_1$$

$$-\frac{E\alpha}{2(1-\nu)}\frac{\left(T_b-T_a\right)}{\ln\left(\frac{b}{a}\right)}\left\{\frac{1-\nu-\nu\sqrt{2(1-\nu)}}{d^{-1-\sqrt{2(1-\nu)}}2\nu\sqrt{2(1-\nu)}}\right\}\left[\ln\left(\frac{b}{d}\right)+\left\{\frac{b^2-d^2}{b^2+d^2(2\nu-1)}\right\}\left\{\ln\left(\frac{b}{a}\right)(2\nu-1)-\nu-\ln\left(\frac{d}{a}\right)\right\}\right]$$

$$+\frac{2\nu b^2}{(2\nu-1)\left\{b^2+d^2(2\nu-1)\right\}}\left\{\frac{1-\nu-\nu\sqrt{2(1-\nu)}}{d^{-1-\sqrt{2(1-\nu)}}2\nu\sqrt{2(1-\nu)}}\right\}\left(k_1\sigma_Y+E\alpha T_a\right), \tag{A.5}$$

$$Q=-Pd^{-2\sqrt{2(1-\nu)}}\left\{\frac{1-\nu+\nu\sqrt{2(1-\nu)}}{1-\nu-\nu\sqrt{2(1-\nu)}}\right\}+\frac{k_1K\left\{1-\nu+\nu\sqrt{2(1-\nu)}\right\}}{2\sqrt{2(1-\nu)}}\int_c^d r_1^{-\sqrt{2(1-\nu)}}\left(\varepsilon_{\mathrm{eq}}^p\right)^n \mathrm{d}r_1$$

$$-\frac{k_1Kd^{-2\sqrt{2(1-\nu)}}\left\{1-\nu+\nu\sqrt{2(1-\nu)}\right\}}{2\sqrt{2(1-\nu)}}\int_c^d r_1^{\sqrt{2(1-\nu)}}\left(\varepsilon_{\mathrm{eq}}^p\right)^n \mathrm{d}r_1$$

$$-\frac{Ea\left(T_b-T_a\right)}{(2\nu-1)\ln\left(\frac{b}{a}\right)}\frac{(1+\nu)}{d^{-1+\sqrt{2(1-\nu)}}\left\{1-\nu-\nu\sqrt{2(1-\nu)}\right\}}. \tag{A.6}$$

The first condition of Equation (A.2) provides the value of C_1 as:

$$C_1=R+\left[\begin{array}{l}\dfrac{2\nu b^2c^{-1+\sqrt{2(1-\nu)}}}{(2\nu-1)\left\{b^2+d^2(2\nu-1)\right\}}d^{1-\sqrt{2(1-\nu)}}\left\{\dfrac{1-\nu+\nu\sqrt{2(1-\nu)}}{2\nu\sqrt{2(1-\nu)}}\right\}\\ -\dfrac{2\nu b^2c^{-1-\sqrt{2(1-\nu)}}}{(2\nu-1)\left\{b^2+d^2(2\nu-1)\right\}}\left\{\dfrac{1-\nu-\nu\sqrt{2(1-\nu)}}{d^{-1-\sqrt{2(1-\nu)}}2\nu\sqrt{2(1-\nu)}}\right\}-\dfrac{1}{(2\nu-1)}\end{array}\right]E\varepsilon_0, \tag{A.7}$$

where:

$$R = Qc^{-1+\sqrt{2(1-\nu)}} - k_1\sigma_Y \ln c - k_1 K \int_a^c \frac{\left(\varepsilon_{eq}^p\right)^n}{r_1} dr_1 + Pc^{-1-\sqrt{2(1-\nu)}} + \frac{k_1\sigma_Y}{(2\nu-1)} + \frac{E\alpha T_a}{(2\nu-1)}$$

$$+ \frac{E\alpha(T_b - T_a)}{(2\nu-1)\ln\left(\frac{b}{a}\right)} \ln\left(\frac{c}{a}\right) - \frac{2E\alpha(T_b - T_a)}{(2\nu-1)^2 \ln\left(\frac{b}{a}\right)} - \frac{E\alpha(T_b - T_a)}{(2\nu-1)\ln\left(\frac{b}{a}\right)}. \tag{A.8}$$

Using the second condition of Equation (A.2) the value of C_2 is obtained as:

$$C_2 = \frac{1-2\nu}{E}\left[-\frac{c^2}{2}(k_1\sigma_Y \ln c + C_1) - 2k_1 K \int_a^c r_1 \left(\varepsilon_{eq}^p\right)^n dr_1 + k_1 K c^2 \int_a^c \frac{\left(\varepsilon_{eq}^p\right)^n}{r_1} dr_1 - 3k_1 K \int_a^c \left\{ r_2 \int_a^{r_2} \frac{\left(\varepsilon_{eq}^p\right)^n}{r_1} dr_1 \right\} dr_2 \right]$$

$$+ \frac{3-2\nu}{4E} c^2 k_1 \sigma_Y + \frac{(1-\nu)}{E} c^2 k_1 K \left(\varepsilon_{eq}^p\big|_{r=c}\right)^n - \frac{1}{2} c^2 \alpha T_a - \frac{\alpha(T_b - T_a)}{2\ln\left(\frac{b}{a}\right)} \left\{ c^2 \ln\left(\frac{c}{a}\right) - \frac{3c^2}{2} \right\} + \frac{c^2}{2}\varepsilon_0. \tag{A.9}$$

The constant axial strain ε_0 is obtained by using free end condition (zero total axial force) given by:

$$\left[\int_a^c \sigma_z^{(\text{plastic zone I})} r dr + \int_c^d \sigma_z^{(\text{plastic zone II})} r dr + \int_d^b \sigma_z^{(\text{elastic zone})} r dr \right] = 0. \tag{A.10}$$

Inserting the expressions of σ_z for plastic zones I, II, and the elastic zone in Equation (A.10), the constant axial strain during first stage of elastic–plastic deformation is obtained as:

$$\varepsilon_0 = \frac{k_1\sigma_Y}{AE}\left\{ \ln c \frac{c^2}{2} - \ln a \frac{a^2}{2} + \frac{1}{4}(c^2 - a^2) + \frac{\nu(d^2 - c^2)}{2\nu - 1} + \frac{\nu d^2 (b^2 - d^2)}{b^2 + d^2(2\nu - 1)} \right\}$$

$$+ \frac{k_1 K}{AE} \int_a^c \left\{ r \int_a^r \frac{\left(\varepsilon_{eq}^p\right)^n}{r_1} dr_1 \right\} dr + \frac{R}{AE}\left(\frac{c^2 - a^2}{2}\right) + \frac{k_1 K}{AE} \int_a^c r \left(\varepsilon_{eq}^p\right)^n dr + \frac{Q}{AE}\left\{ \frac{d^{1+\sqrt{2(1-\nu)}} - c^{1+\sqrt{2(1-\nu)}}}{1+\sqrt{2(1-\nu)}} \right\}$$

$$+ \frac{P}{AE}\left\{ \frac{d^{1-\sqrt{2(1-\nu)}} - c^{1-\sqrt{2(1-\nu)}}}{1-\sqrt{2(1-\nu)}} \right\} + \frac{k_1 K\left\{1 - \nu + \nu\sqrt{2(1-\nu)}\right\}}{2\sqrt{2(1-\nu)}AE} \int_c^d \left[r^{\sqrt{2(1-\nu)}} \int_c^r r_1^{-\sqrt{2(1-\nu)}} \left(\varepsilon_{eq}^p\right)^n dr_1 \right] dr$$

$$
-\frac{k_1K\left\{1-\nu-\nu\sqrt{2(1-\nu)}\right\}}{2\sqrt{2(1-\nu)}AE}\int_c^d\left[r^{-\sqrt{2(1-\nu)}}\int_c^r r_1^{\sqrt{2(1-\nu)}}\left(\varepsilon_{\text{eq}}^p\right)^n\mathrm{d}r_1\right]\mathrm{d}r+\frac{\alpha T_a}{A}\left[\frac{d^2-c^2}{2(2\nu-1)}-\frac{\left(b^2-d^2\right)^2}{2\left\{b^2+d^2(2\nu-1)\right\}}\right]
$$

$$
+\frac{\alpha(T_b-T_a)}{(2\nu-1)\ln\left(\frac{b}{a}\right)A}\left\{\ln\left(\frac{d}{a}\right)\frac{d^2}{2}-\ln\left(\frac{c}{a}\right)\frac{c^2}{2}-\frac{1}{2}\left(d^2-c^2\right)\frac{6\nu+1}{2(2\nu-1)}\right\}+\frac{k_1K}{AE}\int_c^d r\left(\varepsilon_{\text{eq}}^p\right)^n\mathrm{d}r
$$

$$
+\frac{\alpha}{2(1-\nu)}\frac{(T_b-T_a)}{\ln\left(\frac{b}{a}\right)A}\left[\begin{array}{l}-\nu d^2\ln\left(\frac{b}{d}\right)-\left\{\frac{\nu d^2\left(b^2-d^2\right)}{b^2+d^2(2\nu-1)}\right\}\left\{\ln\left(\frac{b}{a}\right)(2\nu-1)-\nu-\ln\left(\frac{d}{a}\right)\right\}\\-2(1-\nu)\left\{\ln\left(\frac{b}{a}\right)\frac{b^2}{2}-\ln\left(\frac{d}{a}\right)\frac{d^2}{2}-\frac{1}{4}\left(b^2-d^2\right)\right\}\end{array}\right],
\tag{A.11}
$$

where:

$$
A=\frac{d^2-c^2}{2(2\nu-1)}-\frac{\left(b^2-d^2\right)^2}{2\left\{b^2+d^2(2\nu-1)\right\}}
$$

$$
-\left(\frac{c^2-a^2}{2}\right)\left\{\frac{2\nu b^2}{(2\nu-1)\left\{b^2+d^2(2\nu-1)\right\}2\nu\sqrt{2(1-\nu)}}\left(\begin{array}{l}\left(\frac{c}{d}\right)^{-1+\sqrt{2(1-\nu)}}\left(1-\nu+\nu\sqrt{2(1-\nu)}\right)\\-\left(\frac{c}{d}\right)^{-1-\sqrt{2(1-\nu)}}\left(1-\nu-\nu\sqrt{2(1-\nu)}\right)\end{array}\right)-\frac{1}{(2\nu-1)}\right\}
$$

$$
-\left\{\frac{d^{1+\sqrt{2(1-\nu)}}-c^{1+\sqrt{2(1-\nu)}}}{1+\sqrt{2(1-\nu)}}\right\}\frac{2\nu b^2}{(2\nu-1)\left\{b^2+d^2(2\nu-1)\right\}}d^{1-\sqrt{2(1-\nu)}}\left\{\frac{1-\nu+\nu\sqrt{2(1-\nu)}}{2\nu\sqrt{2(1-\nu)}}\right\}
$$

$$
+\left\{\frac{d^{1-\sqrt{2(1-\nu)}}-c^{1-\sqrt{2(1-\nu)}}}{1-\sqrt{2(1-\nu)}}\right\}\frac{2\nu b^2}{(2\nu-1)\left\{b^2+d^2(2\nu-1)\right\}}\left\{\frac{1-\nu-\nu\sqrt{2(1-\nu)}}{d^{-1-\sqrt{2(1-\nu)}}2\nu\sqrt{2(1-\nu)}}\right\}.
\tag{A.12}
$$

A.2 Evaluation of Different Integration Constants and Constant Axial Strain ε_0 in the Second Stage of Elastic–Plastic Deformation

Using the relevant equations of stresses and strain from second stage of elastic–plastic deformation in Equation (A.1) provides the values of C_3 and C_4 as:

$$C_3 = -C_4 d^{-2\sqrt{2(1-\nu)}}\left[\frac{1-\nu+\nu\sqrt{2(1-\nu)}}{1-\nu-\nu\sqrt{2(1-\nu)}}\right]+\frac{k_1 K\left\{1-\nu+\nu\sqrt{2(1-\nu)}\right\}}{2\sqrt{2(1-\nu)}}\int_c^d r_1^{-\sqrt{2(1-\nu)}}\left(\varepsilon_{eq}^p\right)^n dr_1$$

$$-\frac{k_1 K d^{-2\sqrt{2(1-\nu)}}\left\{1-\nu+\nu\sqrt{2(1-\nu)}\right\}}{2\sqrt{2(1-\nu)}}\int_c^d r_1^{\sqrt{2(1-\nu)}}\left(\varepsilon_{eq}^p\right)^n dr_1-\frac{E\alpha(T_b-T_a)}{(2\nu-1)\ln\left(\frac{b}{a}\right)}\frac{(1+\nu)}{d^{-1+\sqrt{2(1-\nu)}}\left\{1-\nu-\nu\sqrt{2(1-\nu)}\right\}},\tag{A.13}$$

$$C_4 = \frac{E\alpha(T_b-T_a)}{(2\nu-1)\ln\left(\frac{b}{a}\right)}\left\{\frac{1-\nu-\nu\sqrt{2(1-\nu)}}{d^{-1-\sqrt{2(1-\nu)}}2\nu\sqrt{2(1-\nu)}}\right\}\left\{\ln\left(\frac{d}{a}\right)-\frac{2\nu+1}{2\nu-1}-\frac{(1+\nu)}{1-\nu-\nu\sqrt{2(1-\nu)}}\right\}$$

$$+k_1 K\left\{\frac{1-\nu-\nu\sqrt{2(1-\nu)}}{2\sqrt{2(1-\nu)}}\right\}\int_c^d r_1^{\sqrt{2(1-\nu)}}\left(\varepsilon_{eq}^p\right)^n dr_1-\left\{\frac{1-\nu-\nu\sqrt{2(1-\nu)}}{d^{-1-\sqrt{2(1-\nu)}}2\nu\sqrt{2(1-\nu)}}\right\}\left(\frac{2\nu}{2\nu-1}\right)k_1\sigma_Y\frac{e^2}{2d^2}$$

$$-\frac{E\alpha(T_b-T_a)}{2(1-\nu)\ln\left(\frac{b}{a}\right)}\left\{\frac{1-\nu-\nu\sqrt{2(1-\nu)}}{d^{-1-\sqrt{2(1-\nu)}}2\nu\sqrt{2(1-\nu)}}\right\}\left[\frac{1}{(2\nu-1)}\left\{\ln\left(\frac{d}{a}\right)+\frac{1}{2}-\frac{e^2}{2d^2}\right\}-\ln\left(\frac{d}{a}\right)+\frac{1}{2}-\frac{e^2}{2d^2}\right],\tag{A.14}$$

Using the first condition of Equation (A.2), the constant C_1 is obtained as:

$$C_1 = N+\frac{E\varepsilon_0}{(1-2\nu)},\tag{A.15}$$

where:

$$N = C_3 c^{-1+\sqrt{2(1-\nu)}}+C_4 c^{-1-\sqrt{2(1-\nu)}}+\frac{k_1\sigma_Y}{(2\nu-1)}+\frac{E\alpha T_a}{(2\nu-1)}+\frac{E\alpha(T_b-T_a)}{(2\nu-1)\ln\left(\frac{b}{a}\right)}\left\{\ln\left(\frac{c}{a}\right)-\frac{2\nu+1}{2\nu-1}\right\}$$

$$-k_1\sigma_Y\ln c-k_1 K\int_a^c\frac{\left(\varepsilon_{eq}^p\right)^n}{r_1}dr_1,\tag{A.16}$$

The constant C_2 is still given by Equation (A.9). To find out the constant C_5, the boundary condition, at $r=f$, $\sigma_r^{(\text{plastic zone III})} = \sigma_r^{(\text{plastic zone IV})}$ is used. This provides:

$$C_5 = M+\frac{E\varepsilon_0}{(1-2\nu)},\tag{A.17}$$

where:

$$M=\frac{Ea}{(2\nu-1)}T_a+\frac{Ea(T_b-T_a)}{2(1-\nu)\ln\left(\frac{b}{a}\right)}\left[\frac{1}{(2\nu-1)}\left\{\ln\left(\frac{d}{a}\right)+\frac{1}{2}-\frac{e^2}{2d^2}\right\}-\ln\left(\frac{e}{a}\right)\right]$$

$$\left\{\frac{1}{2\nu-1}\left(1+\frac{e^2}{2d^2}\right)+\frac{1}{2}-\ln\left(\frac{f}{e}\right)+\ln f\right\}k_1\sigma_Y-k_1K\int_e^f\frac{\left(\varepsilon_{\rm eq}^p\right)^n}{r_1}dr_1. \tag{A.18}$$

The boundary condition, at $r=e$, $\varepsilon_\theta^{p(\text{plastic zone IV})}=0$, provides the value of C_7 as:

$$C_7=-\frac{e^2a(T_b-T_a)}{2(1-\nu)\ln\left(\frac{b}{a}\right)}\left(\nu^2-1\right)-e^2\left(1-\nu^2\right)\frac{k_1\sigma_Y}{E}. \tag{A.19}$$

At $r=f$, $\varepsilon_\theta^{p(\text{plastic zone III})}=\varepsilon_\theta^{p(\text{plastic zone IV})}$. This provides the constant C_6 as:

$$C_6=\frac{af^2(T_b-T_a)}{2(1-\nu)\ln\left(\frac{b}{a}\right)}\left(\nu^2-1\right)+\left(1-\nu^2\right)f^2\frac{k_1\sigma_Y}{E}+\frac{4\nu^2}{E}k_1K\left(\frac{2}{\sqrt{3}}\right)^n\int_e^f\left\{r_2\int_e^{r_2}\frac{\left(\varepsilon_{\rm eq}^p\right)^n}{r_1}dr_1\right\}dr_2$$

$$+\frac{2\nu^2}{E}k_1K\int_e^f r_1\left(\varepsilon_{\rm eq}^p\right)^n dr_1+\left(1-\nu^2\right)f^2\frac{k_1K}{E}\left(\varepsilon_{\rm eq}^p\right)^n\Big|_{r=f}-\frac{2\nu^2k_1K}{E}f^2\int_e^f\frac{\left(\varepsilon_{\rm eq}^p\right)^n}{r_1}dr_1+C_7$$

$$-\frac{(1-2\nu)}{E}\left[\frac{f^2}{2}\left(C_7-k_1\sigma_Y\ln f\right)\right]-\frac{3-2\nu}{4E}f^2k_1\sigma_Y-\frac{1-\nu}{E}f^2k_1K\left(\varepsilon_{\rm eq}^p\right)^n\Big|_{r=f}-\frac{f^2}{2}aT_a$$

$$-\frac{a(T_b-T_a)}{2\ln\left(\frac{b}{a}\right)}f^2\left\{\ln\left(\frac{f}{a}\right)-\frac{3}{2}\right\}+\frac{f^2}{2}\varepsilon_0. \tag{A.20}$$

The constant axial strain ε_0 in the second stage of elastic–plastic deformation is obtained by using the following free end condition:

$$\left[\int_a^c\sigma_z^{(\text{plastic zone I})}r\,dr+\int_c^d\sigma_z^{(\text{plastic zone II})}r\,dr+\int_d^e\sigma_z^{(\text{elastic zone})}r\,dr+\int_e^f\sigma_z^{(\text{plastic zone IV})}r\,dr+\int_f^b\sigma_z^{(\text{plastic zone III})}r\,dr\right]=0. \tag{A.21}$$

Thus, Equation (A.21) provides:

$$\varepsilon_0 = \frac{k_1\sigma_Y}{BE}\left[\begin{array}{l}\ln c\left(\frac{c^2}{2}\right)-\ln a\left(\frac{a^2}{2}\right)+\frac{1}{4}\left(c^2-a^2\right)+\frac{\nu}{2\nu-1}\left(d^2-c^2\right)+\frac{\nu}{2\nu-1}\left(1+\frac{e^2}{2d^2}\right)\left(f^2-d^2\right)\\ -\nu\left\{\ln\left(\frac{f}{e}\right)f^2-\frac{1}{2}\left(f^2-e^2\right)\right\}-\frac{1}{4}\left(b^2-f^2\right)-\ln b\left(\frac{b^2}{2}\right)+\ln f\left(\frac{f^2}{2}\right)\end{array}\right]$$

$$+\frac{N}{2BE}\left(c^2-a^2\right)+\frac{C_5}{BE}\left\{\frac{d^{1+\sqrt{2(1-\nu)}}-c^{1+\sqrt{2(1-\nu)}}}{1+\sqrt{2(1-\nu)}}\right\}+\frac{C_6}{BE}\left\{\frac{d^{1-\sqrt{2(1-\nu)}}-c^{1-\sqrt{2(1-\nu)}}}{1-\sqrt{2(1-\nu)}}\right\}+\frac{\alpha T_a}{2(2\nu-1)B}\left(f^2-c^2\right)$$

$$+\frac{\alpha\left(T_b-T_a\right)}{(2\nu-1)\ln\left(\frac{b}{a}\right)B}\left\{\ln\left(\frac{d}{a}\right)\frac{d^2}{2}-\ln\left(\frac{c}{a}\right)\frac{c^2}{2}-\frac{3}{4}\left(d^2-c^2\right)-\frac{d^2-c^2}{2\nu-1}\right\}$$

$$+\frac{\alpha\left(T_b-T_a\right)}{2(1-\nu)\ln\left(\frac{b}{a}\right)B}\left[\begin{array}{l}\frac{\nu}{(2\nu-1)}\left\{\ln\left(\frac{d}{a}\right)+\frac{1}{2}-\frac{e^2}{2d^2}\right\}\left(f^2-d^2\right)-\left\{\ln\left(\frac{e}{a}\right)e^2-\ln\left(\frac{d}{a}\right)d^2-\frac{1}{2}\left(e^2-d^2\right)\right\}\\ +\nu\left\{\ln\left(\frac{f}{e}\right)f^2-\frac{1}{2}\left(f^2-e^2\right)\right\}-\left\{\ln\left(\frac{f}{a}\right)f^2-\ln\left(\frac{e}{a}\right)e^2-\frac{1}{2}\left(f^2-e^2\right)\right\}\end{array}\right]$$

$$+\frac{M}{2BE}\left(b^2-f^2\right)+k_1K\int_a^c r\left(\varepsilon_{\text{eq}}^p\right)^n \mathrm{d}r+\frac{k_1K\left\{1-\nu+\nu\sqrt{2(1-\nu)}\right\}}{2BE\sqrt{2(1-\nu)}}\int_c^d\left[r^{\sqrt{2(1-\nu)}}\int_c^r r_1^{-\sqrt{2(1-\nu)}}\left(\varepsilon_{\text{eq}}^p\right)^n \mathrm{d}r_1\right]\mathrm{d}r$$

$$-\frac{k_1K\left\{1-\nu-\nu\sqrt{2(1-\nu)}\right\}}{2BE\sqrt{2(1-\nu)}}\int_c^d\left[r^{-\sqrt{2(1-\nu)}}\int_c^r r_1^{\sqrt{2(1-\nu)}}\left(\varepsilon_{\text{eq}}^p\right)^n \mathrm{d}r_1\right]\mathrm{d}r$$

$$+\frac{k_1K}{BE}\int_c^d r\left(\varepsilon_{\text{eq}}^p\right)^n \mathrm{d}r-\frac{2\nu k_1K}{BE}\int_e^f\left\{r\int_e^r\frac{\left(\varepsilon_{\text{eq}}^p\right)^n}{r_1}\mathrm{d}r_1\right\}\mathrm{d}r-\frac{\nu k_1K}{BE}\int_e^f r\left(\varepsilon_{\text{eq}}^p\right)^n \mathrm{d}r-\frac{k_1K}{BE}\int_f^b r\left(\varepsilon_{\text{eq}}^p\right)^n \mathrm{d}r$$

$$-\frac{k_1K}{BE}\int_f^b\left\{r\int_f^r\frac{\left(\varepsilon_{\text{eq}}^p\right)^n}{r_1}\mathrm{d}r_1\right\}\mathrm{d}r+\frac{k_1K}{BE}\int_a^c\left\{r\int_a^r\frac{\left(\varepsilon_{\text{eq}}^p\right)^n}{r_1}\mathrm{d}r_1\right\}\mathrm{d}r, \tag{A.22}$$

where:

$$B=\left\{\frac{b^2-a^2}{2(2\nu-1)}\right\}. \tag{A.23}$$

6

Combining Thermal Autofrettage with Various Other Processes

6.1 Introduction

The underlying principle of an autofrettage process is to subject a thick-walled cylindrical/spherical vessel to a state of partial or full plastic deformation. In the case of partial plastic deformation, the autofrettaged specimen comprises an inner plastic zone that restrains the tendency of an outer elastic zone to contract internally to regain its original undeformed position. This induces compressive residual stresses in the vicinity of the inner wall. In the case of full plastic deformation, the material at the inner wall suffers a greater degree of strain hardening than that at the outer wall. Here, at the inner wall, compressive residual stresses are induced due to non-uniform straining. In both cases, tensile residual stresses are induced at the vicinity of the outer wall to attain a state of equilibrium. While the compressive residual stresses are beneficial for strengthening against internal load, the tensile residual stresses reduce the strength of the autofrettaged specimen when there are external surface flaws like cracks (Parker, 1981). Tensile residual stresses are also detrimental for the fatigue strength, and may promote stress corrosion cracking. Seifi and Babalhavaeji (2012) demonstrated the adverse effect of the tensile residual stresses in reducing the bursting pressure of a thick-walled autofrettaged cylinder with a surface crack using the FEM, as well as experiments. Four types of surface crack orientations were considered. These were an axial crack, a circumferential crack, and two angular cracks inclined at 30° and 60°, respectively, with respect to the axis of the cylinder. Of all of these types of crack orientations, the effect of the axial crack was the most significant in reducing the bursting pressure by about 30%. The effect of the circumferential crack was the least significant and caused about 6% reduction in bursting pressure. A few researchers suggested techniques to reduce the harmful effect of tensile residual stresses in the outer wall of a typical hydraulically autofrettaged cylinder. These techniques include a post mild heat treatment to reduce the outer wall tensile residual stresses (Franklin and Morrison, 1960), external wire winding (Sedighi and Jabbari, 2013) and shot peening of the outer wall (Koh et al., 1997; Koh and Stephens, 1991) to induce compressive residual stress at the outer wall of an autofrettaged cylinder.

Recently a technique has been proposed and analyzed by Shufen and Dixit (2018) to combine the thermal autofrettage process with a post-heat treatment to induce compressive residual stress at the outer wall of the thick-walled cylinder. In this technique, a thermally autofrettaged thick-walled cylinder is subjected to a temperature gradient; the outer wall is heated above the lower critical temperature, which is 727 °C for most steels (Callister and Rethwisch, 2009), whilst the inner wall is kept well below the lower critical temperature. Subsequent cooling of the cylinder induces a compressive residual stress field in the vicinity of the outer wall, while preserving the autofrettage-induced residual stresses in the vicinity of the inner wall. This technique will be described in detail in this chapter.

This chapter is also devoted to the theoretical analysis of thermal autofrettage combined with other frettaging methods. These are a combined hydraulic and thermal autofrettage process and a thermal autofrettage process combined with shrink-fitting. As discussed in detail in Chapter 5, one of the main advantages in thermal autofrettage is that the process is very simple and can be cost effective. As described by Kamal et al. (2016), the experimental setup of thermal autofrettage does not require any complex or moving machine parts. However, despite its simplicity and economical aspect, the maximum increase in the pressure-bearing capacity of a thermally autofrettaged specimen is less than that of a hydraulically autofrettaged specimen. There is also a restriction on the maximum allowable temperature difference to which the specimen can be subjected during the thermal loading stage, because the property of the material begins to change at elevated temperatures. Nevertheless, sometimes the thermal autofrettage process may be preferred to other autofrettage processes due to the simplicity and cost-effectiveness of the former. Depending on the need, the thermal autofrettage process may be augmented with other processes to enhance its overall effectiveness.

The chapter is organized as follows. Section 6.2 briefly introduces the microstructure. Section 6.3 explains alloys and phase diagrams. Section 6.4 explains the phase transformation in steel and its relevance to autofrettage. Section 6.5 describes the concept of the heat treatment coupled with the thermal autofrettage process. In Section 6.6, the detailed procedure for the FEM modeling of the heat treatment coupled thermal autofrettage process is presented. Section 6.7 presents the typical results of the process. Section 6.8 and 6.9 present the analysis of the thermal autofrettage process combined with hydraulic autofrettage and shrink-fitting, respectively. Section 6.10 concludes the chapter.

6.2 A Brief Introduction to Microstructure

When a metal undergoes solidification from its liquid phase in a molten state during casting, the motion of the free-flowing atoms in the metal gradually slows down with the decrease in temperature. With the continued decrease in temperature, a particular atom gets restricted in its motion and starts to bond with other atoms, thus initiating the formation of the solid phase. Bonded atoms do keep vibrating with small amplitude but do not slide past other atoms. A group of atoms get arranged in a particular periodic manner, which is called crystal. The beginning of the formation of a new phase (liquid to solid in this case) is called nucleation, and the initial group of atoms that start this process is called the nucleus. Subsequently, other surrounding atoms start to pattern themselves around this nucleus, forming a well-ordered three-dimensional arrangement of atoms called the crystalline lattice. The smallest group of atoms that is repeated to give the overall crystalline lattice is called a crystal. Since the crystal grows in all three directions, it is generally convenient to designate the directions in which the atoms are arranged in each crystal. A vector connecting two atoms in a particular crystal that signifies the direction of the growth of the crystal and the arrangement of atoms in a crystal is called its crystallographic direction. Likewise, the same nucleation and growth phenomenon takes place at other locations in the bulk of the solidifying metal where similar clusters of atoms position themselves according to another set of crystallographic directions to give another crystalline lattice. A fully-grown cluster of atoms with a particular set of crystallographic orientations gives rise to a structure termed as a grain. As crystallization continues with further solidification, the differently oriented grains impinge at each other and form an interface. This interface is called the grain boundary. The totality of the grains in the polycrystalline material is called the microstructure.

There are several methods of carrying out grain refinement, i.e., grains can be reduced in size. One method is the adding of some alloying element during melting. The other method is by carrying out heat treatment, which is accomplished by the proper design of heating and cooling cycles. The third method is by carrying out excessive plastic deformation; for example, severe plastic deformation can cause a considerable reduction in grain size. A refined microstructure leads to an enhancement of strength. The Hall–Petch relation provides a quantitative relation between the grain size and the yield stress of a material. It is also possible to carry out grain coarsening. The process of replacing deformed grains by a new set of defect-free grains is called recrystallization. Recrystallization invariably leads to a reduction in the strength of the metal but enhances ductility. Recrystallization occurs after attaining a threshold temperature. The threshold temperature decreases with an increase in prior plastic work and holding time at the threshold temperature. One can define recrystallization temperature as the temperature at which 50% new stress-free grains are formed in one hour. For most of the metals, a typical value of recrystallization temperature is 0.4 times the melting point temperature in K.

Figure 6.1a shows the two-dimensional representation of the formation of a grain boundary at the junction of two grains comprising groups of atoms

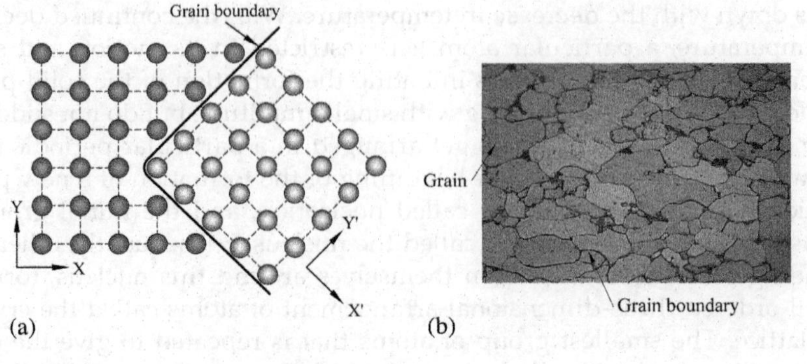

FIGURE 6.1
Illustration of grain and grain boundary: (a) a two-dimensional schematic representation, (b) as observed in an optical microscope with a magnification of X50. Part b modified with permission from Dutta et al. (2017). Copyright Springer (2017).

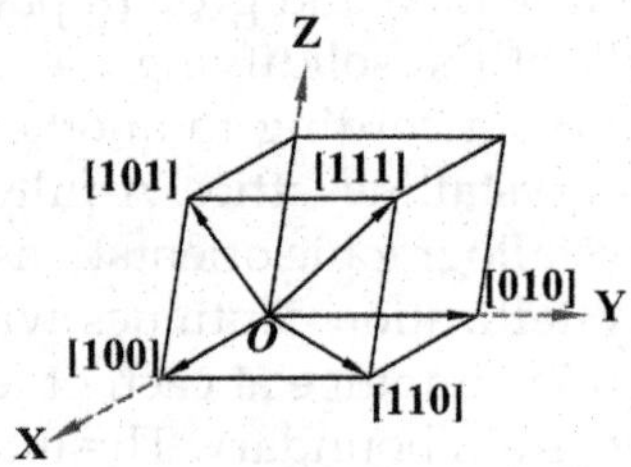

FIGURE 6.2
Crystallographic directions designated with Miller's indices.

that grow in two different directions. Figure 6.1b shows the microstructure of an annealed AH36 steel strip after laser-assisted bending. Details are available in the reference (Dutta et al., 2017). This micrograph is taken at a magnification of 50 and pertains to the base (undeformed and unheated) region. It clearly shows the grains and grain boundaries. The designation of crystallographic directions in a typical crystal is shown in Figure 6.2. The directions are expressed as unit vectors and enclosed in a square bracket without any space or commas. These sets of numbers are called Miller's indices.

6.3 Alloys and Phase Diagram

In brief, a phase is the part of an alloy system that has similar physical and chemical properties. Under conditions of constant external conditions such as temperature, pressure, and composition, the phase properties in an alloy remain in a state of equilibrium. In general, the pressure remains unchanged and only the temperature and composition are taken into account. The interrelationship between the phases at equilibrium in an alloy system with a given composition and temperature can be mapped into what is popularly called the phase diagram. A brief description on the iron–carbon phase diagram is presented here. Figure 6.3 shows the schematic of an iron–carbon phase diagram. Temperature is plotted along the vertical axis and the carbon composition is plotted along the horizontal axis. Depending on the content of carbon, iron carbon alloys may be of two types—steel (carbon concentration less than 2%), or cast iron (carbon concentration more than 2%). The discussion of the iron carbon alloy system for the present chapter will be solely based on steel; the portion of the phase diagram pertaining to cast iron will not be discussed.

By chemical definition, steel is an alloy of pure iron containing a carbon concentration of less than 2% by weight. Depending on the temperature of the alloy system, steel can exist in different allotropic forms. Each form comprises different types of phases in the microstructure. For all compositions, an iron–carbon steel system cools down from the liquid phase to provide the allotropic form called austenite, a solid phase called γ phase. For a composition of 2.1% carbon by weight, a 100% solid phase is formed at about 1147 °C. As the carbon percentage in steel reduces, its solidus temperature (temperature below which full solidification occurs) increases as apparent in Figure 6.3. When austenite cools down below a particular threshold

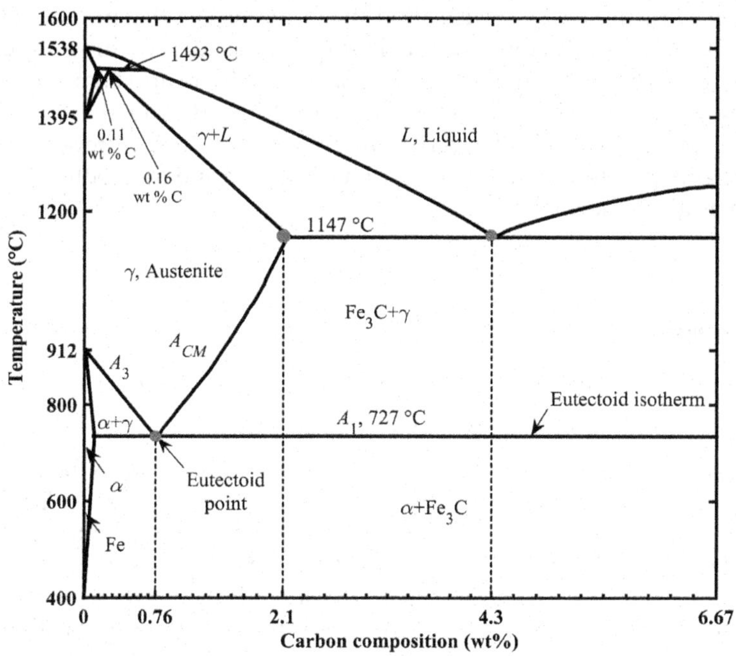

FIGURE 6.3
A typical iron–carbon diagram.

temperature, a new phase starts appearing in the microstructure. This new phase may be proeutectoid cementite or proeutectoid ferrite depending on whether the carbon concentration is more or less than 0.76%, respectively. Ferrite is pure iron and cementite is a stable compound of iron and carbon with the chemical formula Fe_3C. Ferrite is generally referred to as the α phase in the iron carbon diagram. The significance of the prefixed word "proeutectoid" will be understood later.

Steel with 0.76% carbon concentration is called eutectoid steel. Steel with more than 0.76% carbon concentration is called hypereutectoid steel and that with less than 0.76% carbon is called hypoeutectoid steel. The threshold temperature for the formation of proeutectoid ferrite decreases with the increase in the carbon concentration; the variation of this temperature with the carbon concentration is represented by the line A_3. On the other hand, the threshold temperature for the formation of cementite increases with the increase in the carbon concentration and the variation is represented by the line A_{CM}. The temperatures represented by points on the lines A_3 and A_{CM} are called upper critical temperatures. Upper critical temperature is the temperature above which the full austenization occurs during heating. The lines A_3 and A_{CM} meet at the point where the carbon concentration is 0.76% at a temperature of 727 °C. This point that vertically demarcates the hypo and hyper eutectoid steels in the iron carbon diagram is called the eutectoid point. The horizontal line passing through the eutectoid point with a temperature of 727 °C is called the eutectoid isotherm. The temperature 727 °C is called the lower critical temperature; below this temperature, austenite decomposes into two stable phases of ferrite and cementite. These (eutectoid) ferrite and cementite phases are formed after the temperature has reached below the eutectoid temperature and are different than the proeutectoid ferrite and cementite phases that appeared before reaching the eutectoid isotherm.

6.4 Phase Transformation Kinetics and Temperature Time Transformation (TTT) Diagram

Heat treatment is a process whereby a cold worked metal is heated above a certain temperature and then cooled down at a desired rate. This objective of heat treatment is to tailor the microstructure of the material to bring out various desirable properties to suit the operating conditions to which it will be subjected. For example, a fast rate of cooling increases the hardness and tensile strength at the cost of a reduced ductility whereas a slow rate of cooling increases the ductility of the metal. In the heat treatment of typical steels, the workpiece is heated above the austenization temperature and then cooled down in a suitable cooling medium. The austenization temperature is the minimum temperature for the formation of austenite in the steel with the particular carbon composition. For eutectoid steel with 0.76% carbon concentration, this is equal to the lower critical temperature 727 °C. In hypoeutectoid and hypereutectoid steels, the austenization temperature for the steel with the particular composition is the upper critical temperature. Various types of heat treatment according to the type of cooling used are explained in (Callister and Rethwisch, 2009).

The iron carbon diagram provides information about the evolution of phases at various temperatures for a particular composition, but it does not convey the rate at which the phases transform with the variation of temperature with time. The aspect that describes the rate at which a parent phase transforms into its microstructural constituents at different temperatures is called phase transformation kinetics, and it is of significant importance to the study of heat treatment processes. The chapter will focus on the phase transformation kinetics of steel. Based on how the atoms move in an alloy system, a typical steel can transform by either a diffusional mode or displacive mode. A diffusional transformation is where there is breakage of the atomic bonds and each atom moves and rearranges to form a new microstructural lattice; the composition of phase changes due to diffusion. On the other hand, a displacive transformation is where ordered groups of atoms move without breakage of the atomic bonds; there is no breakage of the bond. In a typical heat treatment of a steel, the material is heated above the austenization temperature and cooled at a suitable rate to obtain one or more of the constituents, which are as follows—proeutectoid ferrite (in case of hypoeutectoid steels), proeutectoid cementite (in case of hypereutectoid steels), pearlite, bainite, and martensite. Except martensite, which arises due to displacive transformation, all of the other constituents are formed by diffusional transformation. Based on the variation of temperature with time, diffusional transformation may be classified into either isothermal or anisothermal. First, the kinetics of isothermal diffusional transformation are explained, which is followed by its extension to the non-isothermal case.

6.4.1 Kinetics of Isothermal Diffusional Phase Transformation

The mathematical model describing the kinetics of isothermal diffusional transformation is governed by the following equation (Callister and Rethwisch, 2009):

$$X_k(t) = X_k^{\max}\left\{1-\exp(-b_k t^{n_k})\right\}, \tag{6.1}$$

where $X_k(t)$ is the volume fraction of a kth phase constituent that grows at the rate defined by the constants n_k and b_k. The mathematical expression given in Equation (6.1) is called the Johnson–Mehl–Avrami–Kolmogorov (JMAK) equation. The rate constants n_k and b_k are functions of the temperature. The parameter $X_k^{\max}$ denotes the maximum volume fraction of the kth phase that can be obtained.

In all types of steel, namely hypoeutectoid, eutectoid, and hypereutectoid steel, $X_k^{\max}$ for phase constituents pearlite and bainite is equal to unity. This means that, a typical steel in an austenitic state can transform into a microstructure comprising either 100% pearlite or 100% bainite depending on the cooling rate. In addition, austenite may also decompose to give proeutectoid ferrite (in hypoeutectoid steels) or cementite (in case of hypereutectoid steels) when the temperature is above the eutectoid isotherm but below the upper critical temperature.

The variation of $X_k^{\max}$ for any proeutectoid phase at different temperatures T is computed as follows (Phadke et al., 2004):

$$X_k^{\max} = \begin{cases} 0, & T > Ac_3 \\ X_k^{\text{eq}} \dfrac{Ac_3 - T}{Ac_3 - Ac_1}, & Ac_1 \geq T \geq Ac_3 \\ X_k^{\text{eq}}, & T < Ac_1 \end{cases} \tag{6.2}$$

where Ac_3 is the upper critical temperature and X_k^{eq} is the volume fraction of the proeutectoid phase at equilibrium. The equilibrium volume fraction X_k^{eq} is computed from the iron carbon diagram using the lever rule, which is stated as follows for proeutectoid ferrite (Callister and Rethwisch, 2009):

$$X_k^{\text{eq}} = \frac{0.76 - \text{C\%}}{0.76 - 0.022}, \tag{6.3}$$

where C% is the carbon content of the steel.

Partial differentiation of Equation (6.1) with respect to time t provides:

$$\frac{\partial X_k(t)}{\partial t} = X_k^{\max} b_k n_k t^{(n_k - 1)} \exp\left(-b_k t^{n_k}\right), \tag{6.4}$$

which indicates that, for a particular set of the rate constants n_k and b_k, the rate of transformation is slow at the beginning, when t is as small as both power and the exponential terms are small. With an increase in t, the power term increases rapidly as compared to the exponential term and rate of transformation increases rapidly. With further increase in t, the exponential term becomes the dominant term and the rate slows down.

When Equation (6.1) is plotted in a 2D Cartesian system with a logarithm of time along the horizontal axis and volume fraction along the vertical axis, an S-shaped curve is obtained as shown in Figure 6.4a for two different sets of rate constants n_k and b_k. The characteristic trend of the rate effect governed by Equation (6.2) is observed in Figure 6.4a where the curves 1–2 and 3–4 start off with almost zero slopes, which then increase sharply after some interval and then decreases again approaching a zero slope with further

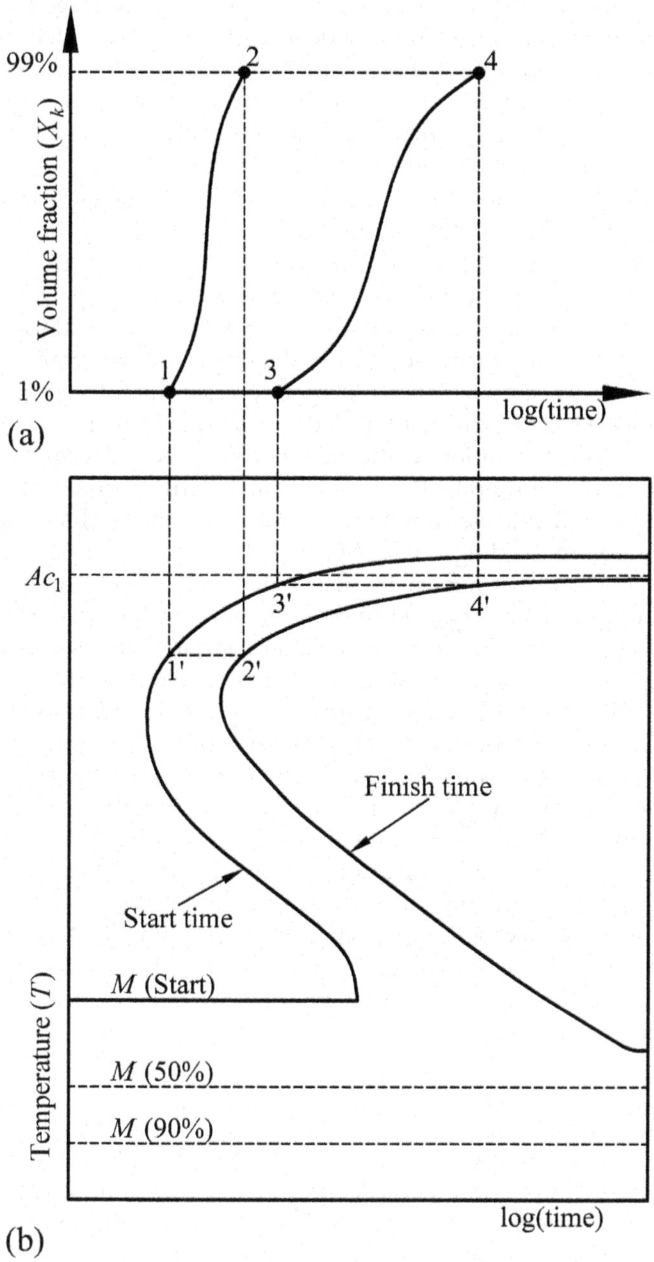

FIGURE 6.4
Schematic of typical (a) JMAK equation governed phase transformation curves and (b) corresponding temperature time transformation (TTT) diagram obtained by projection of start and finish times of several phase transformation curves under isothermal condition for the case of eutectoid steel.

increase in time. The finish time tends to an infinitely large value as the temperature approaches the lower critical temperature Ac_1. Hence, Equation (6.1) exhibits an asymptotic behavior, both at the start and towards the finish of the transformation. This is why it becomes necessary to identify a unique volume fraction at which the start and finish of the transformation are defined. Thus, the start and finish of the transformation is generally set when the volume fraction has reached 1% and 99%, respectively, of X_k^{max}. This is labelled in Figure 6.4a. When the start and finish times of several JMAK curves for various sets of rate constants n_k and b_k are projected in a time–temperature two-dimensional Cartesian space, a pair of C-shaped curves are obtained as shown in Figure 6.4b, which is popularly called the temperature time transformation (TTT) diagram. TTT diagram depicted in Figure 6.4b is for a eutectoid steel. As mentioned earlier, the phase transformation is of two types—diffusional and displacive. TTT diagram deals with both types of transformation. If the steel is cooled below a certain temperature with a high cooling rate, there is no time for diffusional transformation to occur and the displacive transformation to martensite phase commences. This is indicated in the diagram by *M* (start). Lines indicating the completion of 50% and 90% transformation are also indicated.

The TTT diagram for a typical hypoeutectoid and hypereutectoid steel is depicted in Figure 6.5b. There are three characteristic curves—(i) a start time curve for the proeutectoid phase, (ii) a finish time curve for the proeutectoid phase that coincides with the start time curve for the eutectoid phase, and (iii) a finish time curve for the eutectoid phase. These are obtained by projecting the start and finish times from the transformation curves projected from the time volume fraction plot of Figure 6.5a. This TTT diagram is similar to the TTT diagram of the eutectoid steel of Figure 6.4, except for the addition of an extra start time curve for the proeutectoid phase. Like in case of the eutectoid steel, the start time and finish time curves become asymptotic as the temperature approaches the upper critical temperature Ac_3 and the lower critical temperature Ac_1, for proeutectoid and eutectoid phases, respectively. Points corresponding to 1% and 99% volume fractions signify the start and finish of the transformation of the proeutectoid phase.

6.4.2 Kinetics of Anisothermal Diffusional Phase Transformation

The kinetics of anisothermal transformation are based on the assumption that the overall transformation consists of a series of small, successive isothermal transformations. The kinetics of a typical anisothermal transformation of a particular *k*th constituent are depicted in Figure 6.6.

The anisothermal temperature variation is shown in the cooling curve, which is approximated by a series of small discreet isotherms where the temperature is assumed to remain constant for a duration of time Δt. Each temperature point at the beginning of the isothermal transformation is mapped in the TTT diagram. Subsequently, each mapped temperature point from the

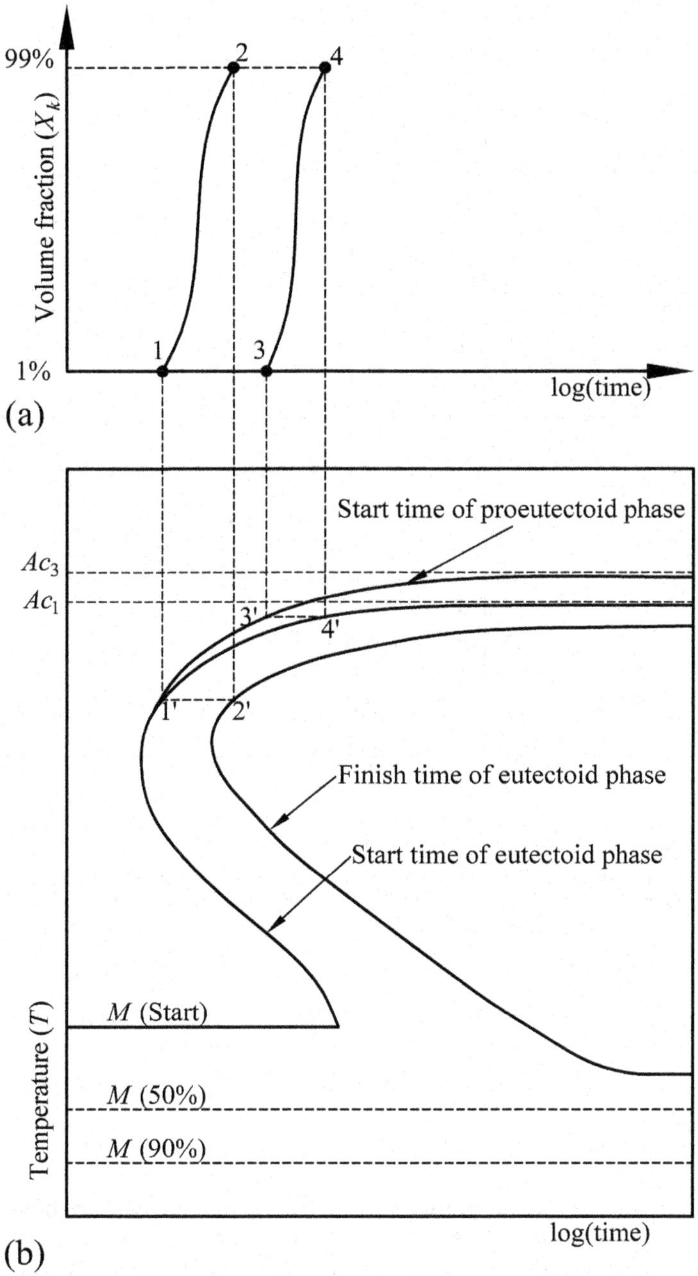

FIGURE 6.5
Schematic of typical (a) JMAK equation governed phase transformation curves and (b) corresponding temperature time transformation (TTT) diagram obtained by projection of start and finish times of several phase transformation curves under isothermal condition for the case of hypoeutectoid or hypereutectoid steel.

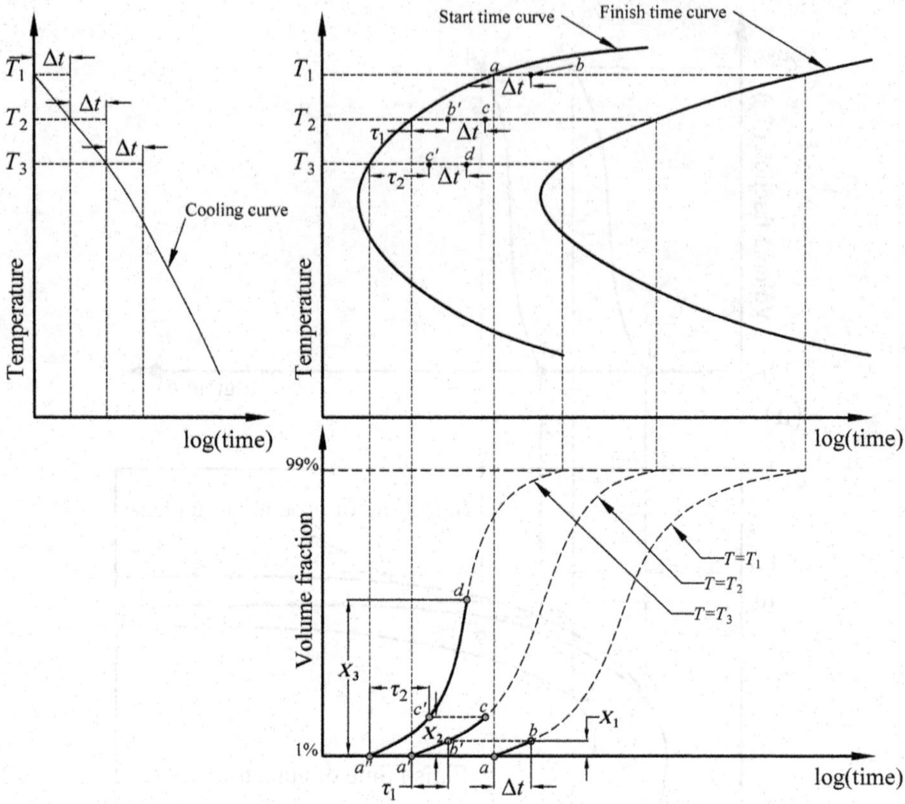

FIGURE 6.6
Schematic of incremental transformation along multiple isothermal transformation curves under anisothermal condition based on the fictitious time concept. With permission from Shufen and Dixit (2018). Copyright Elsevier.

TTT diagram is then projected onto the time–volume fraction plane that provides the characteristic JMAK curve governing the transformation with the rate constants n_k and b_k existent at the prescribed temperature under isothermal condition.

The transformation begins along the JMAK curve for an isothermal temperature T_1 along the path *ab* till a volume fraction X_1 up to time Δt. At the end of Δt, the temperature changes to T_2; the transformation behavior is now governed by the isothermal JMAK curve defined for T_2. To establish the continuity of transformation, it must be ensured that the transformation along the JMAK curve for temperature T_2 compensates for the volume fraction X_1 that has already been transformed in the previous time increment. To facilitate this, a fictitious time parameter is used, which is computed as follows (de Oliveira et al., 2010):

$$\tau = \left\{ \frac{-1}{b_k} \ln\left(1 - \frac{X_k(t)}{X_k^{\max}}\right) \right\}^{\frac{1}{n_k}}. \tag{6.5}$$

Equation (6.5) is simply a rearranged form of Equation (6.1) that gives the total time elapsed till the end of the previous time increment, during which the *k*th constituent has attained the volume fraction $X_k(t)$. The fictitious time parameter accounts for the transformation that has already taken place. It transforms the actual time of transformation to the equivalent time of transformation at the current temperature. The rate constants b_k and n_k are inversely computed from Equation (6.1) using the volume fractions 0.01 and 0.99 at the start and finish times t_s and t_f, respectively. Hence,

$$n_k = \frac{\ln\left\{ \dfrac{\ln\left(1 - \dfrac{0.01}{X_k^{\max}}\right)}{\ln\left(1 - \dfrac{0.99}{X_k^{\max}}\right)} \right\}}{\ln\left(\dfrac{t_s}{t_f}\right)}, \tag{6.6}$$

$$b_k = \frac{1}{t_s^{n_k}} \left\{ -\ln\left(1 - \frac{X_k(t_s)}{X_k^{\max}}\right) \right\}. \tag{6.7}$$

The points *a* and *b* in the JMAK curve for T_1 are projected to the points *a′* and *b*, respectively, in the JMAK curve for T_2. The time corresponding to the path *a′b′* is the fictitious time τ_1 over which the new time Δt must be incremented so that the transformation takes place along the path *a′b′c* through the point *b′*. Similarly, for the next time increment, the fictitious time τ_2 is computed for the JMAK curve at the temperature T_3, and the volume fraction is computed along the path *a′c′d*. The solid path in each of the JMAK curves depicted in Figure 6.5 are the paths where transformation actually take place.

The JMAK equation for the kinetics of anisothermal transformation can thus be written as follows:

$$X_k^{t+\Delta t} = X_k^{\max} X_\gamma \left\{ 1 - \exp\left(b_k (\tau + \Delta t)^{n_k} \right) \right\}, \tag{6.8}$$

where X_γ is the volume fraction of austenite that was available at the commencement of the transformation of each subsequent k^{th} phase constituent as follows:

$$X_\gamma = 1 - \sum_{i \neq k} X_i, \tag{6.9}$$

where X_i is the volume fraction of each preceding i^{th} constituent whose transformation has already occurred.

6.4.3 Kinetics of Displacive or Diffusion-Less Transformation of Steel

In typical steels, austenite undergoes displacive transformation to give martensite (a form of steel), which is very hard. The mathematical expression for computing the volume fraction of martensite is computed only as a function of temperature. Once the temperature reaches the threshold for martensitic transformation, which is generally about 225 °C (Callister and Rethwisch, 2009), all of the austenite available up to that point steadily transforms into martensite until it reaches room temperature. In brief, martensite is a microstructural form of steel that is formed due to the rapid cooling or quenching of the austenite at such a fast rate that the carbon atoms have no time to diffuse out of the crystalline lattice to form cementite (Fe_3C). The phenomena of martensite formation cannot be inferred from the iron-carbide diagram, which is an equilibrium phase diagram. The volume fraction of martensite X_M is computed (Koistinen and Marburger, 1959) as follows:

$$X = X_\gamma\left[1-\exp\left\{-0.011\left(T_{MS}-T_j\right)\right\}\right], \tag{6.10}$$

where T_{MS} is the martensitic temperature and T_j is the current temperature.

6.5 Concept of Thermal Autofrettage with Heat Treatment

A conventional thermal autofrettage process comprises two steps—a loading step where a radial temperature gradient is setup across the wall of the cylinder and an unloading step where the temperature gradient is removed by slowly cooling down the cylinder to room temperature. This induces residual stresses, which are compressive at the vicinity of the inner wall and tensile at the vicinity of the outer wall. As a means to negate the harmful outer wall tensile stresses, Shufen and Dixit (2018) suggested a technique where the thermal autofrettage process is combined with a post-heat treatment. Here the cylinder is subjected to two additional process steps after the thermal autofrettage *namely* a reloading step where the thermally autofrettaged cylinder is again subjected to a temperature gradient that involves temperatures above the recrystallization temperature and a quenching step where the cylinder is quenched from both the inner and outer walls. The schematic of the process is depicted in Figure 6.7. In the reloading step, the inner wall of the thermally autofrettaged cylinder is heated to a temperature T_{in} that is greater than the lower critical temperature Ac_1, and the outer wall is kept at a temperature T_{out} that is lesser than Ac_1. In the quenching step, the cylinder is cooled down from the both the inner and outer walls. This induces a residual stress field that is compressive at the vicinity of the inner wall as well as the outer wall.

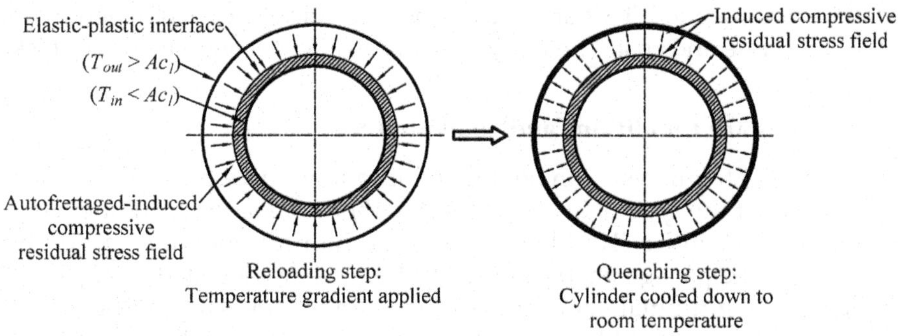

FIGURE 6.7
Heat treatment of thermally autofrettaged cylinder.

Shufen and Dixit (2018) carried out a thermal–mechanical–metallurgical FEM analysis in commercial package ABAQUS®. The metallurgical aspect of the process in the FEM model was incorporated through user material subroutine UMATHT. The details of the modeling are explained in the following subsections.

6.6 FEM Implementation of Thermal Autofrettage with Heat Treatment

This section presents the methodology for FEM implementation of thermal autofrettage coupled with heat treatment. The section is organized in the following subsections. Section 6.6.1 describes the model. The Section 6.6.1 describes the boundary conditions. Section 6.6.3 explains the load step sizing with a brief description of the concept. Section 6.6.4 presents material behavior. The mathematical models governing the physics of the three involved process, namely metallurgical, thermal, and mechanical are presented in Section 6.6.4, 6.6.5, and 6.6.6, respectively.

6.6.1 Model Description

At the outset, a model of the problem needs to be defined. In general, if the length of the cylinder is six times the thickness of the cylinder, it can be modeled as a generalized plane–strain model (Kamal et al., 2017). In most of the cases, it is possible to exploit the symmetry of the cylinder and an axisymmetric model can be developed as depicted in Figure 6.8. The axisymmetric model simplifies the analysis of a three-dimensional cylindrical geometry to that of a simple two-dimensional rectangular shell representing a longitudinal section. This approach yields significant reduction of the computational time and cost. The mesh generation also becomes an easier task.

6.6.2 Boundary Conditions and Load Step Size

The overall FEM analysis can be divided into four successive analysis steps on the basis of the type of analysis and the thermal loading conditions subjected to the cylinder. The thermal and mechanical boundary conditions subjected to the part geometry of the FEM model are depicted in Figure 6.8. Due to the longitudinal symmetry of the cylinder specimen, the FEM part model has been reduced to half the length of the cylinder. This condition is applied using the mechanical boundary condition $u_z = 0$ at the mid section (represented by AB in Figure 6.8) of the cylinder. In all of the analysis' steps, the ends of the cylinder are insulated. The insulation boundary condition is expressed as follows:

$$\frac{\partial T}{\partial r}(0,t) = 0, \tag{6.11}$$

which is assigned to the edges AB and CD as depicted in Figure 6.8. Other thermal boundary conditions exclusive to each analysis step are described as follows:

***First and Second Step (Thermal Autofrettage)*:** The first two steps are the heating and cooling processes of a conventional thermal autofrettage process.

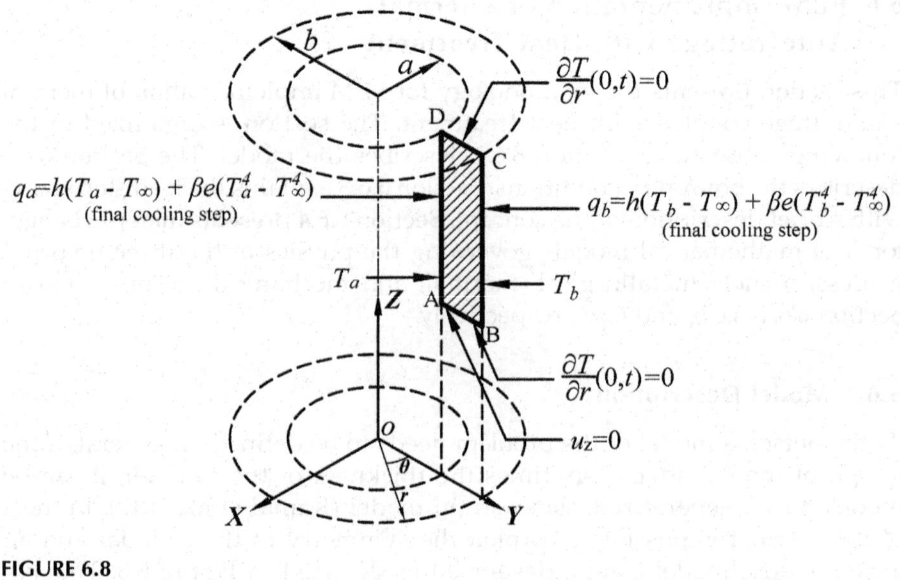

FIGURE 6.8
Axisymmetric model of the thick-walled cylinder. With permission from Shufen and Dixit (2018). Copyright Elsevier.

In the first step, the inner wall edge (AD in Figure 6.8) is assigned an ambient temperature, say $T_a = 20$ °C, while the outer wall edge BC is assigned a higher temperature $T_b = T_a + \Delta T$, where ΔT is the requisite temperature difference for the desired level of autofrettage. In the second step, the temperature of the outer wall BC is gradually lowered to $T_b = 20$ °C, while retaining the temperature of the inner wall AD. The distribution of autofrettage-induced residual stresses obtained at the end of these two steps are propagated as predefined fields to the third step. Only the final states of the field variables are of interest to steps three and four; hence a steady state analysis is sufficient for executing steps one and two.

Third Step (Austenization): In this step, the temperature of the outer wall edge BC is assigned a temperature higher than the lower critical temperature. For steel, one can take $T_a \approx 725$ °C temperature for the outer wall and a much lower temperature at the inner wall. However, the difference is kept such that there is no reyielding of the inner wall. This is essential so that residual stresses generated after the thermal autofrettage process are not mitigated. The stress state and the temperature distribution at the end of step three due to reheating are propagated to step four. Like in the case of the first two steps, only the steady state values of the field variables were required, and a steady state analysis was carried out.

Fourth Step (Cooling): The phase transformation kinetics are incorporated in this step, hence it is required to carry out a transient analysis. Transient analysis takes the results at the end of step three as the initial condition. The cylinder is cooled at a fast rate from both the inner and outer walls simultaneously. A combined convection and radiation boundary condition is assigned to both the inner and outer walls, which is mathematically expressed as follows:

$$q_a = h(T_a - T_\infty) + \beta e\left(T_a^4 - T_\infty^4\right), \tag{6.12}$$

$$q_b = h(T_b - T_\infty) + \beta e\left(T_b^4 - T_\infty^4\right), \tag{6.13}$$

where q_a, q_b are the heat fluxes at the inner and outer wall edges AD and BC, respectively, h is the convective heat transfer coefficient, T_∞ is the ambient temperature (room temperature), e is the thermal emissivity, and $\beta = 5.67 \times 10^{-8}$ W/m^2 is the Stefan–Boltzmann constant.

6.6.3 Load Step Size

As explained in Chapter 2, FEM is essentially a numerical technique for solving differential equations. It has been successfully applied for solving complicated nonlinear problems. In nonlinear problems, FEM solves the governing differential equation progressively by applying the total external load/displacement in small incremental steps in either an explicit (forward difference) or implicit (backward difference) regime. For each increment, a convergent solution is sought based on satisfying the equilibrium equation up to a certain desired tolerance. The solver moves on to the next increment only if the solution based on the current load increment converges. A very large increment can cause difficulty in convergence and a very small increment can lengthen the computational time. Hence, the selection of a proper load step size is important. The magnitude of each incremental load per step is called the load step size, and it is assigned as a fraction of the total load amplitude that is to be applied. For example, a load step size of 0.01 per increment for a total load amplitude of 1 means that the total load will be applied in 100 steps. This is identical to a case of assigning a load step size of 0.03 for a load amplitude 3.

For the first three steps, fixed load step size may be chosen. For example, the temperature difference ΔT can be applied in 100 load steps. This will ensure an accuracy of 1% over the full range in determining the temperature difference at which yielding starts. By increasing the number of load steps, the accuracy of estimation can be enhanced, but the computational time will also increase. At each load step, equivalent plastic strain can be noted. The load step at which equivalent plastic strain becomes nonzero, indicates the commencement of yielding. For step four, i.e., the transient analysis of heat treatment, the simulations can be run till the cylinder cools down to room temperature. Usually for this problem, the time increments vary from 1×10^{-6} s and 1 s. The increments can be fixed, or adaptive, based on the consideration of accuracy and stability.

6.6.4 Material Behavior

The material behavior needs to be defined using temperature as well as microstructure-dependent properties. The rate of transformation of a phase constituent with the temperature is obtained from a TTT diagram of the material. The start and end curves in the TTT diagram are approximated by piecewise polynomial functions. Due to the asymptotic nature of the finish time curve, the lower critical temperature may be offset to a slightly lower value. This is because, as the temperature approaches the lower critical value, the value of t_f becomes very large and produces computational difficulty.

6.6.4.1 Implementation of Metallurgical Analysis Module

By default, the material definition library in most FEM packages does not include features that readily allow for a definition of the material properties according to the microstructure. However, such special material behavior can be incorporated through user-defined subroutines. In general computer terminology, a subroutine is a modular program that contains a specific set of instructions, which can be called by a main program whenever it is required to perform a desired set of operations. As an example, the mathematical expressions described in Section 6.4 governing phase transformation kinetics explained in Section 6.4 can be implemented in the user subroutine, UMATHT, which is a subroutine for incorporating user-defined thermal material properties in ABAQUS®. The subroutine is encoded in FORTRAN language. For each incremental step, the main code in ABAQUS® calls the UMATHT subroutine at each integration point. The volume fractions of the microstructural constituents are stored in the solution-dependent variables (SDVs). The SDVs are global variables that can be utilized to store the values of field variables in each incremental analysis step. Depending on the number of field variables required, the number of SDVs must be allocated in the memory. The SDVs are stored in the state variable (STATEV) array defined in the subroutine.

For a typical steel, the volume fractions of austenite, pearlite, bainite, and martensite were stored in arrays SDV1, SDV2, SDV3, and SDV4, respectively. The initial condition can be supplied by using the SDVINI user subroutine, which can be called at the very beginning (Figure 6.9). After that, the SDVs to which the initial volume fractions of the phases are assigned are passed into the UMATHT subroutine. Representing the volume fractions of austenite,

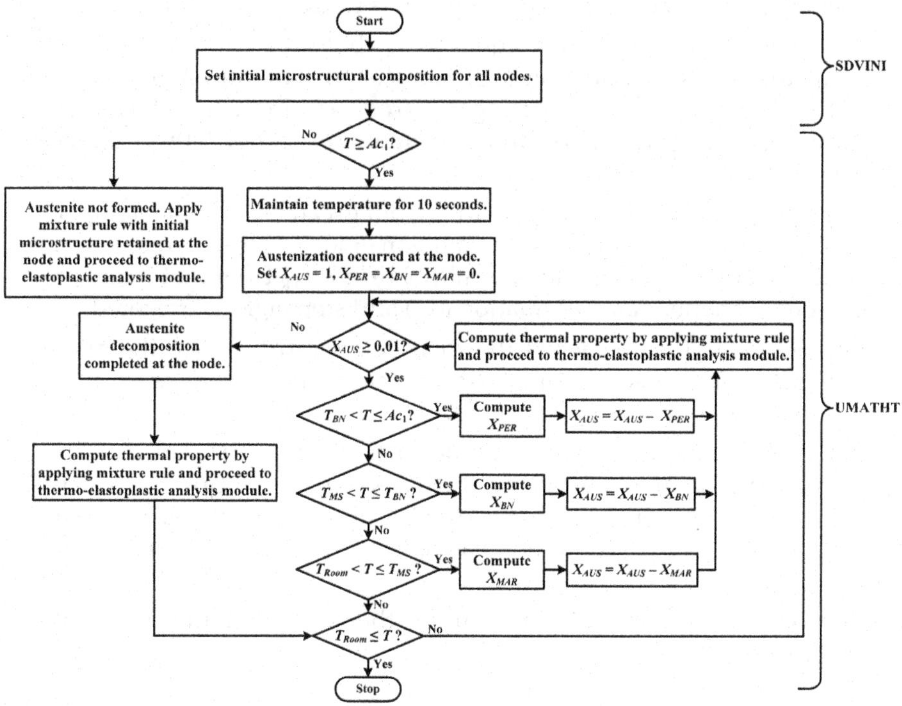

FIGURE 6.9
Flowchart of the algorithm applied for phase transformation kinetics. With permission from Shufen and Dixit (2018). Copyright Elsevier.

ferrite, pearlite, bainite, and martensite by X_{AUS}, X_{FER}, X_{PER}, X_{BN}, and X_{MAR}, respectively, the pseudo code for the algorithm used to simulate the phase transformation kinetics at a particular node is given as follows:

Step One: The subroutine SDVINI is called at the node to initialize a microstructural state. The SDVs storing these volume fractions are passed into the UMATHT subroutine in Step two.

Step Two: Nodal temperature T is checked. If T is greater than Ac_1, then assuming that the node remains at the temperature for more than 10 s, a fully austenitic microstructure with $X_{AUS} = 1$ is assigned at the particular node. Otherwise step eight is executed.

Step Three: The available volume fraction of austenite X_{AUS} is checked. If it is greater than 1%, Step four is executed. Otherwise, the austenite decomposition is considered complete and step eight is executed.

Step Four: If T is less than, or equal to, the lower critical temperature Ac_1, but greater than T_{BN}, X_{PER} is calculated using Equation(6.8) and the current volume fraction of X_{AUS} is updated by subtracting X_{BN} from that in the previous increment using Equation (6.9) and step eight is executed. Otherwise step five is executed.

Step Five: If T is less than or equal to T_{BN} but greater than T_{MS}, X_{BN} is calculated using Equation (6.8) and the current volume fraction of X_{AUS} is updated by subtracting X_{BN} from that in the previous increment using Equation (6.9) and step eight is executed. Otherwise step six is executed.

Step Six: If T is less than or equal to T_{MS} but greater than room temperature, X_{MAR} is calculated using Equation (6.10) and the current volume fraction of X_{AUS} is updated by subtracting X_{MAR} from that in the previous increment using Equation (6.9) and step eight is executed.

Step Seven: If T has reached the room temperature T_{Room}, the microstructure has reached a metastable state and step nine is executed. Otherwise step eight is executed.

Step Eight: Thermal properties at the particular node are calculated by applying the mixture rule on the basis of the resulting microstructural composition, and the UMATHT subroutine for thermal analysis is executed. Execute the elastoplastic analysis by incrementing the time and go to step three.

Step Nine: Stop simulation by calling the exit command XIT in the subroutine.

6.6.4.2 Implementation of Thermal Analysis Module

After the computation of the volume fractions, the linear mixture law given in Equation (A.1) is applied to calculate the thermal properties, such as specific heat and thermal conductivity. The thermal behavior is governed by Fourier's law of heat conduction, which is expressed as follows:

$$\rho c\frac{\partial T(r,t)}{\partial t}=\nabla.\{k\nabla T(r,t)\}+Q(r,t), \tag{6.14}$$

where c is the specific heat at constant volume, ∇ is the gradient operator, $T(r, t)$ is the temperature as a function of spatial coordinates r and time t, k is the thermal conductivity and Q is the rate of internal heat generated per unit volume. The specific heat is defined in the subroutine as:

$$\frac{\partial u}{\partial T}=c, \tag{6.15}$$

where u is the internal energy per unit mass. The latent heat generated in the material during phase transformation causes a change in the internal energy and is included in the expression for the change in internal energy as follows:

$$\mathrm{d}u=\frac{\partial u}{\partial T}\Delta T+\left(\beta\frac{\Delta T}{|\Delta T|}\right)\frac{\Delta H}{\rho}\Delta X_k, \tag{6.16}$$

where ΔT is the incremental change in temperature, ΔH is the enthalpy of formation per unit volume, ΔX_k is the incremental volume fraction of the kth constituent and β is a constant. The value of β is either 0 or 1 depending on the magnitude of ΔT. When ΔT is 0, β is 0 to ensure the that term $\Delta T/|\Delta T|$ does not become undefined. When ΔT is not 0, β is 1. The values of ΔH for austenite–pearlite and austenite–martensite transformations are (1.56×10^9–$1.5\times10^6\,T$) J/m^3, and 640×10^6 J/m^3, respectively. These values have also been used by (Wang et al., 1997).

6.6.4.3 Implementation of the Mechanical Analysis Module

Using index notations, the model for material plasticity is governed by the von Mises yield criterion and the subsequent flow stress yielding is based on Ziegler's linear kinematic hardening rule, which is expressed as follows (Dixit and Dixit, 2008):

$$\sigma_Y=\sqrt{\frac{3}{2}\left(\sigma'_{ij}-\mathrm{d}\alpha_{ij}\right)\left(\sigma'_{ij}-\mathrm{d}\alpha_{ij}\right)}, \tag{6.17}$$

$$\mathrm{d}\alpha_{ij}=\frac{H'}{\sigma_Y}\left(\sigma_{ij}-\alpha_{ij}\right)\mathrm{d}\varepsilon^p_{\text{eq}}, \tag{6.18}$$

where σ'_{ij} is the deviatoric stress tensor, $\mathrm{d}\alpha_{ij}$ is the incremental back stress tensor, σ_{ij} is the stress tensor, H' is the kinematic hardening modulus obtained as the slope of stress and equivalent plastic strain curve, and $d\varepsilon_{eq}^{p}$ is the incremental equivalent plastic strain, which is defined as:

$$d\varepsilon_{eq}^{p} = \sqrt{\frac{2}{3} d\varepsilon_{ij}^{p} d\varepsilon_{ij}^{p}}. \tag{6.19}$$

This takes care of the Bauschinger effect. The total strain increment is defined as the sum of the following components:

$$d\varepsilon_{ij} = d\varepsilon_{ij}^{e} + d\varepsilon_{ij}^{p} + d\varepsilon_{ij}^{th}, \tag{6.20}$$

where $d\varepsilon_{ij}^{eq}$ is the incremental elastic strain, $d\varepsilon_{eq}^{p}$ is the incremental plastic strain, and $d\varepsilon_{ij}^{th}$ is the incremental thermal strain, which are expressed as:

$$\begin{cases} d\varepsilon_{ij}^{e} = \frac{1}{E}\left\{(1+\nu)d\sigma_{ij} - \nu d\sigma_{kk}\delta_{ij}\right\} \\ d\varepsilon_{ij}^{p} = d\lambda \frac{\partial f}{\partial \sigma_{ij}}, \\ d\varepsilon_{ij}^{th} = (\alpha dT)\delta_{ij}, \end{cases} \tag{6.21}$$

where

ν is the Poisson's ratio,

δ_{ij} is the Kronecker delta function,

$d\lambda$ is the plastic multiplier,

f is the von Mises yield function,

α is the coefficient of thermal expansion,

$d\sigma_{ij}$ is the incremental stress tensor,

$d\sigma_{kk}$ is the trace of the incremental stress tensor, and

dT is the incremental change in temperature.

The effect of dilatational strain resulting from phase changes and the strain due to transformation induced plasticity are neglected.

6.7 Mesh Sensitivity Analysis

To optimize the computational time and cost, a mesh sensitivity analysis is generally carried out to select the most efficient mesh that can provide the desired accuracy with the least computational time. The optimum mesh may be decided using a criterion in which the maximum values of the radial, hoop, and axial stresses converge towards approximate constant values during the loading step in a typical thermal autofrettage process. Usually the mesh will be finer in the inner and the outer portion, because these regions are prone to plastic deformation. The intermediate region remains elastic and a coarse mesh can be used here.

6.8 Results of the Heat Treatment Coupled Thermal Autofrettage Process

This section discusses the results of the FEM analysis for the heat treatment coupled thermal autofrettage process. Some typical results have been presented by Shufen and Dixit (2018). The autofrettage specimen was a 120 mm long cylinder with an inner radius of 10 mm and an outer radius of 30 mm made up of 1080 steel. The required properties of this steel are presented in Appendix A. It was assumed that the cylinder was initially composed of 55% pearlite and 45% bainite in its microstructure. Here, additional results are presented based on a new parameter for the same specimen. The discussion is based on the results obtained in the thermal autofrettage step, the reheating step and the final cooling step.

6.8.1 Results at the End of the Thermal Autofrettage Step

The cylinder was subjected to an initial thermal autofrettage with a temperature difference of 325 °C, with an inner wall temperature of 20 °C and an outer wall temperature of 345 °C. The first yielding of the cylinder commenced at the inner wall when the temperature difference reached 229 °C. With further increase in temperature difference, yielding kept on spreading to adjacent portion. This generated a forward moving inner plastic zone. . Increasing temperature gradient kept on increasing the compressive stresses at the outer wall. When the temperature difference reached 323 °C, the outer wall of the cylinder commenced yielding due to the compressive stresses. The distribution of stresses in the cylinder after the thermal loading step and the residual stresses after the unloading step are depicted in Figure 6.10a,b, respectively. The cylinder comprises an inner plastic zone I up to a radius of 11.41 mm, and an elastic zone II almost up to the outer wall. The maximum hoop and axial compressive residual stresses at the inner wall of the cylinder are 231.94 MPa and 260.17 MPa, respectively.

6.8.2 Results at the End of the Austenization Step Due to Reheating

The thermally autofrettaged cylinder was subjected to a radial temperature gradient that austenized the outer wall. The outer wall was heated to 725 °C

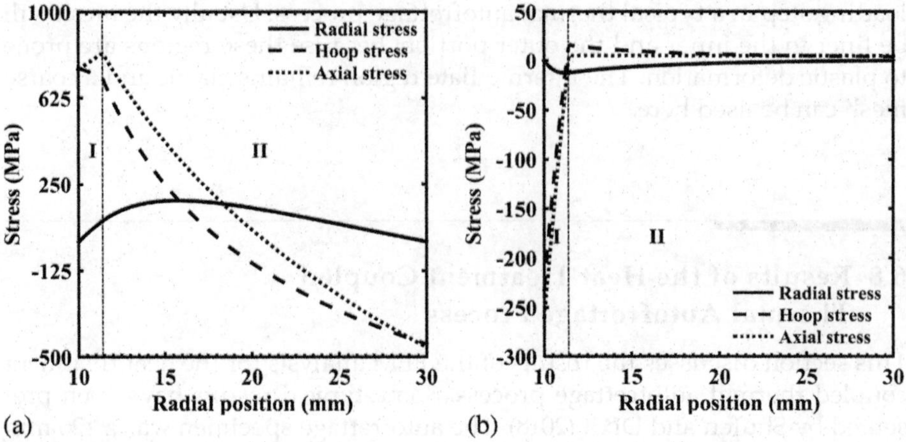

FIGURE 6.10
Radial distribution of (a) stresses after loading step and (b) residual stresses after unloading step in the cylinder thermally autofrettaged with wall temperatures $T_a = 20$ °C and $T_b = 345$ °C.

and the temperature of the inner wall was based on a criterion that the resulting temperature gradient did not further yield the thermally autofrettaged cylinder. This minimum temperature difference was about 60 °C. Thus, the inner wall was heated to 700 °C. The stress distribution after the reheating step is depicted in Figure 6.11. As can be inferred, there is no further yielding in the cylinder and it remains in the same deformed state as obtained from the preceding thermal autofrettage process step.

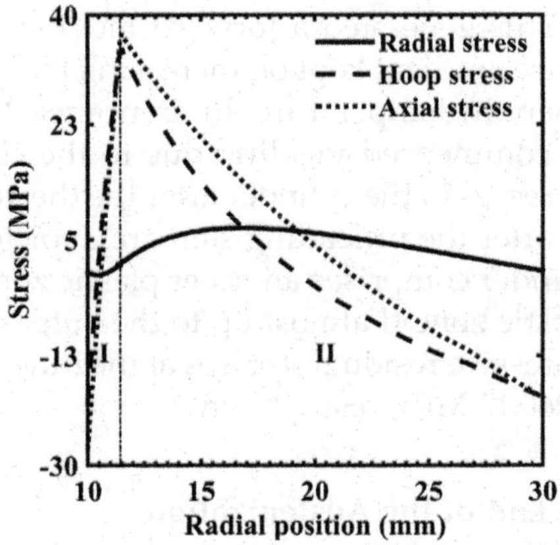

FIGURE 6.11
Distribution of stresses at the end of the reloading step with the temperature boundary conditions $T_a = 700$ °C, $T_b = 725$ °C.

6.8.3 Results at the End of the Final Cooling Step

The austenized cylinder was simultaneously cooled from both the inner and outer walls by subjecting it to the heat transfer boundary conditions given in Equations (6.12) and (6.13). A thermal emissivity e of 0.3 was used as the typical value. Two cases of cooling rates due to $h=300$ W/(m^{2}K) and $h=500$ W/(m^{2}K) were considered. The variation of the hoop and axial residual stresses at the outer surface with time and temperature in each case of cooling rate are depicted in Figure 6.12. The behavior of residual stresses, time,

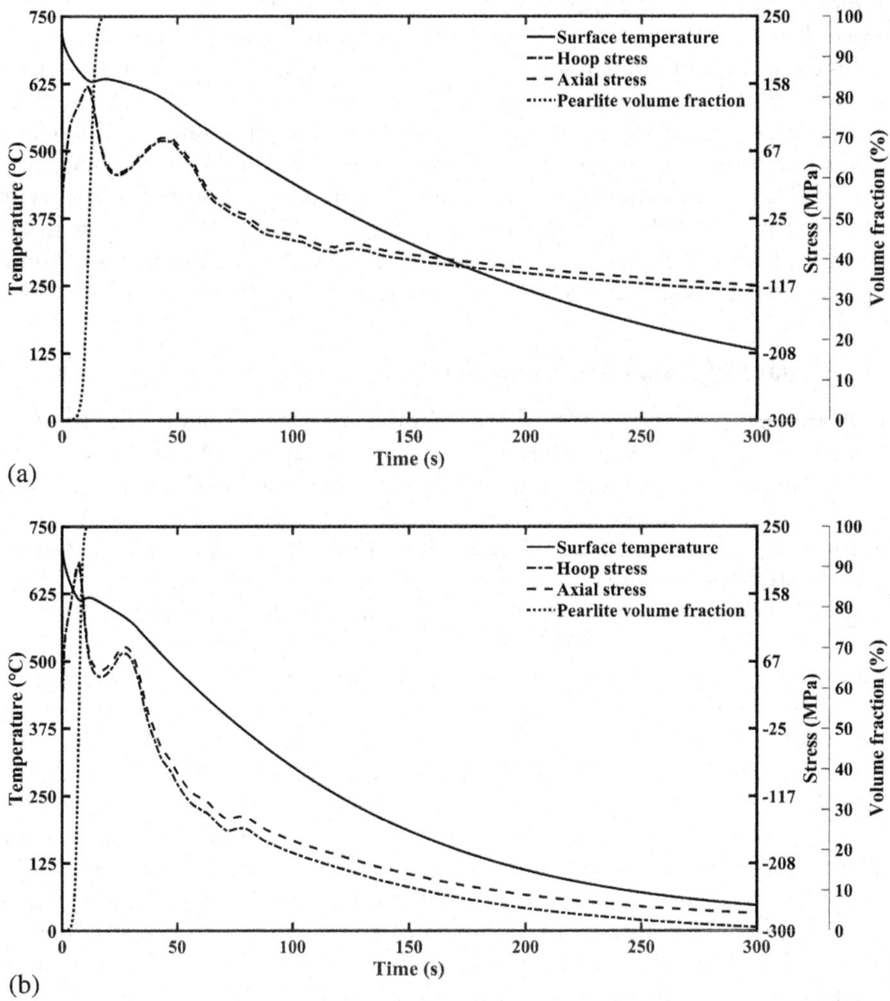

FIGURE 6.12
Variation of temperature, pearlite volume fraction, hoop, and axial stresses with time at the outer wall of the cylinder for (a) $h=300$ W/(m^{2}K) and (b) $h=500$ W/(m^{2}K).

and temperature may be divided into two stages—tensile and compressive. These two stages are explained in the following subsections.

6.8.3.1 Stages of Tensile Behavior

When the austenized outer wall is rapidly cooled, the surface temperature of the outer wall drops instantly. This leads to an immediate contraction of the outer wall surface. The underneath material resists this contraction due to which tensile stresses are generated at the outer wall surface. With further cooling, the temperature of the outer wall surface reaches the lower critical temperature 715 °C. At this temperature, pearlite starts to form as can be inferred from the transformation JMAK curves in Figure 6.12a,b.

This causes a sudden increase in temperature due to the release of latent heat as shown by the temperature variation curve in Figure 6.12a,b. The momentary increase in temperature causes an expansion of the outer wall surface, which counteracts the contraction process and decreases the tensile hoop and axial stresses at the outer wall. This is also evident from Figure 6.12a,b. With time, the rate of pearlite transformation decreases leading to a decrease in the release of latent heat. Afterwards, the contraction of the outer surface dominates and the tensile stresses again start to increase up to some time.

6.8.3.2 Stage of Compressive Behavior

With further cooling, the rest of the cylinder material underneath the outer wall begins to gradually contract. Because of the larger volume, the contraction of the underlying material pulls the already cooled and hardened outer wall surface. This decreases the tensile residual stresses at the outer surface. As the temperature approaches room temperature, the outer initial tensile residual stresses become compressive.

The distribution of residual stresses at the end of the final cooling step with the cooling rates provided by $h=300$ W/(m^{2}K) and $h=500$ W/(m^{2}K) are depicted in Figure 6.13a,b, respectively. The comparison of the maximum compressive residual stresses induced in the cylinder for these two cases are tabulated in Table 6.2. For the case of $h=300$ W/(m^{2}K), the percentage decrease in the maximum hoop and axial residual stresses at the inner wall are 6.5% and 4.18%, respectively. For $h=500$ W/(m^{2}K), the percentage decrease in the maximum hoop and axial residual stresses at the inner wall were 17.79% and 12.39%, respectively. This was because of plastic deformation of the outer wall that created the outer plastic zone III, which decreased the size of the elastic zone II, available at the end of the thermal autofrettage process. As a result, the volume of material that suppressed the inner plastic zone decreased, which also reduced the inner wall compressive residual stresses.

To determine the loss of compressive residual stresses at the inner wall, the cylinder was pressurized with 600 MPa. The pressure-bearing capacity of the cylinder without autofrettage was obtained as 372 MPa.

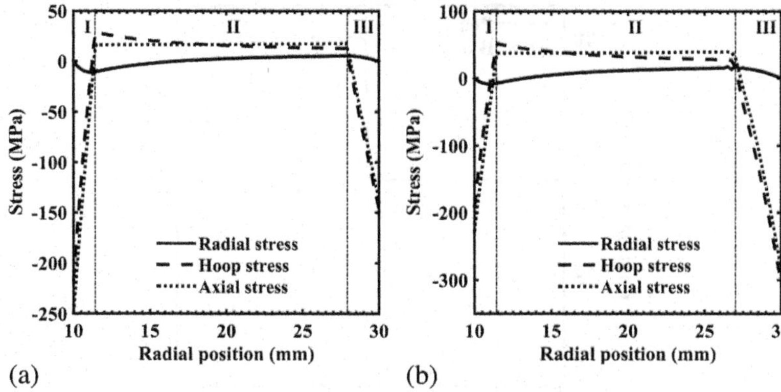

FIGURE 6.13
Distribution of residual stresses due to elastic recovery in the unloading step for various cases of cooling rates given by (a) $h = 300$ W/(m²K) and (b) $h = 500$ W/(m²K).

TABLE 6.1
Comparison of Residual Stresses and Increase in Pressure-Bearing Capacity at Different Cooling Rates

	Maximum compressive residual stress (MPa)				
	Inner wall		Outer wall		
h W/(m².K)	Hoop	Axial	Hoop	Axial	Increase in pressure-bearing capacity (%)
300	216.86	249.29	147.59	139.27	20.97%
500	190.67	227.94	304.77	285.94	20.16%

This increased to 465 MPa after thermal autofrettage, an increase of 25%. Coupled heat treatment slightly decreased the pressure-bearing capacity as evident from Table 6.1; it became 20.97% for cooling with $h = 300$ W/m².K and 20.16% for cooling with $h = 500$ W/m².K. The distribution of microstructural phases in the cylinder after the process is depicted in Figure 6.14. The microstructure in the cylinder retains the initially assumed 55% pearlitic and 45% bainitic state up to a radius of about 20 mm while the remainder of the cylinder comprises a 100% pearlitic state.

6.9 Combined Hydraulic and Thermal Autofrettage

Although the thermal autofrettage is a simpler process of inducing beneficial residual stresses in the pressure vessels, it does not enhance the pressure capacity to the extent possible by a hydraulic autofrettage process. However,

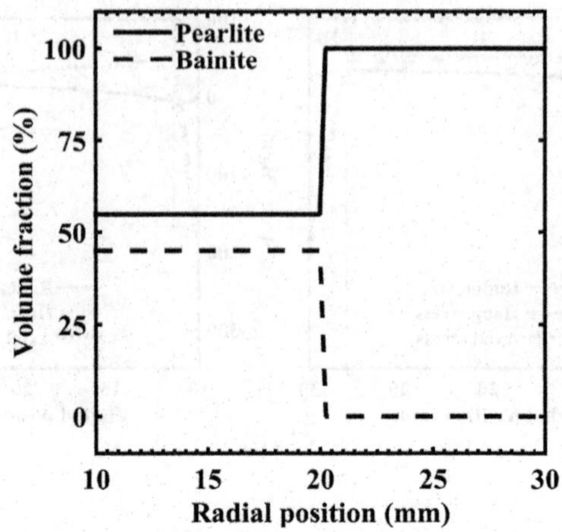

FIGURE 6.14
Distribution of microstructural constituents for the cooling rates corresponding to $h = 300$ W/(m^{2}K) and $h = 500$ W/(m^{2}K).

the hydraulic autofrettage will require a high-pressure capacity hydraulic power pack, which increases the initial, operating, and maintenance costs. A good trade-off can be to combine the two processes. This section presents the theoretical FEM analysis of a hydraulic autofrettage process combined with the thermal autofrettage process. The general procedure for the FEM analysis is explained in the following subsections. Most of these general steps are similar to those already explained in Section 6.6.

6.9.1 Model Description and Boundary Conditions

This study can also be based on the axisymmetric model already explained in Section 6.6.1. The typical model is shown in Figure 6.15, along with the boundary conditions. The inner and outer radii are denoted by a and b, respectively. The symbol r denotes an intermediate radial position. The edges AD and BC represent the inner and outer wall, respectively. The edge AD is subjected to an autofrettage pressure p and temperature $T_a = 20$ °C whereas the edge BC is subjected to temperature $T_b = 20$ °C $+ \Delta T$, where ΔT is the required temperature difference. The edges AB and CD remain traction-free and insulated.

The FEM analysis can be divided into two analysis steps—a loading step and an unloading step. The combined thermal and pressure loading–unloading operation of the cylinder during the autofrettage analysis can be implemented by using two successive linearly ramped amplitude functions—one with a positive slope applied during the loading step, and then one with a negative slope applied during the unloading step. The concept of load step sizing explained in Section 6.6.3 is also applicable in this analysis. A small

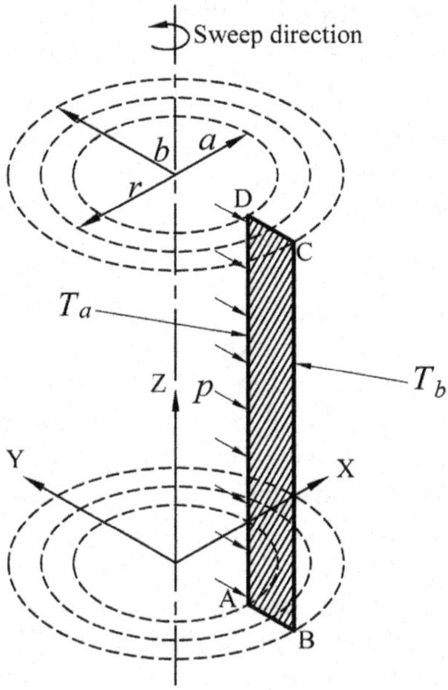

FIGURE 6.15
Schematic of axisymmetric part with boundary conditions. With permission from (Shufen and Dixit, 2017). Copyright American Society of Mechanical Engineers (ASME).

step size will enhance the accuracy of estimated threshold load combination at which the cylinder commences yielding.

6.9.2 Material Behavior and Mesh Sensitivity Analysis

The material must be defined using temperature-dependent properties. This has also been explained in Section 6.6.4. A linear kinematic hardening model governed by Ziegler can be considered as the material behavior to incorporate the Bauschinger effect. The basis for the selection of the optimum mesh is also similar to that explained in Section 6.7.

6.9.3 Typical Results of Thermal Autofrettage Combined with Hydraulic Autofrettage

This section presents and discusses the results for the combined hydraulic–thermal autofrettage process. Some results have been already presented by Shufen and Dixit (2017) for two materials—aluminum and steel alloy SS304. Here, a new set of results is presented for AH36 mild steel, whose temperature-dependent properties are taken from (Zhang et al., 2004). The section presents and discusses the results of the hydraulic–thermal autofrettage process. A 120 mm long cylinder with an inner radius of 10 mm and an outer radius of 30 mm is taken as the typical autofrettage specimen. The sets of results for the combined autofrettage process is presented based on two cases—one based on temperature-dependent mechanical and thermal properties, and another based on constant properties corresponding to an ambient temperature of 20 °C.

For the case considering temperature-dependent properties, the yield pressure of the cylinder in hydraulic autofrettage came out to be 176 MPa. The yield temperature difference in thermal autofrettage came out to be 146 °C. At the temperature difference of 224 °C, the outer wall also starts yielding. When the temperature dependency is not considered and the properties corresponding to a temperature of 20 °C are taken, the inner wall started to yield at 148 °C, and the outer wall started to yield at 315 °C. This value is about 41% greater than that considering temperature-dependent properties. Usually, an attempt is made to avoid outer wall yielding. Ignoring the property variation with temperature significantly overestimates the allowable temperature difference. The minimum temperature difference due to which compressive yielding occurs after unloading is 295 °C, which is 5.36% more than that obtained using temperature-dependent properties.

The yield pressure and yield temperature difference were combined to carry out the autofrettage of the cylinder. The distribution of stresses after loading and residual stresses after unloading are found to be almost identical in both of the cases i.e., considering temperature-dependent and temperature-independent properties. Figure 6.16a shows the stress distribution after loading and Figure 6.16b shows the residual stresses after unloading. The radius of the elastic–plastic interface is 13.5 mm. The compressive residual hoop stress is the maximum at the inner wall with a magnitude of 353 MPa.

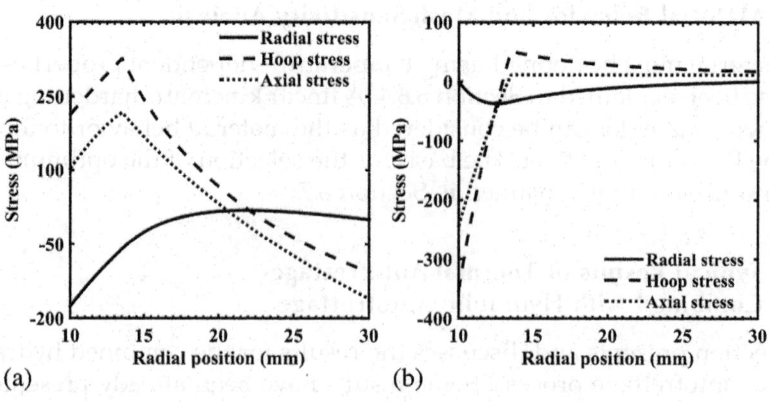

FIGURE 6.16
Distribution of (a) stresses after loading and (b) residual stresses after unloading of the AH36 steel cylinder autofrettaged with the yield pressure and yield temperature difference for both the cases i.e., considering temperature-dependent and temperature-independent properties.

6.9.4 Increase in Pressure-Bearing Capacity of Combined Autofrettaged Cylinder

Based on the results obtained in Section 6.9.3, combined autofrettage is carried out using various possible pressure and temperature difference loading combinations. A series of combined autofrettage processes are carried out using thermal and pressure loading conditions combined in the following manner. The hydraulic pressure is gradually varied from zero to the maximum allowable limit that just avoids the reverse yielding of the inner wall of the cylinder during unloading in a hydraulic autofrettage process. The minimum autofrettage pressure at which the cylinder yields due to compressive residual induced after unloading is 353 MPa for both cases, considering and neglecting temperature-dependent properties. The minimum thermal autofrettage temperature difference at which the cylinder yields due to compressive residual stress induced after cooling is 280 °C for the case with temperature-dependent properties and 295 °C when temperature dependency is neglected.

Figure 6.17 depicts the increase in pressure-bearing capacities for a combination of the autofrettage pressure and temperature difference for the AH36 mild steel specimen. The properties of AH36 mild steel are listed in Tables B.1 and B.2 in Appendix B. Figure 6.17a depicts the results considering temperature-dependent properties while Figure 6.17b depicts a similar distribution by considering the properties at 20 °C. The difference in the results based on two cases of considering and neglecting temperature-dependent properties lies mainly in the pressure that can be applied in combination with a particular temperature difference. As an example, for a temperature difference of 220 °C, the maximum pressure that can be applied without causing reverse yielding is 105 MPa by neglecting temperature-dependent properties and 95 MPa by considering temperature-dependent properties. Both the cases result in about 70.44% increase in the pressure-bearing

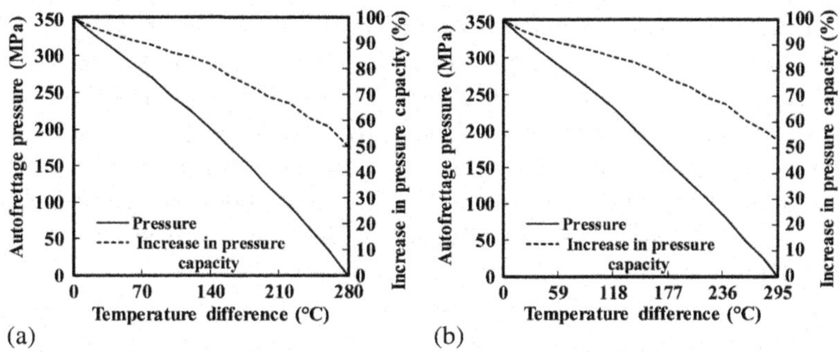

FIGURE 6.17
Percentage increase in pressure-bearing capacities of AH36 cylinder autofrettaged with various combinations of pressure and temperature difference considering (a) temperature-dependent properties and (b) temperature independent properties.

capacity. The decrease in the maximum pressure in case of considering temperature-dependent properties is because of lower flow stress at the higher temperature. At elevated temperature, the material yields at lower stress. Hence, highly realistic results can be obtained only by considering temperature-dependent properties.

The curves depicted in Figure 6.17 clearly highlight the significance of the thermal load. For the case considering temperature-dependent properties a hydraulic pressure of 175 MPa does not increase the pressure-bearing capacity as it is not sufficient for carrying out hydraulic autofrettage. When this pressure applied with a temperature difference of 160 °C, it results in 80.53% (with respect to the original pressure-bearing capacity of 176 MPa) increase in the pressure-bearing capacity. Similarly, considering the temperature-independent properties, a hydraulic pressure of 180 MPa results in only 1.86% in the pressure-bearing capacity, but when applied with a temperature difference of 160 °C, it causes 80.53% increase in the pressure-bearing capacity. Figure 6.17 may be useful for a design strategy where, for a desired increase in the pressure-bearing capacity of a cylinder, the feasible combination of thermal and pressure loading has to be identified.

6.10 Thermal Autofrettage with Shrink-Fitting and Its Effect on Fatigue Life

This section presents the theoretical analysis of increasing the fatigue life performance of a thermally autofrettaged cylinder by shrink-fitting. Autofrettaged cylinders are generally shrunk-fit to increase the pressure-bearing capacity as well as to maximize the fatigue life. This may also reduce the Bauschinger effect. Literature includes the studies carried out by (Abdelsalam and Sedaghati, 2013; Bhatnagar, 2013; Jahed et al., 2005; Lee et al., 2009; Parker, 2000; Parker and Kendall, 2003; Seifi, 2018). The theoretical analysis of a shrink-fitted thermally autofrettaged cylinder has been carried out by (Kamal and Dixit, 2016). The researchers considered a thick-walled thermally autofrettaged cylinder that is shrink-fitted with an outer cylinder made up of the same material. A contact pressure at the interface of the two cylinders due to the shrink fit. The enhancement in the maximum pressure-bearing capacity and fatigue lifetime through shrink-fit is presented in the following subsections.

6.10.1 Enhancement in the Maximum Pressure-Bearing Capacity

This Subsection is based on the work carried out by (Kamal and Dixit, 2016). A 20 mm thick cylinder having an inner radius 10 mm and outer radius 30 mm was taken as the typical case of the autofrettage specimen. This cylinder was initially thermally autofrettaged and then shrink-fitted to an outer cylinder with an outer radius of 40 mm. Both of the cylinders were made of stainless steel SS304. The distribution of the residual stresses in the thermally autofrettaged cylinder after performing shrink-fitting was given by the superposition of the residual stresses due to the shrink-fit with those due to thermal autofrettage. The residual stresses obtained in the outer layer were solely due to shrink-fitting. A design strategy was developed to improve the maximum pressure-bearing capacity by using suitable combinations of temperature difference (T_b–T_a) and shrink-fit allowance δ. The pressure-bearing capacity of the thermally autofrettaged cylinder with shrink-fit was at the maximum for the combination of temperature difference (T_b–T_a) = 103 °C and shrink-fit allowance δ = 0.027 mm. The maximum pressure-bearing capacity of the shrink-fitted thermally autofrettaged cylinder was 159 MPa. This was compared with the case of a single monobloc cylinder with the same thickness i.e., having an inner radius of 10 mm and an outer radius of 40 mm. After the cylinder was thermally autofrettaged with a maximum temperature difference of 122 °C without shrink-fitting, the pressure-bearing capacity was 133.90 MPa. Therefore, the maximum pressure-bearing capacity increased by 65.88% in case of the shrink-fitted thermally autofrettaged and 39.34% in case of the thermally autofrettaged monobloc cylinder without shrink-fit. This depicted the significance of the shrink-fitting. The distribution of the net residual stresses in the shrink-fitted thermally autofrettaged as well as in outer monobloc cylinder is depicted in Figure 6.18a,b, respectively. As observed, the residual hoop stress in the inner autofrettaged cylinder is

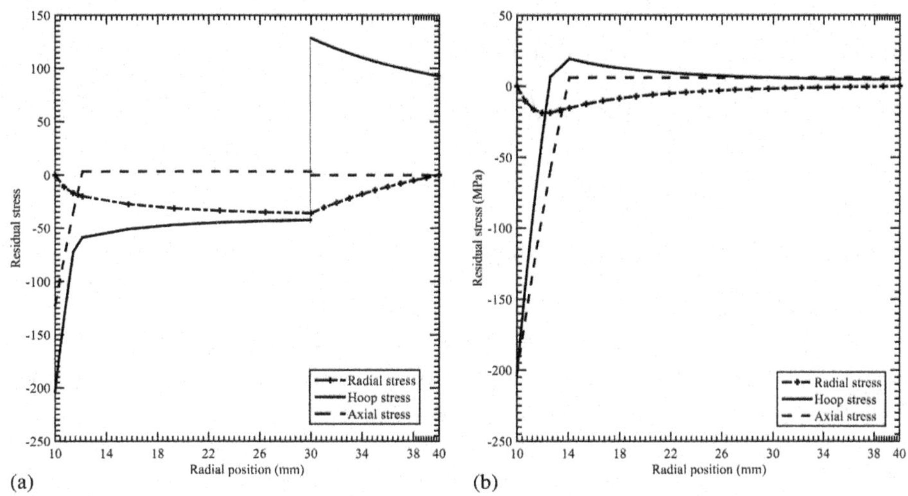

FIGURE 6.18
Distribution of residual in (a) shrink-fitted thermally autofrettaged SS304 cylinder and (b) thermally autofrettaged monobloc SS304 cylinder. With permission from (Kamal and Dixit, 2016). Copyright ASME.

compressive throughout the thickness. Hence, shrink-fitting also results in additional increases in the fatigue life performance of the compound cylinder than in the case of a thermally autofrettaged monobloc cylinder.

6.10.2 Enhancement in Fatigue Life Performance

The fatigue life of the shrink-fitted thermally autofrettaged is estimated numerically using Paris' law (Paris and Erdogan, 1963). Paris' law is a power law expression that relates the stress intensity factor range to the crack growth rate during a state of cyclic loading. Mathematically, it is expressed as:

$$\frac{dl}{dN} = A(\Delta K_I)^m, \tag{6.22}$$

where l is the crack length, N is the fatigue life in cycles, A and m are constants of the material, and ΔK_I is the range of mode I stress intensity factor. The expression for K_I for a straight-fronted longitudinal crack having a crack-depth ratio of $l/w<0.25$ (l is the crack depth and w is the wall thickness) due to an applied pressure p is given by (Underwood, 1972):

$$K_I = 1.12\sigma_\theta\sqrt{\pi l} + 1.13p\sqrt{\pi l} + 1.12\sigma_{\theta R}\sqrt{\pi l}, \tag{6.23}$$

where σ_θ denotes the hoop stress and $\sigma_{\theta R}$ denotes the residual hoop stress. For the case of cyclic pressurization, the range of stress intensity, ΔK_I can be expressed as:

$$\Delta K_I = K_{Imax} - K_{Imin}, \tag{6.24}$$

where K_{Imax} and K_{Imin} are given by:

$$K_{Imax} = 1.12\sigma_{\theta max}\sqrt{\pi l} + 1.13p_{max}\sqrt{\pi l} + 1.12\sigma_{\theta R}\sqrt{\pi l}, \tag{6.25}$$

$$K_{Imin} = 1.12\sigma_{\theta min}\sqrt{\pi l} + 1.13p_{min}\sqrt{\pi l} + 1.12\sigma_{\theta R}\sqrt{\pi l}. \tag{6.26}$$

where p_{max} and p_{min} are the maximum and minimum pressure in the cycle, respectively. It is to be noted that K_{Imin} may result into a negative value for large compressive residual stresses. In that case, K_{Imin} is set to zero that provides $\Delta K_I = K_{Imax}$. Thus, from Equation (6.25) and defining the hoop stress range $\Delta\sigma_\theta$ using Lamé hoop stress solution for a pressure range Δp, the non-dimensional stress intensity factor can be expressed as:

$$\frac{\Delta K_I}{\Delta K_o} = 1.12\left\{\begin{bmatrix}\dfrac{\left(1+\dfrac{b^2}{r^2}\right)}{\left(W^2-1\right)} + 1.009 + \dfrac{\sigma_{\theta R}}{\Delta p}\end{bmatrix}\right\}, \tag{6.27}$$

where $\Delta K_o = \Delta p\sqrt{\pi l}$ and $W = b/a$. In case of non-autofrettaged cylinder, $\sigma_{\theta R}$ is zero. The crack depth ratio (l/w) is defined as:

$$\frac{l}{w} = \frac{r-a}{b-a}, \tag{6.28}$$

the non-dimensional stress intensity factor, $\frac{\Delta K_I}{\Delta K_o}$ can be expressed as function of (l/w) from Equation (6.27). Thus, expressing $\frac{\Delta K_I}{\Delta K_o} = f(l/w)$, Paris' equation (Equation 6.22) can be rearranged to obtain the following equation:

$$\frac{dl}{dN} = A(\pi)^{\frac{m}{2}}(\Delta p)^m (l)^{\frac{m}{2}} \{f(l/w)\}^m. \tag{6.29}$$

Integration of Equation (6.8) provides the fatigue life as:

$$N_f = \frac{w^{1-\frac{m}{2}}}{A(\pi)^{\frac{m}{2}}(\Delta p)^m} \int_{(l/w)_i}^{(l/w)_f} \frac{d(l/w)}{(l/w)^{\frac{m}{2}}\left[f(l/w)\right]^m}. \tag{6.30}$$

The lower and upper limit of integration $(l/w)_i$ and $(l/w)_f$ in Equation (6.30) are the initial and final crack depth ratios, respectively. Thus, the fatigue life of an internally pressurized thick-walled cylinder internal working pressure can be evaluated using Equation (6.30). The integral can be evaluated using a numerical technique, for example Simpson's 1/3rd rule.

For the estimation of fatigue life of the shrink-fitted thermally autofrettaged SS304 cylinder, the distribution of ($\Delta K/\Delta K_\theta$) was first obtained as a function of crack depth ratio for different safe working pressures including the maximum working pressure. The distribution is depicted in Figure 6.19. As observed, the value of ($\Delta K/\Delta K_\theta$) increases up to the elastic–plastic radius of elastic–plastic interface after which it decreases gradually. Also, the same trend is observed in the increase of the stress intensity factor with the working pressure.

For the evaluation of the fatigue life, the initial crack depth ratio was taken as 0.001 (Rees, 1990) and the final crack depth ratio was taken as 0.25, following the ASME pressure vessel code (ASME, 2007).

In this range of crack depth ratio, the values of $f(l/w)$ are taken from Figure 6.19. for the present case. For SS304 the material constants are taken as $A = 5.5 \times 10^{-12}$ and $m = 3.25$ (Barsom and Rolfe, 1999). The fatigue life of the shrink-fitted thermally autofrettaged SS304 cylinder was evaluated from Equation (6.30) for various working pressures. The resulting fatigue lives at each case of working pressure were also computed for monobloc cylinders with and without thermal autofrettage, and compared with those of the shrink-fitted thermally autofrettaged cylinder. These comparisons are

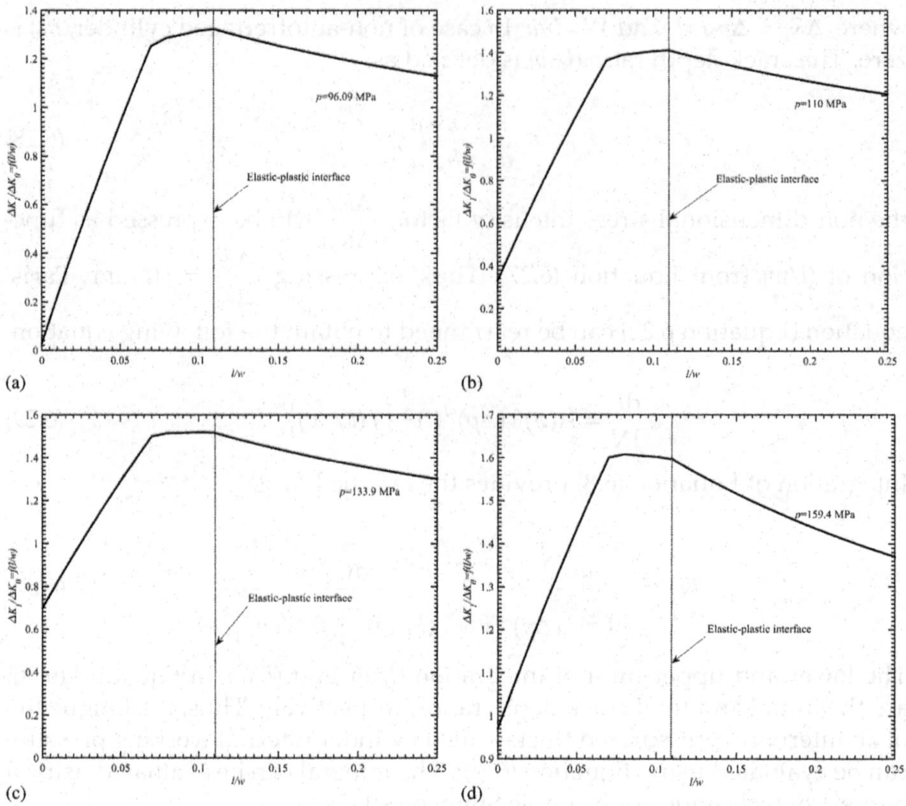

FIGURE 6.19
Stress intensity factors for a straight-fronted longitudinal crack in the shrink-fitted thermally autofrettaged cylinder for different working pressures: (a) $p=96$ MPa, (b) $p=110$ MPa, (c) $p=134$ MPa, (d) $p=159$ MPa (Kamal and Dixit, 2019).

TABLE 6.2

Fatigue Lives of Cylinders for Different Range of Working Pressures (Kamal and Dixit, 2019)

	Fatigue lives (Cycles)		
Working pressure (MPa)	**Shrink-fitted thermally autofrettaged**	**Thermally autofrettaged monobloc cylinder**	**Non-autofrettaged monobloc cylinder**
96	241,38,54,31	241,20,85,28	13,679
110	33,63,670	33,06,248	–
134	1,88,934	1,85,362	–
159	39,674	–	–

depicted in Table 6.2. As observed, there is a considerable improvement in the fatigue life performance in the shrink-fitted thermally autofrettaged cylinder for a particular safe working pressure.

6.11 Conclusion

This chapter presented the theoretical analysis of combining the conventional thermal autofrettage with heat treatment and other frettaging techniques—hydraulic autofrettage and shrink-fitting. In the heat treatment coupled thermal autofrettage process, a thermally autofrettaged thick-walled cylinder was subjected to a secondary temperature gradient where the outer wall is austenized by heating above the lower critical temperature while the temperature of the inner wall was kept a lower temperature. Then the cylinder was simultaneously cooled from both the inner and outer walls. In the combined thermal and hydraulic autofrettage process, a thick-walled cylinder was subjected to both a pressure loading and a temperature gradient during the loading step of a typical autofrettage process. This helped in economizing the process in terms of the pressure requirement for a desired level of increase in the pressure-bearing capacity. In the analysis of a shrink-fitted thermally autofrettaged cylinder, a cylinder initially subjected to thermal autofrettage was shrink-fitted to an outer cylinder made of the same material as the autofrettaged cylinder. It was depicted that the static and dynamic performance of a thermally autofrettaged cylinder can be improved by performing shrink-fitting. The chapter introduced the reader to the basics of metallurgy such as microstructure, the phase diagram of iron carbon alloy (especially steel), and the kinetics of isothermal and anisothermal phase transformation. This was followed by the description of the FEM modeling procedure for incorporating phase transformation kinetics in FEM package ABAQUS. Typical results for the heat treatment coupled thermal autofrettage process were presented. After this, the FEM modeling procedure for the combined thermal and hydraulic autofrettage process was explained and typical results were discussed. Finally, the theoretical analysis of the shrink-fitted thermally autofrettaged cylinder was explained along with the presentation of typical results.

Acknowledgments

This chapter is mainly based on the following articles:

(1) Kamal, S.M. and Dixit, U.S., (2016), A study on enhancing the performance of thermally autofrettaged cylinder through shrink-fitting, *Journal of Manufacturing Science and Engineering*, **138**, 094501–094501-5. https://doi.org/10.1115/1.4033083.

(2) Shufen, R. and Dixit, U.S., (2017), A finite element method study of combined hydraulic and thermal autofrettage process, *Journal of Pressure Vessel Technology*, **139**, 041204–041204-9. https://doi.org/10.1115/1.4036143.

(3) Shufen, R. and Dixit, U.S., (2018), An analysis of thermal autofrettage process with heat treatment, *International Journal of Mechanical Sciences*, **144**, 134–145. https://doi.org/10.1016/j.ijmecsci.2018.05.053.

Authors are grateful to the publishers of these articles.

References

Abdelsalam, O.R. and Sedaghati, R., (2013), Design optimization of compound cylinders subjected to autofrettage and shrink-fitting processes, *Journal of Pressure Vessel Technology*, **135**, 021209–021209-11. https://doi.org/10.1115/1.4007960

ASME, (2007), ASME Boiler and Pressure Vessel Code, 2007, "'Rules for Construction of High Pressure Vessels,'" Division 3, Section 8, Article KD-4 74–76.

Barsom, J.M. and Rolfe, S.T., (1999), *Fracture and Fatigue Control in Structures: Applications of Fracture Mechanics*, 3rd edition. ASTM International, West Conshohocken, PA.

Bhatnagar, R.M., (2013), Modelling, validation and design of autofrettage and compound cylinder, *European Journal of Mechanics – A/Solids*, **39**, 17–25. https://doi.org/10.1016/j.euromechsol.2012.09.013

Callister, W.D. and Rethwisch, D.G., (2009), *Materials Science and Engineering: An Introduction*, 8th edition. John Wiley and Sons, Hoboken, NJ.

de Oliveira, W.P., Savi, M.A., Pacheco, P.M.C.L. and de Souza, L.F.G., (2010), Thermomechanical analysis of steel cylinders quenching using a constitutive model with diffusional and non-diffusional phase transformations, *Mechanics of Materials*, **42**, 31–43. https://doi.org/10.1016/j.mechmat.2009.09.006

Dixit, P.M. and Dixit, U.S., (2008), *Modeling of Metal Forming and Machining Processes: by Finite Element and Soft Computing Methods*, 2008 edition. Springer, London, UK.

Dutta, P.P., Kalita, K., Dixit, U.S. and Liao, H., (2017), Magnetic-force-assisted straightening of bent mild steel strip by laser irradiation, *Lasers in Manufacturing and Materials Processing*, **4**, 206–226. https://doi.org/10.1007/s40516-017-0047-x

Franklin, G.J. and Morrison, J.L.M., (1960), Autofrettage of cylinders: prediction of pressure/external expansion curves and calculation of residual stresses, *Proceedings of the Institution of Mechanical Engineers*, **174**, 947–974. https://doi.org/10.1243/PIME_PROC_1960_174_069_02

Jahed, H., Farshi, B. and Karimi, M., (2005), Optimum autofrettage and shrink-fit combination in multi-layer cylinders, *Journal of Pressure Vessel Technology*, **128**, 196–200. https://doi.org/10.1115/1.2172957

Kamal, S.M., Borsaikia, A.C. and Dixit, U.S., (2016), Experimental assessment of residual stresses induced by the thermal autofrettage of thick-walled cylinders, *The Journal of Strain Analysis for Experimental Design*, **51**, 144–160. https://doi.org/10.1177/0309324715616005

Kamal, S.M. and Dixit, U.S., (2016), A study on enhancing the performance of thermally autofrettaged cylinder through shrink-fitting, *Journal of Manufacturing Science and Engineering*, **138**, 094501–094501-5. https://doi.org/10.1115/1.4033083

Kamal, S.M. and Dixit, U.S., (2019), Enhancement of fatigue life of thick-walled cylinders through thermal autofrettage combined with shrink-fit, in *Strengthening and Joining by Plastic Deformation, Lecture Notes on Multidisciplinary Industrial Engineering*, edited by U.S. Dixit and R.G. Narayanan, Springer, Singapore, pp. 1–30.

Kamal, S.M., Dixit, U.S., Roy, A., Liu, Q. and Silberschmidt, V.V., (2017), Comparison of plane-stress, generalized-plane-strain and 3D FEM elastic–plastic analyses of thick-walled cylinders subjected to radial thermal gradient, *International Journal of Mechanical Sciences*, **131–132**, 744–752. https://doi.org/10.1016/j.ijmecsci.2017.07.034

Koh, S.K., Lee, S.I., Chung, S.H. and Lee, K.Y., (1997), Fatigue design of an autofrettaged thick-walled pressure vessel using CAE techniques, *International Journal of Pressure Vessels and Piping*, **74**, 19–32. https://doi.org/10.1016/S0308-0161(97)00066-5

Koh, S.K. and Stephens, R.I., (1991), Stress analysis of an autofrettaged thick-walled pressure vessel containing an external groove, *International Journal of Pressure Vessels and Piping*, **46**, 95–111. https://doi.org/10.1016/0308-0161(91)90071-9

Koistinen, D.P. and Marburger, R.E., (1959), A general equation prescribing the extent of the austenite-martensite transformation in pure iron-carbon alloys and plain carbon steels, *Acta Metallurgica*, **7**, 59–60. https://doi.org/10.1016/0001-6160(59)90170-1

Lee, E.-Y., Lee, Y.-S., Yang, Q.-M., Kim, J.-H., Cha, K.-U. and Hong, S.-K., (2009), Autofrettage process analysis of a compound cylinder based on the elastic-perfectly plastic and strain hardening stress-strain curve, *Journal of Mechanical Science and Technology*, **23**, 3153–3160. https://doi.org/10.1007/s12206-009-1009-9

Paris, P. and Erdogan, F., (1963), A critical analysis of crack propagation laws, *Journal of Basic Engineering*, **85**, 528–533. https://doi.org/10.1115/1.3656900

Parker, A.P., (1981), Stress intensity and fatigue crack growth in multiply-cracked, pressurized, partially autofrettaged thick cylinders, *Fatigue & Fracture of Engineering Materials & Structures*, **4**, 321–330. https://doi.org/10.1111/j.1460-2695.1981.tb01129.x

Parker, A.P., (2000), Bauschinger effect design procedures for compound tubes containing an autofrettaged layer, *Journal of Pressure Vessel Technology*, **123**, 203–206. https://doi.org/10.1115/1.1331281

Parker, A.P. and Kendall, D.P., (2003), Residual stresses and lifetimes of tubes subjected to shrink fit prior to autofrettage, *Journal of Pressure Vessel Technology*, **125**, 282–286. https://doi.org/10.1115/1.1593074

Phadke, S., Pauskar, P. and Shivpuri, R., (2004), Computational modeling of phase transformations and mechanical properties during the cooling of hot rolled rod, *Journal of Materials Processing Technology*, **150**, 107–115. https://doi.org/10.1016/j.jmatprotec.2004.01.027

Rees, D.W.A., (1990), Autofrettage theory and fatigue life of open-ended cylinders, *The Journal of Strain Analysis for Engineering Design*, **25**, 109–121. https://doi.org/10.1243/03093247V252109

Sedighi, M. and Jabbari, A.H., (2013), Investigation of residual stresses in thick-walled vessels with combination of autofrettage and wire-winding, *International Journal of Pressure Vessels and Piping*, **111–112**, 295–301. https://doi.org/10.1016/j.ijpvp.2013.09.003

Seifi, R., (2018), Maximizing working pressure of autofrettaged three layer compound cylinders with considering Bauschinger effect and reverse yielding, *Meccanica*, **53**, 2485–2501. https://doi.org/10.1007/s11012-018-0834-2

Seifi, R. and Babalhavaeji, M., (2012), Bursting pressure of autofrettaged cylinders with inclined external cracks, *International Journal of Pressure Vessels and Piping*, **89**, 112–119. https://doi.org/10.1016/j.ijpvp.2011.10.018

Shufen, R. and Dixit, U.S., (2017), A finite element method study of combined hydraulic and thermal autofrettage process, *Journal of Pressure Vessel Technology*, **139**, 041204–041204-9. https://doi.org/10.1115/1.4036143

Shufen, R. and Dixit, U.S., (2018), An analysis of thermal autofrettage process with heat treatment, *International Journal of Mechanical Sciences*, **144**, 134–145. https://doi.org/10.1016/j.ijmecsci.2018.05.053

Underwood J.H., (1972), Stress Intensity Factors for Internally Pressurized Thick-Walled Cylinders, ASTM STP 513, Part 1, pp. 59–70.

Wang, K.F., Chandrasekar, S. and Yang, H.T.Y., (1997), Experimental and computational study of the quenching of carbon steel, *Journal of Manufacturing Science and Engineering*, **119**, 257–265. https://doi.org/10.1115/1.2831102

Zhang, L., Reutzel, E.W. and Michaleris, P., (2004), Finite element modeling discretization requirements for the laser forming process, *International Journal of Mechanical Sciences*, **46**, 623–637.

Appendix A

Properties of 1080 Steel

The temperature and microstructure dependent properties of 1080 steel are given in Tables A.1–A.6. It is a high carbon steel containing about 0.8% carbon by weight, which corresponds to eutectoid composition. The Poisson's ratio of this steel is about 0.3. The material being made up of various microstructural constituents, the magnitude of each mechanical and thermal property can be computed as an aggregate of that of each constituent existent at the particular temperature using the following linear law of mixtures:

$$P = \sum_{k=1}^{N} X_k P_k, \tag{A.1}$$

where P is the aggregate property, P_k is the property of a kth constituent, X_k is the volume fraction of the kth constituent, and N is the number of constituents present in the alloy. TTT diagram of 1080 steel is depicted in Figure A.1.

TABLE A.1

Temperature-Dependent Young's Modulus of Elasticity. With Permission from Shufen and Dixit (2018) Copyright Elsevier

Temperature (°C)	Young's modulus of elasticity, E (GPa)
12	221
461	165
522	153
570	136
695	62
883	20.45
947	11.5
1000	10

TABLE A.2

Temperature-Dependent Yield Stress and Hardening Modulus of 1080 steel. With Permission from Shufen and Dixit (2018) Copyright Elsevier

Temperature (°C)	σ_Y (MPa)	Hardening modulus, H (MPa)
10	714	25323
554	291	10937
600	215	9010
645	129	6013
720	59	3433
850	40	2231
1000	40	1403

TABLE A.3

Temperature and Microstructure Dependent Specific Heat. With permission from Shufen and Dixit (2018) Copyright Elsevier

	Specific heat, c (J/kg.°C)		
Temperature (°C)	Austenite	Pearlite	Martensite
10	560	479	560
242	564	556	564
388	568	617	568
626	574	738	574
690	585	774	585
1200	679	774	679

TABLE A.4

Temperature and Microstructure Dependent Thermal Conductivity. With Permission from Shufen and Dixit (2018) Copyright Elsevier

	Thermal conductivity, k (W/m.°C)		
Temperature (°C)	**Austenite**	**Pearlite**	**Martensite**
15	11.25	50	41
520	19.5	35	41
790	24	30.5	41
1200	31		

TABLE A.5

Temperature-Dependent Coefficient of Thermal Expansion of 1080 Steel. With Permission from Shufen and Dixit (2018) Copyright Elsevier

Temperature (°C)	Coefficient of thermal expansion, α (/°C)$\times 10^{-6}$
10	15.2
557	15.8
750	21

TABLE A.6

Temperature and Microstructure Dependent Density. With Permission from Shufen and Dixit (2018) Copyright Elsevier

	Density, ρ (kg/m^3)	
Temperature (°C)	**Austenite/Martensite**	**Pearlite**
10	8005	7853
438	7788	7725
783	7618	7592
1200	7405	7592

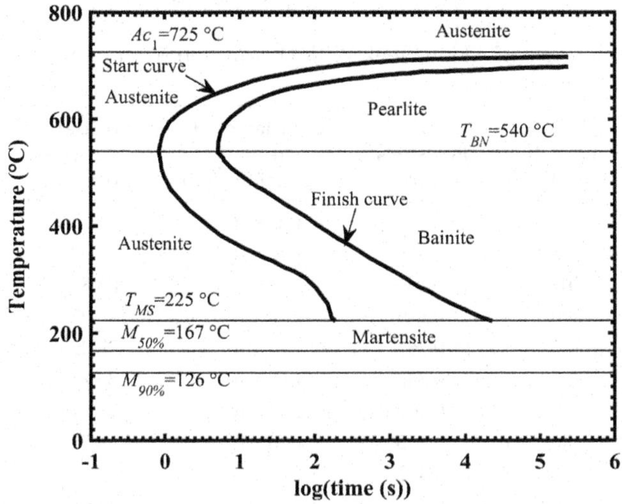

FIGURE A.1
Time–temperature transformation (TTT) diagram of 1080 steel. With permission from (Shufen and Dixit, 2018). Copyright Elsevier.

Appendix B

Properties of AH36 Steel

TABLE B.1
Thermal Properties of AH36 Steel

Temperature (°C)	Thermal conductivity (W/m.°C)	Specific heat (kJ/kg.°C)
0	51.9	486
100	51.1	486
200	48.6	498
300	44.4	515
400	42.7	536
500	39.4	557
600	35.6	586
700	31.8	619
800	26	691
900	26.4	695
1000	27.2	691
3000	120	700

With Permission from Zhang et al. (2004) Copyright Elsevier.

TABLE B.2

Mechanical Properties of AH36 Steel. With Permission from Zhang et al. (2004) Copyright Elsevier.

Temperature (°C)	Elastic modulus (GPa)	Poisson's ratio	Coefficient of thermal expansion (/°C)	Initial yield stress (MPa)	Flow stress at 0.1 plastic strain (MPa)
20	206	0.296	11.7×10^{-6}	344.64	422.64
100	203	0.311	11.7×10^{-6}	331.93	409.93
200	201	0.33	12.2×10^{-6}	308.3	386.3
300	200	0.349	12.8×10^{-6}	276.07	342.57
400	165	0.367	13.3×10^{-6}	235.22	290.22
500	120	0.386	13.8×10^{-6}	185.77	230.77
600	60	0.405	14.4×10^{-6}	127.71	162.71
700	40	0.423	14.8×10^{-6}	68.55	96.05
800	30	0.442	14.8×10^{-6}	64.35	84.35
900	20	0.461	14.8×10^{-6}	45.65	60.65
100	10	0.48	14.8×10^{-6}	11.32	21.32
300	10	0.48	14.8×10^{-6}		

7

Rotational Autofrettage

7.1 Introduction

Rotational autofrettage is a recent development in the area of autofrettage technology. The principle behind any autofrettage process is the generation of beneficial compressive residual stresses at and around the inner surface of a thick-walled cylinder through the plastic deformation of the wall by employing a certain loading at the inner side, before subsequently unloading it. Upon loading the cylinder in service by working mechanical or thermal loads, the compressive residual stress at the inner side offsets the tensile hoop stress due to the working load. This in turn enhances the elastic strength and service life of the cylinder. When this beneficial effect is achieved by rotating the cylinder at a sufficiently high angular velocity about its own axis, and then bringing it back to rest by gradually reducing the angular velocity to zero, this process is known as rotational autofrettage. The idea of rotational autofrettage was first conceived by Zare and Darijani (2016). They carried out a theoretical analysis of the process considering the plane–strain condition. The analysis was based on the Tresca yield criterion with the incorporation of the Bauschinger effect devoid of strain hardening. In a later paper, Zare and Darijani (2017) extended their analysis in order to take into account the effect of strain hardening. They assumed a linear strain-hardening model for the analysis. The work of Zare and Darijani was intended for the strengthening of long, thick-walled cylindrical vessels. The thick-walled hollow circular disks are also widely used in many industries. For example, the rotating disks used in the automobile and aerospace industries, where they are subjected to very high tensile stress in service. Moreover, the outer wall of a butterfly valve and the fastener holes used in high-pressure pipelines need to be able to withstand high pressures during service. Such components need to be strengthened in order to increase their load-carrying capacity in service. This aspect has been studied by Rees (2011) in the swaging of disks, and by Kamal and Dixit (2015) in thermal autofrettage. The rotational autofrettage process may also be a potential procedure for the strengthening of thick-walled disks. To this end, Kamal (2018) developed a theoretical model for the strengthening of thick-walled hollow circular disks through the method of rotational autofrettage. His analysis was based on the plane–stress assumption, the Tresca yield criterion, and its associated flow rule. The effect of strain hardening during plastic flow was taken into account using Ludwik's hardening law. However, the effect of the Bauschinger effect was neglected, and the unloading process was considered to be entirely elastic. Using the analysis of Kamal (2018), Kamal and Kulsum (2019) carried out a parametric study of the influencing parameters during the rotational autofrettage of an axisymmetric circular disk. The available literature on rotational autofrettage analyses the process theoretically for long cylinders and thin hollow circular disks. No experimental study has been reported to date. Nevertheless, a conceptual design of an experimental setup for rotational autofrettage has been presented in Zare and Darijani (2016). The present chapter is devoted to the theoretical analysis of the process based on the available literature. This chapter is organized as follows. Section 7.2 describes the concept of rotational autofrettage. In Section 7.3, the plane–stress model of Kamal (2018) applicable for thin disks is presented. Section 7.4 presents the plane–strain model of Zare and Darijani (2016), which is suitable for long cylinders. A feasibility study of the process is carried out in Section 7.5. Section 7.6 concludes the chapter.

7.2 The Concept of Rotational Autofrettage

In rotational autofrettage, the beneficial residual stresses are generated in the vicinity of the inner wall of a circular disk or cylinder due to the loading and unloading of a centrifugal force. The centrifugal force occurs due to the rotation of the disk/cylinder at high angular velocity. A simple schematic of the rotational autofrettage setup is shown in Figure 7.1a. The setup

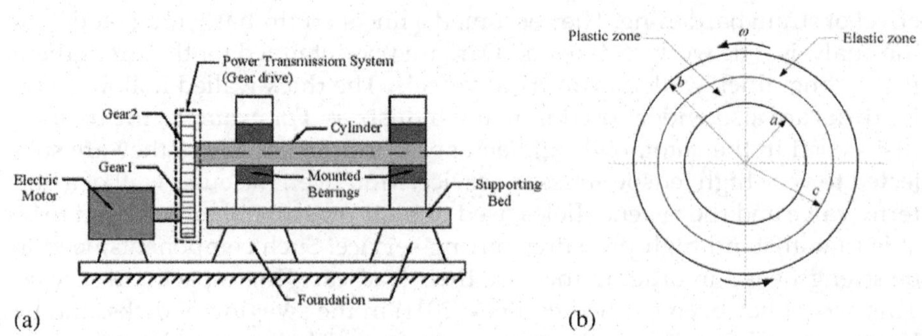

FIGURE 7.1
(a) A schematic setup for rotational autofrettage and (b) cross-section of the disk/cylinder with elastic and plastic zones. Figure 7.1(a) modified with permission from Zare and Darijani (2016). Copyright Elsevier.

was originally conceptualized by Zare and Darijani (2016). The cylinder to be autofrettaged can be rotated by means of an electric motor coupled with a power transmission system such as gear drive. The cylinder is mounted on two suitable bearings as shown in Figure 7.1a. During loading, when the disk/cylinder is rotated at a certain angular velocity ω, the disk/cylinder is subjected to a radial acceleration of $-\mathbf{r}\omega^2$. Here, **r** represents the radius vector of any particle. If a is the inner radius of the disk/cylinder, the material particles at the inner bore surface experience a centrifugal force of magnitude $ma\omega^2$, where m is the mass of the disk/cylinder. As the angular velocity ω is increased gradually, a corresponding quadratic increase in the centrifugal force is achieved. The centrifugal force causes stresses in the disk/cylinder. For the smaller value of ω, the stresses in the disk/cylinder are entirely elastic. If the value of ω is increased gradually, at a certain angular velocity, the yielding initiates at the inner surface of the disk/cylinder for a certain combination of stress state. Upon crossing this threshold velocity, the disk/cylinder experiences partial plastic deformation propagating up to a certain radial position c from the inner bore surface. Thus, an inner plastic zone $a \leq r \leq c$ is formed at the inner bore region of the disk/cylinder. The material beyond the radial position c remains in the elastic state, creating an outer elastic zone $c \leq r \leq b$. Afterwards, the unloading stage is carried out by gradually reducing the angular velocity to zero. At this stage, the disk/cylinder is subjected to compressive residual stresses at and around the inner plastic zone and tensile residual stress towards the outer elastic zone. The cross-section of a typical disk/cylinder with the elastic and plastic zones during the rotational autofrettage procedure is schematically shown in Figure 7.1b.

7.3 Plane–Stress Model of Kamal (2018)

In this section, the plane–stress model of rotational autofrettage developed by Kamal (2018) is presented. A thick-walled circular disk with inner radius a and outer radius b is subjected to an angular velocity ω about its own axis. The cross-sectional dimensions of the disk are shown in Figure 1b. The general concept of rotational autofrettage of a disk/cylinder is discussed in Section 7.2. In order to study the physics of the rotational autofrettage for a hollow circular disk, the plane–stress condition ($\sigma_z = 0$) is appropriate. In the following subsections, the elastic analysis of a thick-walled disk subjected to an angular velocity that is not sufficient to cause yielding at the inner side is carried out, followed by the elasto-plastic analysis of the disk when it is subjected to a plastically deforming angular velocity.

7.3.1 Stresses in the Disk Subjected to a Complete Elastically Deforming Angular Velocity

When the disk is allowed to rotate at a low angular velocity, which is not sufficient enough to cause yielding at the inner surface, the stresses setup throughout the wall thickness of the disk cause only elastic deformation. Under plane–stress condition, the purely elastic response of a body is governed by the following constitutive relations (Chakrabarty, 2006):

$$\varepsilon_r = \frac{1}{E}(\sigma_r - \nu\sigma_\theta), \tag{7.1}$$

$$\varepsilon_\theta = \frac{1}{E}(\sigma_\theta - \nu\sigma_r), \tag{7.2}$$

where ε_r is the radial component of the strain, ε_θ is the hoop strain, σ_r is the radial component of the stress, σ_θ is the hoop stress, ν is the Poisson's ratio, and E is the Young's modulus of elasticity. The axisymmetric equilibrium equation for a disk subjected to a rotational speed ω can be written as (Chakrabarty, 2006):

$$\frac{\mathrm{d}\sigma_r}{\mathrm{d}r} + \frac{\sigma_r - \sigma_\theta}{r} + \rho r\omega^2 = 0, \tag{7.3}$$

where ρ is the density of the material. The strain-displacement relations are given by Equation (5.5) and the compatibility condition is governed by Equation (5.6).

Using the constitutive relations Equations (7.1) and (7.2) in Equation (5.6) along with the equilibrium Equation (7.3), provides the following differential equation:

$$\frac{\mathrm{d}}{\mathrm{d}r}\left\{\frac{1}{r}\frac{\mathrm{d}}{\mathrm{d}r}\left(r^2\sigma_r\right)\right\} + (3+\nu)\rho\omega^2 r = 0. \tag{7.4}$$

The solution of the differential equation represented by Equation (7.4) provides:

$$\sigma_r = A + \frac{B}{r^2} - \left(\frac{3+\nu}{8}\right)\rho\omega^2 r^2, \tag{7.5}$$

where A and B are the constants of integration, which can be evaluated employing boundary conditions. The equilibrium Equation (7.3) along with Equation (7.5) provides:

$$\sigma_\theta = A - \frac{B}{r^2} - \left(\frac{1+3\nu}{8}\right)\rho\omega^2 r^2. \tag{7.6}$$

Using the boundary conditions of vanishing radial stress at the inner and outer radii, the values of constants A and B are evaluated to provide the following expressions of radial and hoop stresses in the elastic disk:

$$\sigma_r = \left(\frac{3+\nu}{8}\right)\rho\omega^2\left(a^2 + b^2 - \frac{a^2b^2}{r^2} - r^2\right), \tag{7.7}$$

$$\sigma_\theta = \left(\frac{3+\nu}{8}\right)\rho\omega^2\left\{a^2 + b^2 + \frac{a^2b^2}{r^2} - \left(\frac{1+3\nu}{3+\nu}\right)r^2\right\}. \tag{7.8}$$

A typical stress distribution in an SS304 steel disk with $b/a=3$, $\sigma_Y=205$ MPa, $\rho=8000$ kg/m^3, and $\nu=0.3$ subjected to an angular velocity of 200 rad/s (≈1910 rpm), shown in Figure 7.2. It is seen that the tensile hoop stress is at the maximum at the inner wall, and that it keeps reducing up to the outer wall. However, it remains tensile throughout the wall. The magnitude of radial stress is much smaller than that of hoop stress; however, the radial stress always remains tensile, except at the inner and outer surfaces where it becomes zero.

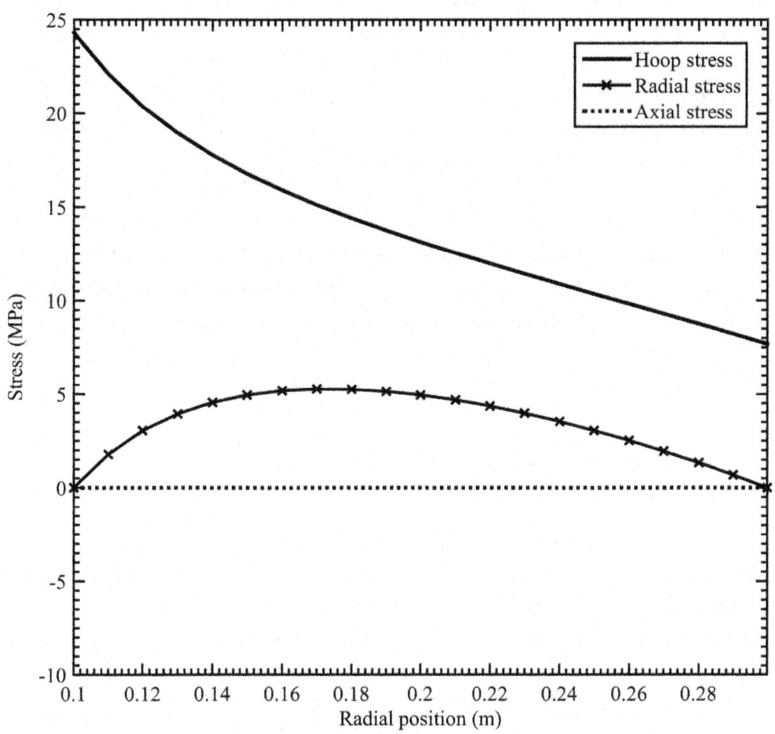

FIGURE 7.2
A typical elastic–stress distribution in an SS304 disk subjected to rotation.

7.3.2 Onset of Yielding

It is observed from Figure 7.2 that when the disk is subjected to angular velocity, the hoop stress becomes the maximum principal stress and the axial stress becomes the minimum principal stress, as it is zero everywhere. Thus, for the initiation of yielding in the disk, the Tresca yield criterion provides:

$$\sigma_\theta - \sigma_z = \sigma_Y, \tag{7.9}$$

where σ_Y is the yield stress in uniaxial tension or compression. As the maximum value of σ_θ appears at the inner radius of the disk, yielding commences at the inner radius upon increasing the angular velocity of rotation. If ω_i is the velocity corresponding to the initiation of yielding at the inner radius, then substituting the value of σ_θ in Equation (7.9) from Equation (7.8) at $r=a$ and incorporating the plane–stress condition ($\sigma_z = 0$), one obtains:

$$\omega_i = \left[\frac{4\sigma_Y}{\rho\left\{(1-\nu)a^2 + (3+\nu)b^2\right\}} \right]^{1/2}. \tag{7.10}$$

Equation (7.10) provides the threshold angular velocity of the disk for the onset of yielding. For any angular velocity in the range $0 \le \omega \le \omega_i$, the stress distribution in the disk is entirely elastic.

In order to get an idea about the order of magnitude of angular velocity needed to start yielding, consider a typical material with $\sigma_Y = 300$ MPa, $\rho = 7800$ kg/m^3, and $\nu = 0.3$. Let the inner radius of the cylinder be 10 mm and the outer radius be 30 mm. Substituting these values in Equation (7.10), $_i$ is obtained as 7114 rad/s. This amounts to 67933 revolutions per minute (RPM). The required angular speed reduces as the size of the cylinder increases in the same proportion. For example, if the dimensions are increased by a factor of 10, i.e., $a = 100$ mm and $b = 300$ mm, then the required RPM will be only 67933/10≈6793 RPM. This aspect is beneficial as the autofrettage of bigger components is possible with a low angular speed.

7.3.3 Elasto-Plastic Analyses during the Loading Stage of the Rotational Autofrettage of Disk

In order to achieve autofrettage, the disk is intended to deform plastically at the inner side. In the case of rotational autofrettage, the disk is loaded with a centrifugal force by subjecting it to an angular velocity $\omega > \omega_i$ to cause plastic deformation. At this velocity, the wall of the disk becomes partially plastic up to a certain intermediate radius c, as long as it is not high enough to cause the yielding of the entire wall. Thus, the wall of the disk is divided into an inner plastic zone $a \le r \le c$ and an outer elastic zone $c \le r \le b$ as shown in Figure 7.1b. The analysis of stress, strain, and the displacement field in the elastic–plastic disk is carried out in the following subsections.

7.3.3.1 The Elastic Zone $c \le r \le b$

In the elastic zone $c \le r \le b$, the stress solutions are still given by Equations (7.5) and (7.6), where the constants A and B are evaluated using the following boundary conditions: $\sigma|_{r=b} = 0$ and $(\sigma_\theta - \sigma_z)|_{r=c} = \sigma_Y$. Thus, the resulting stress distribution are obtained as:

$$\sigma_r = \left(\frac{1+3\nu}{b^2+c^2}\right)\frac{\rho\omega^2c^2}{8}\left(c^2 - \frac{b^2c^2}{r^2}\right) + (3+\nu)\frac{\rho\omega^2b^2}{8}\left\{1 - \frac{c^2}{b^2+c^2} + \frac{b^2c^2}{\left(b^2+c^2\right)r^2} - \frac{r^2}{b^2}\right\}$$
$$+\left(\frac{c^2}{b^2+c^2}\right)\sigma_Y\left(1 - \frac{b^2}{r^2}\right), \tag{7.11}$$

$$\sigma_\theta = \left(\frac{1+3\nu}{b^2+c^2}\right)\frac{\rho\omega^2c^2}{8}\left\{c^2 + \frac{b^2c^2}{r^2} - \frac{\left(b^2+c^2\right)r^2}{c^2}\right\}$$
$$+(3+\nu)\frac{\rho\omega^2b^2}{8}\left\{1 - \frac{c^2}{b^2+c^2} - \frac{b^2c^2}{\left(b^2+c^2\right)r^2}\right\} \tag{7.12}$$
$$+\left(\frac{c^2}{b^2+c^2}\right)\sigma_Y\left(1 + \frac{b^2}{r^2}\right).$$

The substitution of Equations (7.11) and (7.12) in Equation (7.2), along with the use of the second expression of the strain-displacement relation (Equation 5.5), provides the radial displacement u as:

$$u = \left(\frac{r}{E}\right)\left[\begin{array}{l} \left(\frac{1+3\nu}{b^2+c^2}\right)\frac{\rho\omega^2c^2}{8}\left\{(1-\nu)c^2 + \frac{b^2c^2}{r^2} - \frac{\left(b^2+c^2\right)r^2}{c^2} + \frac{\nu b^2c^2}{r^2}\right\} \\ +(3+\nu)\frac{\rho\omega^2b^2}{8}\left\{(1-\nu)\frac{b^2}{b^2+c^2} - (1+\nu)\frac{b^2c^2}{r^2\left(b^2+c^2\right)} + \frac{\nu r^2}{b^2}\right\} \\ +\left(\frac{c^2}{b^2+c^2}\right)\sigma_Y\left\{(1-\nu) + \frac{b^2}{r^2}(1+\nu)\right\} \end{array}\right] \tag{7.13}$$

7.3.3.2 The Plastic Zone $a \le r \le c$

In order to take into account the effect of subsequent yielding or the continued plastic flow of the material in the plastic zone $a \le r \le c$, the Tresca yield criterion can be expressed as:

$$\sigma_\theta - \sigma_z = \sigma_{eq}, \tag{7.14}$$

where σ_{eq} is the equivalent stress and can be represented mathematically by using a suitable hardening function. Here, σ_{eq} is represented by using the hardening function due to Ludwik (Dixit and Dixit, 2008) as:

$$\sigma_{eq} = \sigma_Y + K\left(\varepsilon_{eq}^p\right)^n, \tag{7.15}$$

where K and n are the hardening parameters, known as the hardening coefficient and the strain hardening exponent, respectively. The equivalent plastic strain ε_{eq}^p is given by:

$$\varepsilon_{eq}^p = \left(\frac{2}{3}\varepsilon_{ij}^p \varepsilon_{ij}^p\right)^{1/2}, \tag{7.16}$$

where ε_{ij}^p is the plastic part of the strain tensor. As $\sigma_z = 0$ in the present case, the Tresca yield criterion Equation (7.14) provides the hoop stress distribution in the plastic zone as:

$$\sigma_\theta = \sigma_{eq} = \sigma_Y + K\left(\varepsilon_{eq}^p\right)^n. \tag{7.17}$$

Using Equation (7.17) in the equilibrium Equation (7.3), the following differential equation is obtained:

$$\frac{d}{dr}(r\sigma_r) = \sigma_Y + K\left(\varepsilon_{eq}^p\right)^n - \rho r^2 \omega^2 \tag{7.18}$$

Integrating Equation (7.18) within the limit of the plastic zone from r to $r = c$, the solution for radial stress is obtained as:

$$\begin{aligned}\sigma_r = {}& \frac{c}{r}\left(\frac{c^2 - b^2}{b^2 + c^2}\right)\left\{\sigma_Y + (1 + 3\nu)\frac{\rho\omega^2 c^2}{8}\right\} + \sigma_Y\left(1 - \frac{c}{r}\right) \\ & + (3 + \nu)\left(\frac{\rho\omega^2 b^2}{8}\right)\left(\frac{c}{r}\right)\left(\frac{2b^2}{b^2 + c^2} - \frac{c^2}{b^2}\right) \\ & + \frac{\rho\omega^2}{3}\left(\frac{c^3}{r} - r^2\right) - \frac{K}{r}\int_r^c \left(\varepsilon_{eq}^p\right)^n dr_1,\end{aligned} \tag{7.19}$$

where r_1 is a dummy variable.

The total radial and hoop strain can be expressed as the additive decomposition of their elastic and plastic parts given by:

$$\varepsilon_r = \frac{\mathrm{d}u}{\mathrm{d}r} = \varepsilon_r^e + \varepsilon_r^p, \quad \varepsilon_\theta = \frac{u}{r} = \varepsilon_\theta^e + \varepsilon_\theta^p. \tag{7.20}$$

The superscripts "e" and "p" in Equation (7.20) indicate the elastic and plastic part of the respective strains. The elastic parts of the strains are governed by Equations (7.1) and (7.2). The Tresca associated flow rule provides:

$$\mathrm{d}\varepsilon_r^p = 0, \quad \mathrm{d}\varepsilon_\theta^p = -\mathrm{d}\varepsilon_z^p. \tag{7.21}$$

Equation (7.21) indicates that the radial strain ε_r is wholly elastic. Using Equation (7.20) in Equations (7.1) and (7.2), and with the help of the equilibrium Equation (7.3), a differential equation of the following form is obtained:

$$\frac{\mathrm{d}}{\mathrm{d}r}\left\{\frac{1}{r}\frac{\mathrm{d}}{\mathrm{d}r}(ur)\right\} = \nu\frac{\mathrm{d}\varepsilon_\theta^p}{\mathrm{d}r} - (1-\nu)\frac{\varepsilon_\theta^p}{r} - \left(\frac{1-\nu^2}{E}\right)\rho\omega^2 r. \tag{7.22}$$

The strain compatibility condition in the plastic zone due to Equations (7.20) and (7.21) can be expressed as:

$$r\frac{\mathrm{d}}{\mathrm{d}r}\left(\varepsilon_\theta^e + \varepsilon_\theta^p\right) = \left(\varepsilon_r^e - \varepsilon_\theta^e\right) - \varepsilon_\theta^p. \tag{7.23}$$

Substituting the values of ε_r^e and ε_θ^e from Equations (7.1) and (7.2) in Equation (7.23) one obtains:

$$\frac{1}{E}\frac{\mathrm{d}\sigma_\theta}{\mathrm{d}r} + \frac{1}{E}\frac{\mathrm{d}\sigma_r}{\mathrm{d}r} + \frac{\mathrm{d}\varepsilon_\theta^p}{\mathrm{d}r} + \left(\frac{1+\nu}{E}\right)\rho\omega^2 r = -\frac{\varepsilon_\theta^p}{r}. \tag{7.24}$$

Using the value of ε_θ^p / r from Equation (7.24), Equation (7.22) can be rewritten as:

$$\frac{\mathrm{d}}{\mathrm{d}r}\left\{\frac{1}{r}\frac{\mathrm{d}}{\mathrm{d}r}(ur)\right\} = \nu\frac{\mathrm{d}\varepsilon_\theta^p}{\mathrm{d}r} + (1-\nu)\left\{\frac{1}{E}\frac{\mathrm{d}\sigma_\theta}{\mathrm{d}r} + \frac{1}{E}\frac{\mathrm{d}\sigma_r}{\mathrm{d}r} + \frac{\mathrm{d}\varepsilon_\theta^p}{\mathrm{d}r}\right\}. \tag{7.25}$$

Equation (7.25) is integrated with respect to r to obtain:

$$\frac{1}{r}\frac{\mathrm{d}}{\mathrm{d}r}(ur) = \left(\frac{1-\nu}{E}\right)(\sigma_r + \sigma_\theta) + \varepsilon_\theta^p + C, \tag{7.26}$$

where C is an integration constant. The constant C can be obtained by evaluating the left and right-hand side expressions of Equation (7.26) at $r = c$ taking the corresponding values of u, σ_r, and σ_θ from the elastic zone. The value of the plastic part of the hoop strain ε_θ^p is zero at $r = c$. Thus, one obtains the value of C as zero. Now, substituting the value of C in Equation (7.26) and integrating the resulting expression in the limit from r to $r = c$ in the plastic zone, the following equation is obtained:

$$\left(u\big|_{r=c}\right)c - ur = \left(\frac{1-\nu}{E}\right)\left\{c^2\left(\sigma_r\big|_{r=c}\right) - r^2\sigma_r + \frac{\rho\omega^2}{4}\left(c^4 - r^4\right)\right\} + \int_r^c r_1\varepsilon_\theta^p \mathrm{d}r_1. \quad (7.27)$$

Dividing each term on the left and right-hand side of Equation (7.27) by r^2 and evaluating the corresponding values of $u\big|_{r=c}$ and $\sigma_r\big|_{r=c}$ from the elastic zone, the total hoop strain component ε_θ is obtained as:

$$\frac{u}{r} = \varepsilon_\theta = \frac{1}{E}\left\{\begin{aligned}&\left(\frac{1+3\nu}{b^2+c^2}\right)\left(\frac{\rho\omega^2c^2}{8r^2}\right)\left(b^2c^2 - c^4\right) + \left(\frac{2b^2c^2}{b^2+c^2}\right)\left(\frac{\sigma_Y}{r^2}\right)\\ &+(3+\nu)\left(\frac{\rho\omega^2b^2}{8r^2}\right)\left(\frac{c^4}{b^2} - \frac{2b^2c^2}{b^2+c^2}\right) - (1-\nu)\left(\frac{\rho\omega^2}{4r^2}\right)\left(c^4 - r^4\right)\end{aligned}\right\} \quad (7.28)$$

$$+\left(\frac{1-\nu}{E}\right)\sigma_r - \frac{1}{r^2}\int_r^c r_1\varepsilon_\theta^p \mathrm{d}r_1.$$

Now, the plastic part of the hoop strain can be obtained by subtracting the elastic part ε_θ^e from the total hoop strain given by Equation (7.28). It is to be noted that the elastic part of the hoop strain ε_θ^e is given by Equation (7.2) and that the corresponding values of σ_r, σ_θ in the resulting expression are substituted from the plastic zone. The expression for the plastic part of the hoop strain is then given by:

$$\varepsilon_\theta^p = -\varepsilon_z^p = \frac{1}{E}\left\{\begin{aligned}&\left(\frac{1+3\nu}{b^2+c^2}\right)\left(\frac{\rho\omega^2c^2}{8r^2}\right)\left(b^2c^2 - c^4\right) + \left(\frac{2b^2c^2}{b^2+c^2}\right)\left(\frac{\sigma_Y}{r^2}\right)\\ &+(3+\nu)\left(\frac{\rho\omega^2b^2}{8r^2}\right)\left(\frac{c^4}{b^2} - \frac{2b^2c^2}{b^2+c^2}\right) - (1-\nu)\left(\frac{\rho\omega^2}{4r^2}\right)\left(c^4 - r^4\right)\end{aligned}\right\}$$

$$-\left(\frac{c}{r}\right)\left(\frac{1+3\nu}{E}\right)\left(\frac{\rho\omega^2c^2}{8}\right)\left(\frac{b^2-c^2}{b^2+c^2}\right) - \left(\frac{\sigma_Y}{E}\right)\left(\frac{c}{r}\right)\left(\frac{2b^2}{b^2+c^2}\right) \quad (7.29)$$

$$+\left(\frac{\rho\omega^2}{3E}\right)\left(\frac{c^3}{r} - r^2\right) + \left(\frac{3+\nu}{E}\right)\left(\frac{\rho\omega^2b^2}{8}\right)\left(\frac{c}{r}\right)\left(\frac{2b^2}{b^2+c^2} - \frac{c^2}{b^2}\right)$$

$$-\left(\frac{K}{E}\right)\left(\varepsilon_{\mathrm{eq}}^p\right)^n - \frac{1}{r^2}\int_r^c r_1\varepsilon_\theta^p \mathrm{d}r_1 - \left(\frac{K}{rE}\right)\int_r^c \left(\varepsilon_{\mathrm{eq}}^p\right)^n \mathrm{d}r_1.$$

The equivalent plastic strain $\varepsilon_{\mathrm{eq}}^p$ in Equation (7.29) is provided by Equation (7.16) as:

$$\varepsilon_{\mathrm{eq}}^p = \frac{2}{\sqrt{3}}\,\varepsilon_\theta^p. \quad (7.30)$$

7.3.4 Analysis of Residual Stress Induced in the Disk after Unloading

The unloading of the centrifugal force induced in the loading stage due to rotation is carried out by gradually reducing the angular velocity to zero. At this stage, a significant amount of compressive residual stresses are generated near the inner plastic zone. The residual stress at the outer portion of the disk wall is tensile. During unloading, the stresses in the disk may be either purely elastic or there may be reverse yielding due to compressive residual stresses. However, in the present analysis, the unloading stresses are assumed to be completely elastic and devoid of the Bauschinger effect. Thus, the unloading stresses are given by the following expressions:

$$\sigma_r = -\left(\frac{3+\nu}{8}\right)\rho\omega^2\left(a^2+b^2-\frac{a^2b^2}{r^2}-r^2\right), \tag{7.31}$$

$$\sigma_\theta = -\left(\frac{3+\nu}{8}\right)\rho\omega^2\left\{a^2+b^2+\frac{a^2b^2}{r^2}-\left(\frac{1+3\nu}{3+\nu}\right)r^2\right\}. \tag{7.32}$$

Thus, the residual stresses in the elastic and plastic zones of the disk are obtained by superposing the unloading stresses Equations (7.31) and (7.32) with the corresponding loading stresses developed in Section 7.3.3. The resulting expressions for the residual radial and hoop stresses in the elastic and plastic zones are given as follows:

Plastic Zone $a \le r \le c$:

$$\begin{aligned}(\sigma_r)_{\text{res}} = {} & \frac{c}{r}\left(\frac{c^2-b^2}{b^2+c^2}\right)\left\{\sigma_Y+(1+3\nu)\frac{\rho\omega^2c^2}{8}\right\}+\sigma_Y\left(1-\frac{c}{r}\right)\\ & +(3+\nu)\frac{\rho\omega^2b^2}{8}\frac{c}{r}\left(\frac{2b^2}{b^2+c^2}-\frac{c^2}{b^2}\right)+\frac{\rho\omega^2}{3}\left(\frac{c^3}{r}-r^2\right)\\ & -\frac{K}{r}\int_r^c\left(\varepsilon_{\text{eq}}^p\right)^n \mathrm{d}r_1-(3+\nu)\frac{\rho\omega^2}{8}\left(a^2+b^2-\frac{a^2b^2}{r^2}-r^2\right),\end{aligned} \tag{7.33}$$

$$(\sigma_\theta)_{\text{res}} = \sigma_Y+K\left(\varepsilon_{\text{eq}}^p\right)^n-\frac{\rho\omega^2}{8}\left\{(3+\nu)\left(a^2+b^2+\frac{a^2b^2}{r^2}\right)-(1+3\nu)r^2\right\}. \tag{7.34}$$

Elastic zone $c \le r \le b$:

$$(\sigma_r)_{\text{res}} = \left(\frac{1+3\nu}{b^2+c^2}\right)\frac{\rho\omega^2c^2}{8}\left(c^2 - \frac{b^2c^2}{r^2}\right)$$

$$+(3+\nu)\frac{\rho\omega^2}{8}\left\{\left(a^2 + \frac{b^2c^2}{b^2+c^2}\right)\left(\frac{b^2}{r^2}-1\right)\right\} \quad (7.35)$$

$$+\left(\frac{c^2}{b^2+c^2}\right)\sigma_Y\left(1-\frac{b^2}{r^2}\right),$$

$$(\sigma_\theta)_{\text{res}} = \left(\frac{1+3\nu}{b^2+c^2}\right)\frac{\rho\omega^2c^2}{8}\left\{c^2 + \frac{b^2c^2}{r^2} - \frac{\left(b^2+c^2\right)r^2}{c^2}\right\}$$

$$-(3+\nu)\frac{\rho\omega^2}{8}\left\{\left(a^2 + \frac{b^2c^2}{b^2+c^2}\right)\left(1+\frac{b^2}{r^2}\right)\right\} \quad (7.36)$$

$$+\left(\frac{c^2}{b^2+c^2}\right)\sigma_Y\left(1+\frac{b^2}{r^2}\right)+\frac{\rho\omega^2}{8}(1+3\nu)r^2.$$

The subscript *"res"* in the above equations indicate residual stresses. The generation of the compressive residual stress in the disk indicates that the disk is rotationally autofrettaged.

7.3.5 Evaluation of the Unknown Elastic–Plastic Interface Radius *c*

In order to obtain the complete solution for the stresses in the rotationally autofrettaged disk, the evaluation of the unknown elastic–plastic interface radius c is necessary. Incorporating the boundary condition of vanishing radial stress at the inner radius in Equation (7.19) provides:

$$\frac{c}{a}\left(\frac{c^2-b^2}{b^2+c^2}\right)\left\{\sigma_Y + (1+3\nu)\frac{\rho\omega^2c^2}{8}\right\} + \sigma_Y\left(1-\frac{c}{a}\right)$$

$$+(3+\nu)\left(\frac{\rho\omega^2b^2}{8}\right)\left(\frac{c}{a}\right)\left(\frac{2b^2}{b^2+c^2}-\frac{c^2}{b^2}\right)+\frac{\rho\omega^2}{3}\left(\frac{c^3}{a}-a^2\right)-\left(\frac{K}{a}\right)\int_a^c \left(\varepsilon_{\text{eq}}^p\right)^n dr = 0.$$

(7.37)

The value of c can be evaluated by solving Equation (7.37) for the known values of a, b, ω ($>\omega_i$), and ρ. The integral term in Equation (7.37) can be evaluated numerically using numerical techniques such as the Trapezoidal rule, the Simpson rule, or Gauss quadrature. Moreover, the value of $\varepsilon_{\text{eq}}^p$ in the integrand is evaluated numerically. The complete solution procedure to obtain the value of c is as follows.

Step One: The initial estimate of c is evaluated from Equation (7.37), taking $K=0$ corresponding to a non-hardening case. To solve the resulting non-linear equation, a numerical method such as the bisection method may be employed. The initial values of the plastic hoop strains ε_θ^p are taken as zero at every radial position in the plastic zone, $a \leq r \leq c$.

Step Two: The values of ε_θ^p at every radial position in the plastic zone are now updated using Equation (7.29). For the fixed value of c, the values of ε_θ^p are further updated numerically using the fixed-point iteration method. The process is continued till the convergence in the value of ε_θ^p is achieved. The value of the equivalent plastic strain ε_{eq}^p is updated from Equation (7.30).

Step Three: Using the updated values of ε_{eq}^p obtained in step two, Equation (7.37) is solved to obtain the new estimate of c. If the new estimate of c is same as the previously estimated value of c, the procedure is stopped. Otherwise, starting from step two, the entire procedure is repeated till the convergence in the value of c is achieved.

The value of c and ε_θ^p thus obtained is used to obtain the stress distribution in the elastic and plastic zones of the rotationally autofrettaged disk.

7.3.6 Numerical Simulation

In this section, the rotational autofrettage of a typical disk is assessed by numerically evaluating the stresses using the developed model. For instance, an SS304 disk of inner radius $a=0.1$ m and outer radius $b=0.3$ m is considered. The material properties of SS304 are provided in Section 7.3.1. The hardening parameters K and n for SS304 are taken as 1425 MPa and 0.7, respectively (Kamal et al., 2017). The elastic–plastic stresses in the disk during loading and the residual stresses after unloading are numerically simulated using the equations developed in Sections 7.3.3 and 7.3.4, respectively. Further, the effect of residual stresses on the load-carrying capacity of the disk is studied when the disk is subjected to internal pressure, radial temperature difference, and angular velocity individually. The SS304 disk initiates yielding at an angular velocity (ω_i) of 580.66 rad/s (5545 rpm) according to Equation (7.10). Thus, to achieve rotational autofrettage, the disk is subjected to an angular velocity $\omega=700$ rad/s (6684 rpm) about its own axis. At this velocity, the disk deforms plastically up to an intermediate radial position $c=0.167$ m. Thus, the region $0.1 \leq r \leq 0.167$ within the wall thickness forms the plastic zone and the region $0.167 \leq r \leq 0.3$ forms the elastic zone, where r is measured in m.

7.3.6.1 Elasto-Plastic Stress Distribution in the SS304 Disk during Loading

The loading stress distribution in the plastic and elastic zones of the SS304 disk when it is rotated at $\omega = 700$ rad/s is shown in Figure 7.3. The elasto-plastic stress distribution in the disk shows that the hoop stress throughout the wall thickness of the disk is wholly tensile, providing the maximum value at the inner radius. The hoop stress follows a decreasing trend from the inner radius to the outer radius. The radial stress distribution in the disk is also tensile, vanishing at the extremities. However, the magnitudes of the radial stresses are far less than the hoop stresses everywhere within the disk thickness. The slope of the hoop stress in the plastic zone is observed due to the effect of strain hardening during plastic deformation. In the present case, this slope is small. Using Equations (7.29) and (7.30) together, the maximum equivalent plastic strain in the disk is evaluated to be 0.663×10^{-3} at the inner radius. Consequently, the maximum equivalent stress at the inner radius is obtained as 214.4 MPa. In an equivalent non-hardening disk, the maximum

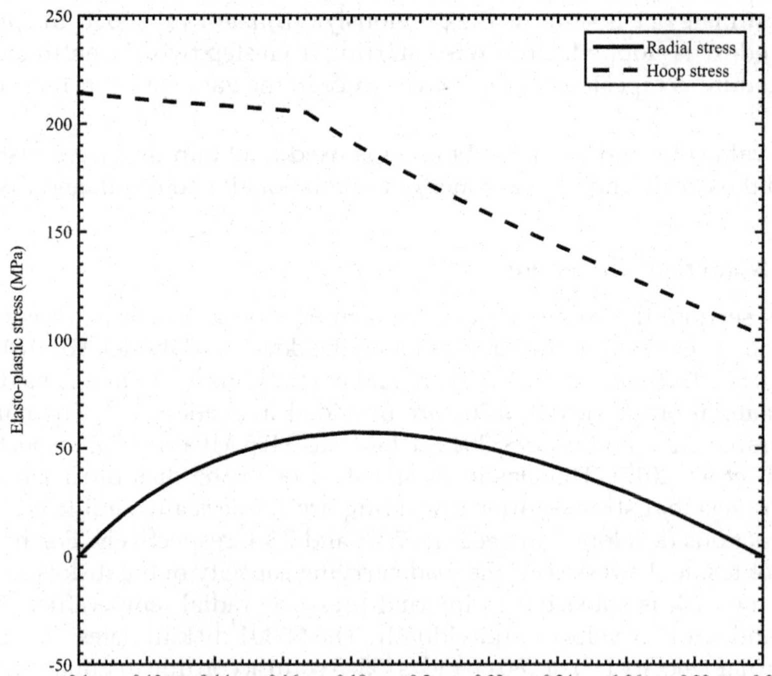

FIGURE 7.3
Elastic–plastic stress distribution in the SS304 disk for $\omega = 700$ rad/sec. With permission from Kamal (2018). Copyright ASME.

equivalent stress at the inner radius is 205 MPa. Thus, the maximum equivalent stress in the disk due to strain-hardening is only 4.6% higher than that of an equivalent non-hardening disk. This shows that the effect of strain-hardening is not significant in the present situation.

7.3.6.2 Residual Stress Distribution in the SS304 Disk after Unloading

The residual stress distribution in the SS304 disk resulting from bringing the angular velocity of the rotating disk to zero from 700 rad/s is shown in Figure 7.4. The distribution is shown as a function of radial position in the positive radial direction. It is observed that a significant amount of compressive residual stresses are generated at the neighborhood of the inner surface of the disk. The highest magnitude of the compressive residual stress occurs at the inner radius and in the present case it is 83.6 MPa. The magnitude of the compressive residual stress gradually decreases from the inner radius to reach the zero-stress level, close to the elastic–plastic interface radius. After crossing the zero-stress level, the residual stress starts increasing in tension, and it reaches the maximum level at the elastic–plastic interface.

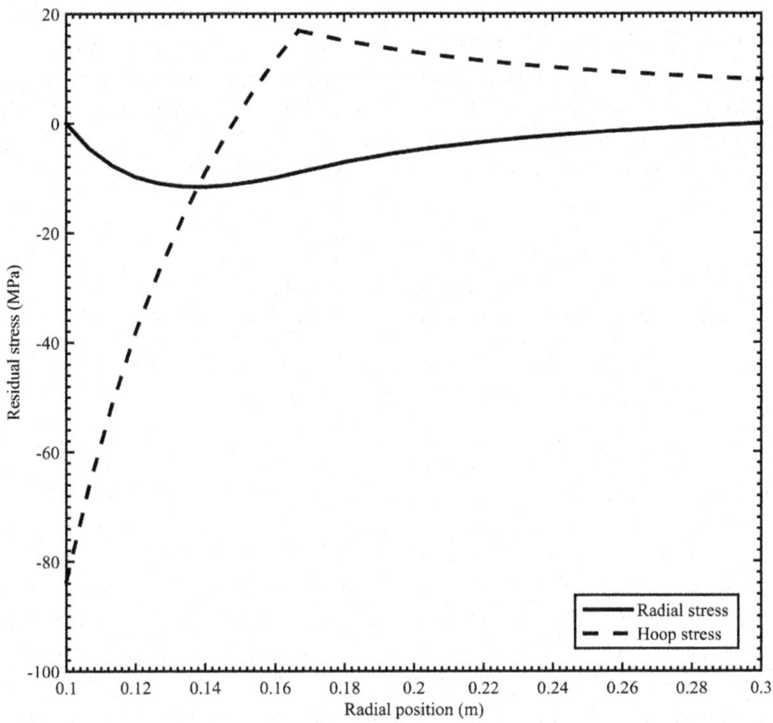

FIGURE 7.4
Residual stress distribution in the SS304 disk. With permission from Kamal (2018). Copyright ASME.

The residual stress from the radius of elastic–plastic interface up to the outer radius remains tensile, following a gradual decreasing trend. As the tensile residual stress at the outer surface of the disk is the smallest in magnitude, the outer surface of the disk is less sensitive to stress corrosion cracking. The residual radial stress within the wall thickness of the disk is entirely compressive, vanishing at the inner and the outer radii. Due to the high magnitude of the compressive residual stress induced at the inner surface of the disk, the load-bearing capacity of the disk will be enhanced when the disk is subjected to subsequent service load. The enhancement in different load bearing capacities of the disk due to rotational autofrettage is discussed in the following subsection.

7.3.7 Effect of Rotational Autofrettage Process on the Load-Carrying Capacity of the SS304 Disk

In this section, the rotationally autofrettaged SS304 disk is considered to be subjected to three different types of individual service loads, namely internal pressure, radial temperature difference, and angular velocity. Under each individual working load, the tensile hoop stress generated in the disk is compensated by the compressive residual hoop stress due to rotational autofrettage to a certain extent. Thus, an increase in the load-carrying capacity is achieved. The net stress distribution in the disk is obtained by superposing the autofrettage residual stress with the corresponding loading stress. It is important to assess the maximum increase in the load-carrying capacity of the disk for each individual loading condition to get an idea about safety during service.

7.3.7.1 Net Stress Distribution in the Autofrettaged SS304 Disk Subjected to Internal Pressure

The stresses in the disk due to internal pressurization are given by Equations (3.3) and (3.4). Thus, the net stress distribution in the elastic and plastic zones of the rotationally autofrettaged SS304 disk is obtained by superposing the respective residual stresses Equations (7.33–7.36) with Equations (3.3) and (3.4). By numerical simulation, it is obtained that the present autofrettaged SS304 disk can be subjected to a maximum working internal pressure of 128.2 MPa. This maximum internal working pressure corresponds to the yield onset pressure of the autofrettaged disk during subsequent loading stage. The yield onset pressure of the corresponding non-autofrettaged SS304 disk is 91.1 MPa. This indicates that the maximum pressure-carrying capacity of the autofrettaged disk is increased by 40.72%, as compared to its non-autofrettaged counterpart. Figure 7.5 shows the net stress distribution in the autofrettaged SS304 disk for the maximum internal pressure of 128.2 MPa. It is observed that the maximum equivalent Tresca stress in the autofrettaged disk under internal pressure appears at the inner radius.

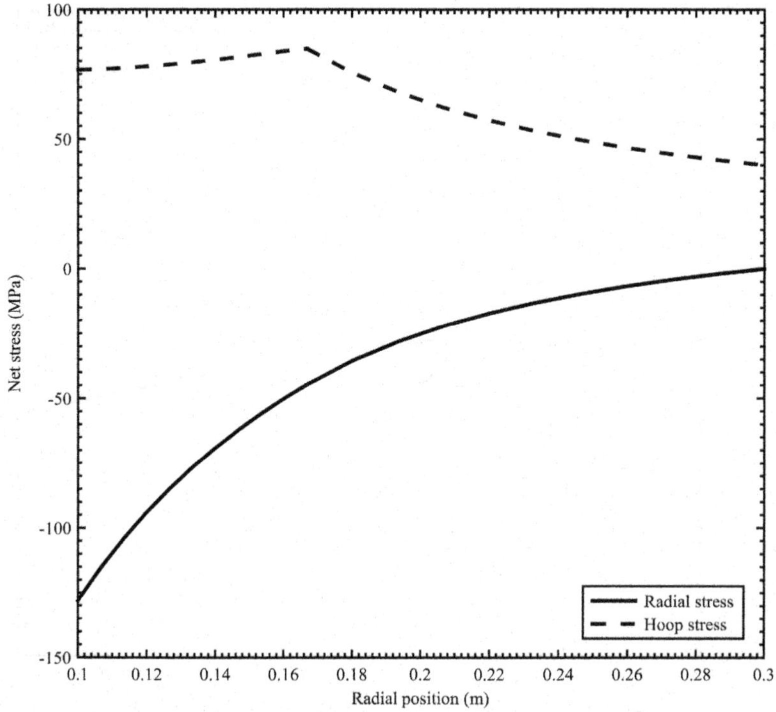

FIGURE 7.5
Net stress distribution in the autofrettaged SS304 disk subjected to 128.2 MPa. With permission from Kamal (2018). Copyright ASME.

7.3.7.2 Net Stress Distribution in the Autofrettaged SS304 Disk Subjected to Radial Temperature Difference

The rotationally autofrettaged SS304 disk is now put under a radial temperature difference (T_b–T_a), where T_a is the inner surface temperature and T_b is the temperature at the outer surface of the disk. At this stage, the overall stress distribution in the autofrettaged SS304 disk is obtained by the superposition of the autofrettage residual stresses with the thermal stresses given by Equations (5.10) and (5.11). The stress distribution in the autofrettaged SS304 disk subjected to a radial temperature difference (T_b–T_a) = 129 °C is shown in Figure 7.6. It is observed that from Figure 7.6 that corresponding to (T_b–T_a) = 129 °C, the disk is at the verge of yielding at the inner radius. Thus, this is the maximum temperature difference that the autofrettaged disk can withstand across the wall thickness during service. The corresponding non-autofrettaged disk can bear a maximum of only 92 °C radial temperature difference, keeping the entire stresses within the elastic limit as per Equation (5.13). This shows that the autofrettaged SS304 disk can withstand 40.22% higher temperature difference than the corresponding non-autofrettaged disk during service.

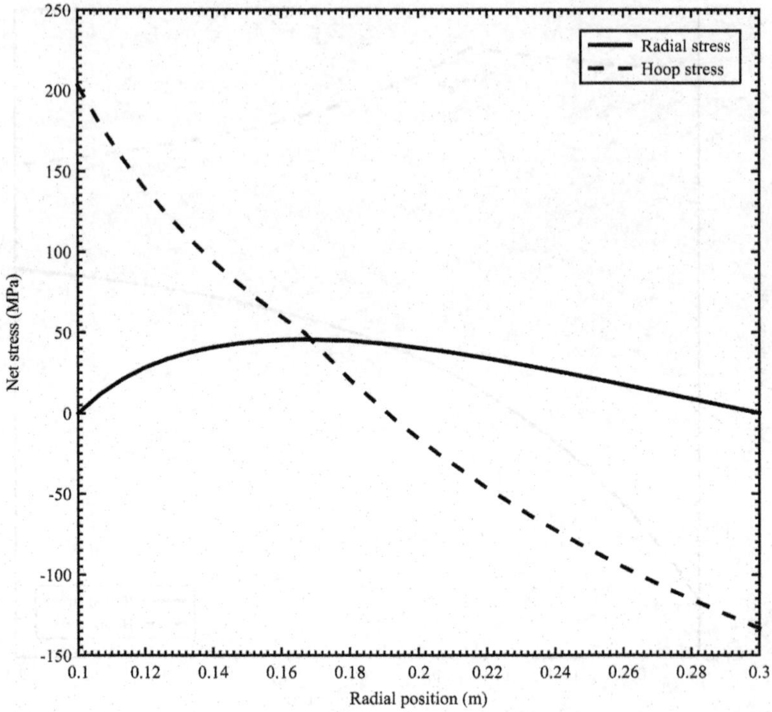

FIGURE 7.6
Net stress distribution in the autofrettaged SS304 disk subjected to $(T_b - T_a)$ =129 °C. With permission from Kamal (2018). Copyright ASME.

7.3.7.3 Net Stress Distribution in the Autofrettaged Disk Subjected to Angular Velocity

The rotationally autofrettaged disks are suitable for use as rotating disks in the automobile and aerospace industries because of the presence of high compressive residual stresses at the inner surface. Keeping this in view, the present rotationally autofrettaged disk is considered to be subjected to angular velocity. The angular velocity in the autofrettaged disk is gradually increased till the yielding initiates at the inner radius. The corresponding yield onset velocity in the rotationally autofrettaged SS304 disk is obtained as 689 rad/s (6579.5 rpm). Equation (7.10) provides the yield onset angular velocity of the equivalent non-autofrettaged SS304 disk as 580.66 rad/s (5545 rpm). Thus, the maximum increase in the rotation of the present autofrettaged rotating disk can be enhanced by 18.66% as compared to a corresponding non-autofrettaged rotating disk. The overall stress distribution in the autofrettaged SS304 disk subjected to the maximum angular velocity of 689 rad/s (6579.5 rpm) is shown in Figure 7.7.

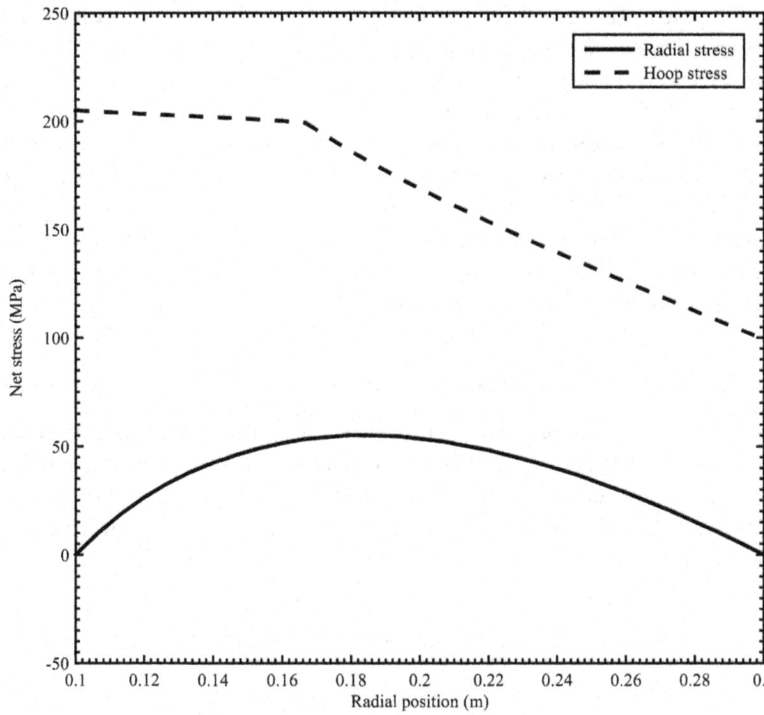

FIGURE 7.7
Net stress distribution in the autofrettaged SS304 disk subjected to $\omega = 689$ rad/s. With permission from Kamal (2018). Copyright ASME.

7.4 Plane–Strain Model of Zare and Darijani (2016)

The theoretical modeling of the rotational autofrettage process was first carried out by Zare and Darijani (2016) applicable for long, thick-walled cylindrical vessels. The model was based on the plane–strain assumption ($\varepsilon_z = 0$). The yielding of the cylinder was assumed to be governed by the Tresca yield criterion without strain hardening. The unloading process was modeled incorporating reverse yielding due to the Bauschinger effect. The Bauschinger effect has been treated in the analysis using a constant Bauschinger effect factor (BEF). However, in general, the BEF is a function of prior plastic strain. In this section, the plane–strain model of Zare and Darijani (2016) is presented without considering the effect of the Bauschinger effect during unloading. Rather, in the present analysis, the unloading process is treated as being completely elastic. The model is presented here with reference to the cross-sectional dimensions shown in Figure 7.1b. Due to the assumption of plane–strain, the axial stress is the intermediate stress throughout the wall thickness of the cylinder. Thus, the incorporation of the Tresca yield criterion provides one inner plastic zone and an outer elastic zone when the cylinder is allowed to rotate about its own axis at a sufficiently high angular velocity. The magnitude of the angular velocity should not be high enough to cause the plastic deformation of the entire wall thickness of the cylinder. It is to be noted that, in the original paper of Zare and Darijani (2016), all of the equations are presented in non-dimensional form. Here, the same equations are presented in dimensional form for convenience. In the following subsections, the analysis of stresses during the different stages of the rotational autofrettage of a cylinder is carried out.

7.4.1 Complete Elastic Analysis

An axisymmetric cylinder of inner radius a and outer radius b is rotated at small angular velocity ω. At this stage, the stresses induced in the cylinder are completely elastic. Hence, the material response is governed by the following constitutive relations (generalized Hook's law) for plane–strain condition:

$$\varepsilon_r = \frac{1}{E}\left\{(1-\nu^2)\sigma_r - \nu(1+\nu)\sigma_\theta\right\}, \tag{7.38}$$

$$\varepsilon_\theta = \frac{1}{E}\left\{(1-\nu^2)\sigma_\theta - \nu(1+\nu)\sigma_r\right\}, \tag{7.39}$$

$$\sigma_z = \nu\left(\sigma_r + \sigma_\theta\right), \tag{7.40}$$

Substituting the constitutive relations in compatibility Equation (5.6) and using equilibrium Equation (7.3) leads to the following differential equation:

$$\frac{d}{dr}\left(r^3\frac{d\sigma_r}{dr}\right) + \left(\frac{3-2\nu}{1-\nu}\right)\rho\omega^2 r^3 = 0. \tag{7.41}$$

Solving Equation (7.41) one obtains:

$$\sigma_r = -\frac{C_1}{2r^2} + C_2 - \frac{(3-2\nu)}{8(1-\nu)}\rho\omega^2 r^2, \tag{7.42}$$

where C_1 and C_2 are integration constants. Substituting Equation (7.42) in Equation (7.3) provides:

$$\sigma_\theta = \frac{C_1}{2r^2} + C_2 - \frac{(1+2\nu)}{8(1-\nu)}\rho\omega^2 r^2. \tag{7.43}$$

Invoking the traction-free boundary conditions: $(\sigma_r)\big|_{r=a\&b}=0$, the constants C_1, C_2 in Equations (7.42) and (7.43) are evaluated to yield the resulting radial and hoop stress distribution in the cylinder as follows:

$$\sigma_r=\frac{(3-2\nu)}{8(1-\nu)}\rho\omega^2\left(a^2+b^2-\frac{a^2b^2}{r^2}-r^2\right), \tag{7.44}$$

$$\sigma_\theta=\frac{(3-2\nu)}{8(1-\nu)}\rho\omega^2\left\{a^2+b^2+\frac{a^2b^2}{r^2}-\left(\frac{1+2\nu}{3-2\nu}\right)r^2\right\}. \tag{7.45}$$

Use of Equations (7.44) and (7.45) in Equation (7.40) furnishes the axial stress distribution as:

$$\sigma_z=\left(\frac{3-2\nu}{1-\nu}\right)\frac{\nu\rho\omega^2}{4}\left\{a^2+b^2-\left(\frac{2}{3-2\nu}\right)r^2\right\}. \tag{7.46}$$

A typical elastic–stress distribution in a typical SS304 cylinder rotating at 200 rad/s is shown in Figure 7.8.

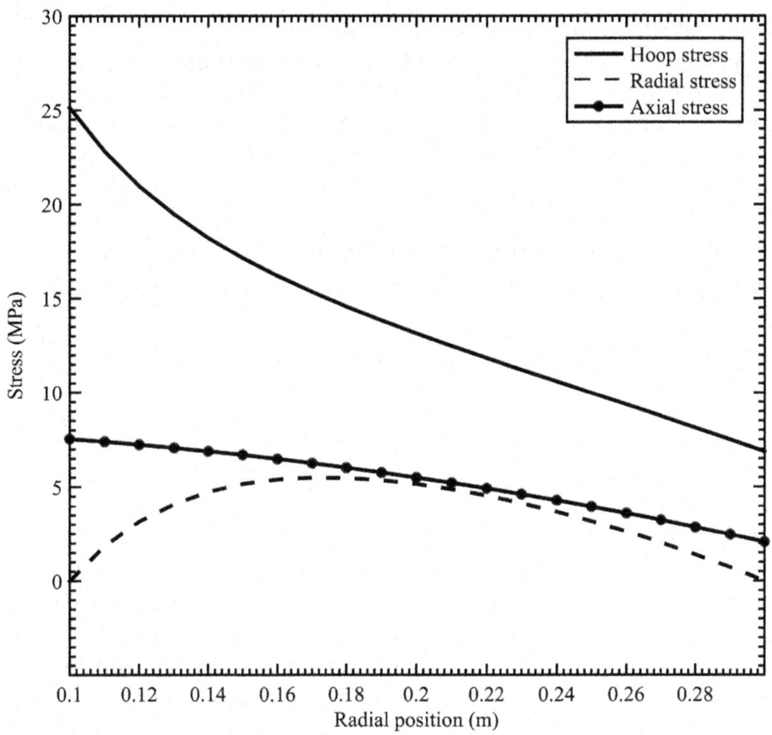

FIGURE 7.8
Typical elastic–stress distribution in an SS304 cylinder subjected to $\omega=200$ rad/s.

7.4.2 Onset of Yielding

The maximum and minimum principal stresses in a cylinder subjected to rotation are the hoop and radial stresses, respectively, as observed from Figure 7.8. If the cylinder is considered to be yielded as per the Tresca yield criterion, then for an increased angular velocity ω, the equivalent stress becomes equal to the yield stress of the material first at the inner radius of the cylinder. Hence,

$$\left(\sigma_\theta - \sigma_r\right)\Big|_{r=a} = \sigma_Y. \tag{7.47}$$

Using the values of σ_r and σ_θ from Equations (7.44) and (7.45) in Equation (7.47), the angular velocity of rotation for the onset of yielding in the cylinder is obtained as:

$$\omega_i = \left[\frac{8(1-\nu)\sigma_Y}{2\rho\left\{b^2(3-2\nu)+a^2(1-2\nu)\right\}}\right]^{1/2}. \tag{7.48}$$

For a typical cylinder specimen having the dimensions $a = 10$ mm and $b = 30$ mm with material properties $\sigma_Y = 300$ MPa, $\rho = 7800$ kg/m³, and $\nu = 0.3$. The angular velocity $_i$ for the onset of yielding calculated using Equation (7.48) is obtained as 6996 rad/s. This amounts to 66807 revolutions per minute (RPM). The required angular speed reduces as the size of the cylinder increases in the same proportion. For example, if the dimensions are increased by a factor of 10, i.e., $a = 100$ mm and $b = 300$ mm, then the required RPM will be only 66807/10≈6680 RPM. It is also to be noted that the angular velocity required for the autofrettage of a cylinder is slightly less than that of the corresponding disk of the same cross-section.

7.4.3 Elastic–Plastic Deformation of the Cylinder Due to Rotation

The minimum angular velocity of rotation at which plastic deformation begins at the inner surface is provided by Equation (7.48). When the cylinder is rotated at a higher velocity than ω_i, the plastic deformation propagates radially outward, up to an intermediate radial position c (Figure 7.1b). The region beyond c up to the outer surface is subjected to elastic deformation. The wall of the cylinder remains partially plastic as long as the angular velocity of rotation is such that $\omega_i < \omega < \omega_o$, where ω_o is the angular velocity corresponding to the complete plastic deformation of the entire wall thickness. The rotational autofrettage of the cylinder is carried out by inducing an angular velocity ω, causing the partial plastic deformation of the cylinder wall during the loading stage. This leads to the formation of an inner plastic zone $a \le r \le c$ and an outer elastic zone $c \le r \le b$, within the wall thickness of the cylinder as shown in Figure 7.1b. In the following subsections, the analyses of stresses in the elastic as well as in the plastic zones of the cylinder are carried out.

7.4.3.1 The Elastic Zone $c \le r \le b$

The stress distributions given by Equations (7.42), (7.43), and (7.40) in the elastic zone $c \le r \le b$ are still valid, provided that the constants of integration become different due to the change of boundary conditions. As the material is assumed to yield as per the Tresca yield criterion, the stress state at the radius of elastic–plastic boundary c leads to:

$$\left(\sigma_\theta - \sigma_r\right)\Big|_{r=c} = \sigma_Y. \tag{7.49}$$

Also, the vanishing of radial stress at the outer surface provides:

$$(\sigma_r)_{r=b} = 0. \tag{7.50}$$

Now, Equations (7.42) and (7.43) are used in the boundary conditions (7.49) and (7.50), and the resulting equations are solved together to obtain the integration constants C_1 and C_2. Then, substituting the values of C_1, C_2, in Equations (7.42) and (7.43) and using Equation (7.40), the radial, hoop, and axial stress distribution in the elastic zone are obtained as:

$$\sigma_r = \left\{\frac{\sigma_Y}{2}c^2 - \left(\frac{1-2\nu}{1-\nu}\right)\frac{\rho\omega^2 c^4}{8}\right\}\left(\frac{1}{b^2} - \frac{1}{r^2}\right) + \left(\frac{3-2\nu}{1-\nu}\right)\frac{\rho\omega^2}{8}\left(b^2 - r^2\right), \tag{7.51}$$

$$\sigma_\theta = \left\{\frac{\sigma_Y}{2}c^2 - \left(\frac{1-2\nu}{1-\nu}\right)\frac{\rho\omega^2 c^4}{8}\right\}\left(\frac{1}{b^2} + \frac{1}{r^2}\right) + \left(\frac{3-2\nu}{1-\nu}\right)\frac{\rho\omega^2}{8}\left\{b^2 - \left(\frac{1+2\nu}{3-2\nu}\right)r^2\right\}, \tag{7.52}$$

$$\sigma_z = \frac{\nu}{b^2}\left\{\sigma_Y c^2 - \left(\frac{1-2\nu}{1-\nu}\right)\frac{\rho\omega^2 c^4}{4}\right\} + \left(\frac{3-2\nu}{1-\nu}\right)\left(\frac{\nu\rho\omega^2}{4}\right)\left\{b^2 - \left(\frac{2}{3-2\nu}\right)r^2\right\}. \tag{7.53}$$

7.4.3.2 The Plastic Zone $a \le r \le c$

For a non-hardening cylinder, the Tresca yield criterion in the plastic zone takes the following form:

$$\sigma_\theta - \sigma_r = \sigma_Y. \tag{7.54}$$

Using Equation (7.54) in equilibrium Equation (7.3) provides:

$$\frac{d\sigma_r}{dr} = \frac{\sigma_Y}{r} - \rho\omega^2 r. \tag{7.55}$$

Integrating equation in the limit from r to $r=c$ within the plastic zone and evaluating the value of σ_r at $r=c$ from Equation (7.51), the solution for radial stress is obtained as:

$$\sigma_r = \left(b^2 - c^2\right)\left\{\left(\frac{3-2\nu}{1-\nu}\right)\frac{\rho\omega^2}{8} - \frac{\sigma_Y}{2b^2} + \left(\frac{1-2\nu}{1-\nu}\right)\frac{\rho\omega^2 c^2}{8b^2}\right\} - \sigma_Y \ln\left(\frac{c}{r}\right) + \frac{\rho\omega^2}{2}\left(c^2 - r^2\right), \tag{7.56}$$

Use of Equation (7.56) in Equation (7.54) provides the hoop stress distribution in the plastic zone as:

$$\begin{aligned}\sigma_\theta = {} & \left(b^2 - c^2\right)\left\{\left(\frac{3-2\nu}{1-\nu}\right)\frac{\rho\omega^2}{8} - \frac{\sigma_Y}{2b^2} + \left(\frac{1-2\nu}{1-\nu}\right)\frac{\rho\omega^2 c^2}{8b^2}\right\} \\ & - \sigma_Y\left\{\ln\left(\frac{c}{r}\right) - 1\right\} + \frac{\rho\omega^2}{2}\left(c^2 - r^2\right).\end{aligned} \tag{7.57}$$

In view of the Tresca associated flow rule, the plastic part of the strains can be expressed as:

$$d\varepsilon_\theta^p = -d\varepsilon_r^p, \quad d\varepsilon_z^p = 0. \tag{7.58}$$

As the total strain component is given by the additive decomposition of the elastic and plastic parts of the respective strain component, according to Equation (7.58), the axial strain is entirely elastic. Thus, using the elastic constitutive relation and invoking the plane–strain condition, the axial stress is still provided by Equation (7.40), even in the plastic zone. Substituting Equations (7.56) and (7.57) in Equation (7.40), the axial stress distribution in the plastic zone is obtained as:

$$\begin{aligned}\sigma_z = {} & \nu\left(b^2 - c^2\right)\left\{\left(\frac{3-2\nu}{1-\nu}\right)\frac{\rho\omega^2}{4} - \frac{\sigma_Y}{b^2} + \left(\frac{1-2\nu}{1-\nu}\right)\frac{\rho\omega^2 c^2}{4b^2}\right\} \\ & - \nu\sigma_Y\left\{2\ln\left(\frac{c}{r}\right) - 1\right\} + \nu\rho\omega^2\left(c^2 - r^2\right).\end{aligned} \tag{7.59}$$

In order to completely obtain the stress distribution in the elastic–plastic cylinder, the value of the elastic–plastic boundary radius c must be known.

The value of c can be obtained by using the boundary condition of vanishing radial stress at the inner radius. This provides:

$$\left(b^2-c^2\right)\left\{\left(\frac{3-2\nu}{1-\nu}\right)\frac{\rho\omega^2}{8}-\frac{\sigma_Y}{2b^2}+\left(\frac{1-2\nu}{1-\nu}\right)\frac{\rho\omega^2c^2}{8b^2}\right\}-\sigma_Y\ln\left(\frac{c}{a}\right)+\frac{\rho\omega^2}{2}\left(c^2-a^2\right)=0. \tag{7.60}$$

Equation (7.60) can be solved numerically for c. For this, the bisection method may be employed. One may also use the FZERO function in MATLAB, which is based on the bisection method to solve Equation (7.60).

7.4.4 Residual Stress Distribution in the Cylinder after the Unloading of Centrifugal Force

The unloading of the induced centrifugal force during the loading stage of the rotational autofrettage process is carried out by gradually reducing the angular velocity to zero. During unloading, the element of a cylinder has the radial acceleration of $r\omega^2$ directed radially outward. During loading, any cylinder element experienced a radially inward acceleration of $-r\omega^2$. Considering the unloading process to be completely elastic without any reverse yielding, the complete elastic unloading stresses are obtained as follows following a similar procedure as in Section 7.4.1.

$$\sigma_r=-\left(\frac{3-2\nu}{1-\nu}\right)\frac{\rho\omega^2}{8}\left(a^2+b^2-\frac{a^2b^2}{r^2}-r^2\right), \tag{7.61}$$

$$\sigma_\theta=-\left(\frac{3-2\nu}{1-\nu}\right)\frac{\rho\omega^2}{8}\left\{a^2+b^2+\frac{a^2b^2}{r^2}-\left(\frac{1+2\nu}{3-2\nu}\right)r^2\right\}. \tag{7.62}$$

$$\sigma_z=-\left(\frac{3-2\nu}{1-\nu}\right)\frac{\nu\rho\omega^2}{4}\left\{a^2+b^2-\left(\frac{2}{3-2\nu}\right)r^2\right\}. \tag{7.63}$$

The residual stresses after unloading are obtained by the superposition of the corresponding loading stresses in the elastic and plastic zones with the unloading stresses given by Equations (7.61–7.63). The resulting expressions of the residual stresses in the plastic and elastic zones of the cylinder are provided as follows:

Plastic zone $a\le r\le c$

$$\begin{aligned}(\sigma_r)_{\text{res}}&=\left(b^2-c^2\right)\left\{\left(\frac{1-2\nu}{1-\nu}\right)\frac{\rho\omega^2c^2}{8b^2}-\frac{\sigma_Y}{2b^2}\right\}-\sigma_Y\ln\left(\frac{c}{r}\right)+\frac{\rho\omega^2}{2}\left(c^2-r^2\right)\\&\quad-\left(\frac{3-2\nu}{1-\nu}\right)\frac{\rho\omega^2}{8}\left(a^2+c^2-\frac{a^2b^2}{r^2}-r^2\right),\end{aligned} \tag{7.64}$$

$$(\sigma_\theta)_{\text{res}} = \left(b^2 - c^2\right)\left\{\left(\frac{1-2\nu}{1-\nu}\right)\frac{\rho\omega^2 c^2}{8b^2} - \frac{\sigma_Y}{2b^2}\right\} - \sigma_Y\left\{\ln\left(\frac{c}{r}\right) - 1\right\} + \frac{\rho\omega^2}{2}\left(c^2 - r^2\right)$$
$$-\left(\frac{3-2\nu}{1-\nu}\right)\frac{\rho\omega^2}{8}\left\{a^2 + c^2 + \frac{a^2b^2}{r^2} - \left(\frac{1+2\nu}{3-2\nu}\right)r^2\right\},$$

(7.65)

$$(\sigma_z)_{\text{res}} = \nu\left(b^2 - c^2\right)\left\{\left(\frac{1-2\nu}{1-\nu}\right)\frac{\rho\omega^2 c^2}{4b^2} - \frac{\sigma_Y}{b^2}\right\} - \nu\sigma_Y\left\{2\ln\left(\frac{c}{r}\right) - 1\right\}$$
$$+\nu\rho\omega^2\left(c^2 - r^2\right) - \left(\frac{3-2\nu}{1-\nu}\right)\frac{\nu\rho\omega^2}{4}\left\{a^2 + c^2 - \left(\frac{2}{3-2\nu}\right)r^2\right\}. \tag{7.66}$$

Elastic zone $c \le r \le b$

$$(\sigma_r)_{\text{res}} = \left\{\frac{\sigma_Y}{2}c^2 - \left(\frac{1-2\nu}{1-\nu}\right)\frac{\rho\omega^2 c^4}{8}\right\}\left(\frac{1}{b^2} - \frac{1}{r^2}\right) - \left(\frac{3-2\nu}{1-\nu}\right)\frac{\rho\omega^2}{8}\left(a^2 + r^2 - \frac{a^2b^2}{r^2} - r^2\right), \tag{7.67}$$

$$(\sigma_\theta)_{\text{res}} = \left\{\frac{\sigma_Y}{2}c^2 - \left(\frac{1-2\nu}{1-\nu}\right)\frac{\rho\omega^2 c^4}{8}\right\}\left(\frac{1}{b^2} + \frac{1}{r^2}\right) - \left(\frac{3-2\nu}{1-\nu}\right)\frac{\rho\omega^2}{8}\left\{a^2 + \frac{a^2b^2}{r^2}\right\}, \tag{7.68}$$

$$(\sigma_z)_{\text{res}} = \frac{\nu}{b^2}\left\{\sigma_Y c^2 - \left(\frac{1-2\nu}{1-\nu}\right)\frac{\rho\omega^2 c^4}{4}\right\} - \left(\frac{3-2\nu}{1-\nu}\right)\frac{\nu\rho\omega^2 a^2}{4}. \tag{7.69}$$

7.4.5 Numerical Simulation

This section presents the numerical evaluation of the stresses during the loading and unloading stages of the rotational autofrettage of a typical cylinder. The material of the cylinder is taken as SS304 steel with the same radial dimensions and material properties as stated in Section 7.3.1. The cylinder initiates yielding at the inner surface when the angular velocity is gradually increased to 571.1 rad/s (5453.6 rpm) according to Equation (7.48). Substituting $c = b$ in Equation (7.60), the angular velocity required to cause complete plastic deformation of the entire wall thickness of the cylinder is obtained as 838.92 rad/s (8011 rpm). Thus, an angular velocity of 750 rad/s (7162 rpm) is selected in order to achieve rotational autofrettage in the cylinder. When the SS304 cylinder is rotated at 750 rad/s (7162 rpm), the cylinder becomes partially plastic up to $c=0.1544$ m. In the following subsections, the stress distributions in the loading stage, the residual stresses after unloading and the overall stress distribution due to internal pressurization in the rotationally autofrettaged cylinder are discussed.

7.4.5.1 Stress Distribution in the SS304 Cylinder during Loading Stage

In the loading stage of the rotational autofrettage of a SS304 cylinder, the stresses are computed using the relevant equations developed in Section 7.4.3, and are shown in Figure 7.9. The trend of the radial and hoop stresses is similar to that of the rotational autofrettage of disks as discussed in Subsection 7.3.6.1. The intermediate principal stress in the axial direction is tensile throughout the wall thickness from the inner to the outer surface of the cylinder.

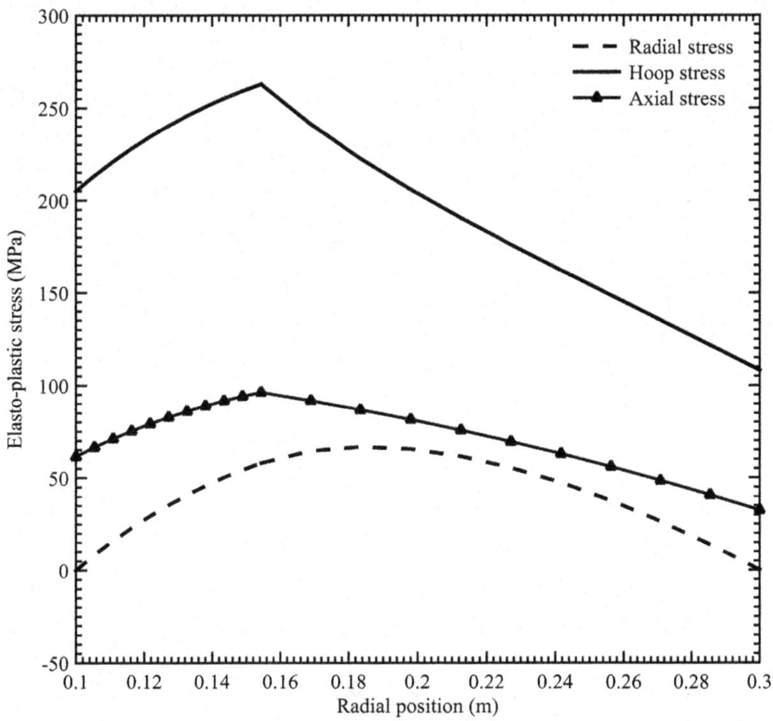

FIGURE 7.9
Elastic–plastic stress distribution in the SS304 cylinder subjected to $\omega=750$ rad/s.

7.4.5.2 Residual Stress Distribution in the SS304 Cylinder after Unloading

The residual stress distribution in the SS304 cylinder after bringing back the rotating cylinder to rest is shown in Figure 7.10. The residual radial and hoop stress follows a similar trend as discussed in Subsection 7.3.6.2. It is observed that a significant amount of compressive residual hoop stress equal to $-0.725\sigma_Y$ is generated at the inner radius of the cylinder. The residual axial stress is also compressive at the neighborhood of the inner surface of the cylinder and becomes slightly tensile towards the outer surface. Due to the large magnitude of the compressive residual hoop stress at the inner surface of the cylinder, the cylinder is now capable of serving against high internal pressure. The pressure-carrying capacity of the rotationally autofrettaged SS304 cylinder will be discussed in Subsection 7.4.5.3. As the residual hoop stress induced at the inner surface of the cylinder is the primary concern during any autofrettage process, a comparison curve of the residual hoop stress between the rotational autofrettage process and the conventional hydraulic autofrettage process (refer to Chapter 3) is shown in Figure 7.11 for the same cylinder. Both autofrettage processes are considered to cause the same amount of overstrain in the cylinder. It is observed that the rotational

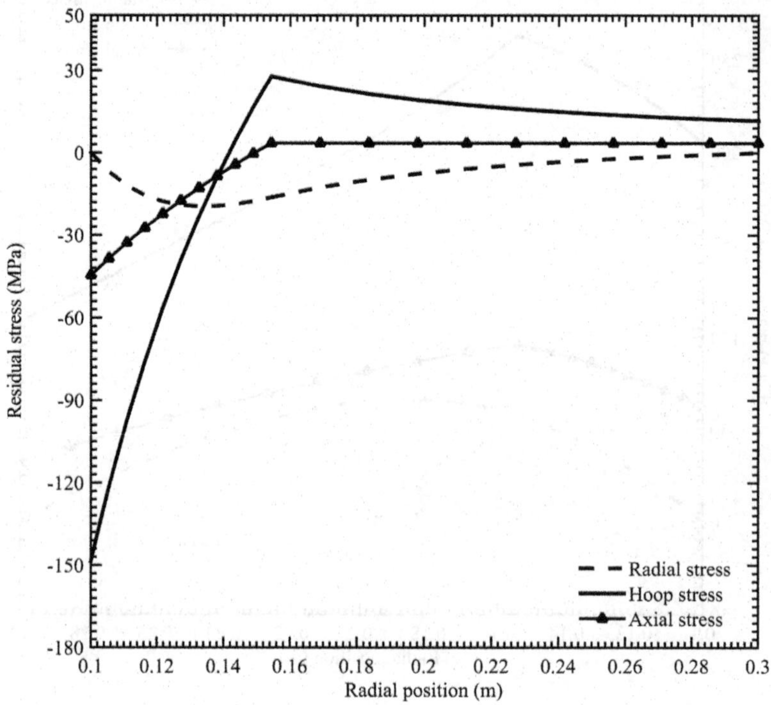

FIGURE 7.10
Residual stress distribution in the SS304 cylinder after unloading.

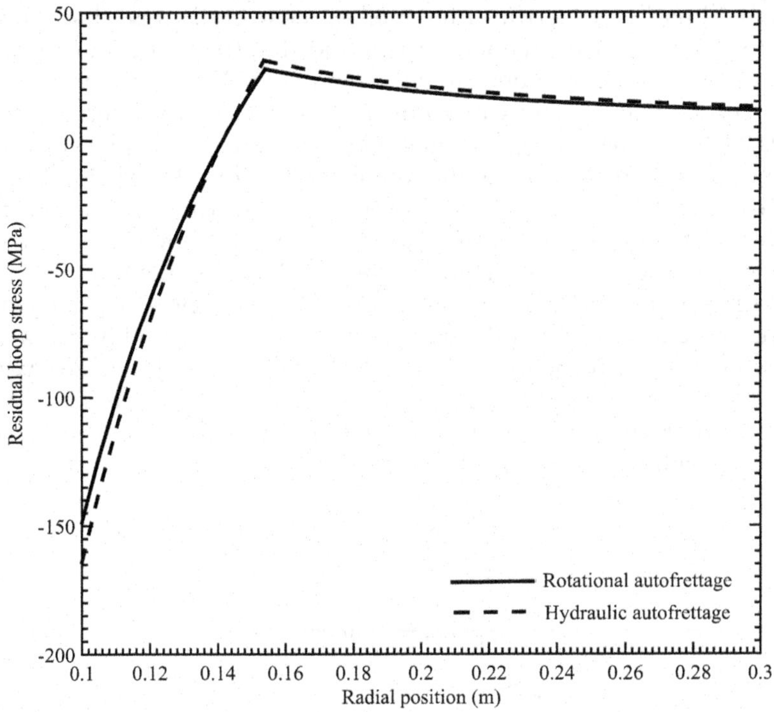

FIGURE 7.11
Comparison of residual hoop stress distribution in the SS304 cylinder subjected to rotational and hydraulic autofrettage for the same level of overstrain.

autofrettage process can produce the residual hoop stress approximately of the same order as that of the conventional hydraulic autofrettage process. Thus, rotational autofrettage is a potential autofrettage procedure to be employed for the strengthening of thick-walled cylindrical vessels.

7.4.5.3 Overall Stress Distribution in the Autofrettaged SS304 Cylinder Subjected to Internal Pressure

The rotationally autofrettaged SS304 cylinder is now considered to be subjected to a high working internal pressure. The overall stress distribution in the autofrettaged cylinder under internal working pressure is obtained by superposing the autofrettage residual stresses with the well-known Lamé's stress solution (Equations 3.3 and 3.4) due to internal pressurization. It is to be noted that the cylinder here is assumed to be open-ended. Thus, no stress is generated in the cylinder in the axial direction due to internal pressure. The highest magnitude of the internal pressure that the autofrettaged cylinder can withstand is that pressure at which the cylinder initiates yielding. In the present case, the yielding of the cylinder is decided based on the Tresca yield criterion. Thus, by numerical simulation of the overall stress distribution in the autofrettaged SS304 cylinder, it is found that the highest magnitude of the pressure that the cylinder can withstand is 157 MPa. The corresponding non-autofrettaged cylinder can withstand the maximum internal pressure of 91.11 MPa only according to Equation (3.13). Thus, in the present case, the maximum pressure capacity of the rotationally autofrettaged SS304 cylinder is increased by 72.3% as compared to the corresponding non-autofrettaged cylinder. The overall stress distribution in the autofrettaged cylinder subjected to the maximum internal pressure of 157 MPa is shown in Figure 7.12. It is observed that at the maximum internal working pressure, the cylinder is at the verge of yielding at the inner surface according to the Tresca equivalent stress ($\sigma_\theta - \sigma_r$). If the cylinder would have been subjected to hydraulic autofrettage by causing the same level of overstrain as that in the rotational autofrettage process, it could withstand 80% higher working pressure than the corresponding non-autofrettaged cylinder.

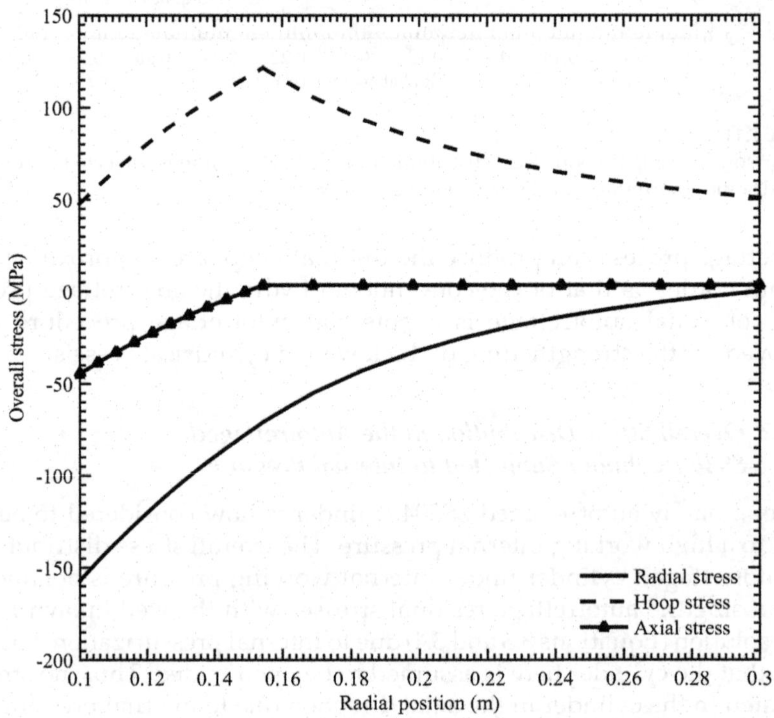

FIGURE 7.12
Overall stress distribution in the rotationally autofrettaged SS304 cylinder subjected to the maximum internal working pressure $p_w = 157$ MPa.

7.5 A Feasibility Study of the Rotational Autofrettage Process

The rotational autofrettage procedure can be accomplished using an experimental setup as shown in Figure 7.1a. The cylinder to be autofrettaged may be mounted on two suitable bearings and made to rotate about its own axis by means of an electric motor as shown in Figure 7.1a. The setup may also be employed for the rotational autofrettage of thick-walled, hollow circular disks. In this section, a feasibility study of the rotational autofrettage process using the setup shown in Figure 7.1a is studied, both for a disk and a cylinder. The theoretical treatments of the rotational autofrettage of a SS304 disk and cylinder as discussed in Sections 7.3.6 and 7.4.5 are considered here for the feasibility study. The study is based on the principle stated in Zare and Darijani (2016). It is assumed that all of the produced energy by the electrical motor is converted to kinetic energy of the disk/cylinder without any energy dissipation. Hence, the energy conservation principle provides:

$$E = \frac{1}{2} I\omega^2 = Pt = T\omega t, s \tag{7.70}$$

where I, ω, P, T, and t represents the moment of inertia, angular velocity, output motor power, output torque of the motor, and time of rotation, respectively. The moment of inertia I of a hollow disk/cylinder is given by the following expression:

$$\begin{aligned} I &= \frac{1}{2} m\left(b^2 + a^2\right) \\ &= \frac{\pi}{2}\rho\left(b^4 - a^4\right)L, \end{aligned} \tag{7.71}$$

where m, ρ, L, a, and b are the mass, density, length, inner radius, and outer radius of the disk/cylinder, respectively.

Taking the length of the SS304 disk as 0.05 m, the moment of inertia of the rotating disk is evaluated from Equation (7.71) as $I = 5.03$ kg m^2. Thus, the energy required to rotate the cylinder at the autofrettage speed of 700 rad/s is $E = 1.23 \times 10^6$ Nm according to Equation (7.70). Similarly, considering the length of the SS304 cylinder as 1 m, the required energy is calculated as $E = 28.27 \times 10^6$ Nm for the autofrettage speed $\omega = 750$ rad/s. Thus, practically we need an electrical motor, which can provide a mechanical energy of 1.23×10^6 Nm for the rotational autofrettage of the present SS304 disk and 28.27×10^6 Nm that for the rotational autofrettage of the cylinder. A common motor is selected purposefully so that the rotational autofrettage operation of both the disk and cylinder can be performed in a single setup. For this, a 30 kW, 10000 rpm high speed permanent magnet motor may be employed. This motor can provide the required energy for the rotational autofrettage of the disk after 41 seconds of operation time. For the rotational autofrettage of the cylinder, the same motor can provide the required energy after an operation time of 942.3 seconds. A simple gear train with a suitable gear ratio may be used for transmitting power from the motor shaft to the disk/cylinder shaft. The required gear ratio can be calculated using the following equation:

$$\frac{T_{\text{motor}}}{T_{\text{disk/cylinder}}} = \frac{N_{\text{disk/cylinder}}}{N_{\text{motor}}}, \tag{7.72}$$

where T indicates the number of gear teeth and N is the rotation in rpm. In the present case, for the rotational autofrettage of the disk, a gear ratio of 0.67 is needed. A gear ratio of 0.72 is needed for the rotational autofrettage of the SS304 cylinder.

7.6 Conclusion

In this chapter, a detailed theoretical analysis of the rotational autofrettage process was carried out. At the beginning, the process was briefly introduced followed by two different theoretical models that have been developed so far. The first theoretical model is based on the plane–stress assumption and suitable for the prediction of stresses and strains in the rotational autofrettage of strain-hardened thick-walled hollow circular disks. The second theoretical model is based on the plane–strain assumption and is applicable for the analysis of rotational autofrettage of non-hardening thick-walled long cylindrical vessels. In both of the theoretical models, the plastic deformation was modeled using the Tresca yield criterion and its associated flow rule. The numerical estimation of stresses was also carried out for a typical SS304 disk and cylinder using the relevant equations. It can be concluded that both SS304 disk and cylinder induces a significant amount of compressive residual hoop stress at the neighborhood of the inner surface of the disk and the cylinder due to rotational autofrettage providing a significant increase in the service load. The service load-carrying capacity of the rotationally autofrettaged disk was assessed by subjecting the disk into three different individual loads: internal pressure, radial temperature difference, and angular velocity. Approximately the same amount of the maximum pressure-carrying capacity and radial temperature difference is achieved in the autofrettaged disk as compared to the non-autofrettaged counterpart. In the present case, it is about 40% corresponding to an overstrain level of 33.5%. However, the maximum increase in the angular velocity (centrifugal load-bearing capacity) is comparatively smaller, which is about 19% in the present case. The rotationally autofrettaged SS304 cylinder can increase the maximum pressure-carrying capacity by 72.3% as compared to its non-autofrettaged counterpart corresponding to an overstrain level of 27%. It is worth mentioning that, if the same cylinder is subjected to the conventional hydraulic autofrettage process causing the same level of overstrain in the cylinder wall, the cylinder can generate residual stresses, which are approximately of the same order as those obtained in the case of rotational autofrettage. Consequently, the maximum achievable increase in the pressure capacity of the cylinder in the rotational autofrettage process is very close to that achieved in the hydraulic autofrettage process. Thus, the rotational autofrettage process could be a potential alternative autofrettage procedure for strengthening both thick-walled disks and cylinders. Until now, no experimental study of the process has been reported. In this chapter, a feasibility study on carrying out the rotational autofrettage operation in practice was carried out for the typical SS304 disk and cylinder. The study showed that rotational autofrettage is a feasible process to be employed in various industries.

Acknowledgments

This chapter is mainly based on the following article:

Kamal, S.M., (2018), Analysis of residual stress in the rotational autofrettage of thick-walled disks, *ASME Journal of Pressure Vessel Technology*, **140**(6), 061402-1–061402-10.

The authors are grateful to the publisher of this article.

References

Chakrabarty, J., (2006), *Theory of Plasticity*, 3rd edition, Butterworth-Heinemann, Burlington, MA.

Dixit, P. M., & Dixit, U. S. (2008), *Modeling of metal forming and machining processes: by finite element and soft computing methods*, Springer Science & Business Media.

Kamal, S.M., (2018), Analysis of residual stress in the rotational autofrettage of thick-walled disks, *ASME Journal of Pressure Vessel Technology*, **140**(6), 061402-1–061402-10.

Kamal, S.M. and Dixit, U.S., (2015), Feasibility study of thermal autofrettage process, in *Advances in Material Forming and Joining*, edited by R.G. Narayanan and U.S. Dixit, Springer, New Delhi, India, pp. 81–107.

Kamal, S.M., Dixit, U.S., Roy, A., Liu Q. and Silberschmidt, V.V., (2017), Comparison of plane-stress, generalized-plane-strain and 3D FEM elastic–plastic analyses of thick-walled cylinders subjected to radial thermal gradient, *International Journal of Mechanical Sciences*, **131–132**, 744–752.

Kamal, S.M. and Kulsum, R., (2019), Parametric study of axisymmetric circular disk subjected to rotational autofrettage, in Proceedings of the 2nd International Conference on Computational Methods in Manufacturing (ICCMM 2019), IIT Guwahati, India (accepted).

Rees, D.W.A., (2011), A theory for swaging of discs and lugs, *Meccanica*, **46**, 1213–1237.

Zare, H.R. and Darijani, H., (2016), A novel autofrettage method for strengthening and design of thick-walled cylinders, *Materials and Design*, **105**, 366–374.

Zare, H.R. and Darijani, H., (2017), Strengthening and design of the linear hardening thick-walled cylinders using the new method of rotational autofrettage, *International Journal of Mechanical Sciences*, **124–125**, 1–8.

8

Selection of an Autofrettage Process and Its Optimization

8.1 Introduction

By now the reader must be well acquainted with the general principle of the autofrettage process along with its various types. Depending on the physics and the experimental procedure associated with their practical implementations, each type of autofrettage process may have its own advantages and disadvantages. Based on these differentiating attributes, it is the responsibility of the engineers designing and manufacturing autofrettage equipment to identify and provide the industrialist with a strategy to select the most suitable autofrettage process depending on the situation. For example, in the manufacturing of gun barrels, the inner wall of the barrel must be provided with helical grooves. The function of these helical grooves is to provide a swirl of hot gases that propels the bullet through the barrel, which in turn spins the bullet. The spinning movement ensures that the bullet travels in a straight path thanks to the presence of gyroscopic action. The process of machining helical grooves in the inner wall of gun barrels for this purpose is called rifling. Also, since gun barrels must be autofrettaged, a question arises as to whether the autofrettage or the rifling operation should be performed first while fabricating a gun barrel. As pointed out by Bihamta et al. (2007), the mandrel driving force in a swage autofrettage process can be reduced by using a mandrel with straight or helical grooves, it is rather interesting that rifling and autofrettage may be done in a single operation with the swage autofrettage process using a well-designed mandrel. Thus, for the manufacture of gun barrels, swage autofrettage is the most convenient process.

As explained in the earlier chapter, the mechanism of inducing compressive residual stresses in an autofrettaged cylinder may be of two types—one where there is partial plastic deformation of the cylinder up to an intermediate radial position, and another where there is complete plastic deformation up to the outer wall. In the first case, the compressive residual stresses are induced due to the inward force exerted by the outer elastic zone on the inner plastic zone due to its tendency to contract internally to regain its original undeformed position. In the second case, the material at the vicinity of the inner wall undergoes more strain than that at the outer wall. Here, the inner wall compressive residual stresses are induced due to non-uniform straining. On the basis of these two types of mechanisms it is necessary to characterize a particular type of autofrettage. The amount of autofrettage in a vessel is generally quantified as (a) the level of autofrettage and (b) the percentage of autofrettage.

The level of autofrettage in a vessel is defined by overstrain, which is the percentage of the thickness of the plastic zone. It is expressed as:

$$\delta = 100\left(\frac{c-a}{b-a}\right), \tag{8.1}$$

where δ represents the overstrain, a and b are the inner radius and outer radius of the cylindrical or spherical vessel and c is the radius of the elastic–plastic interface. A zero overstain means that there is no plastic deformation after the loading phase of the autofrettage of a pressure vessel. A 100% overstrain means that a pressure vessel is completely plastic after the loading phase of the autofrettage process. On the other hand, the percentage of autofrettage is the fraction of the maximum possible compressive residual stresses that can be achieved by an autofrettage process for the particular vessel with the specific material and dimension (Faupel, 1955). A 100% autofrettage means that the maximum possible residual stress has been achieved. It is to be carefully noted that the level of autofrettage and the percentage of autofrettage are entirely different and must not be confused with one another. A 100% autofrettage is possible without 100% overstrain. For example, in a typical open-ended perfectly-plastic specimen (following the von Mises yield criterion) with an inner radius of 10 mm and an outer radius of 30 mm, which is made up of a material with a yield strength of 205 MPa, the required pressure to cause a 100% autofrettage is 210 MPa with a plastic zone up to a radius of 15.53 mm i.e., 27.66% overstrain with a maximum compressive hoop residual stress of about 205 MPa at the inner wall. If the specimen follows the Tresca yield criterion, the elastic–plastic interface radius is 17.46 mm i.e., 37.3% overstrain with a maximum compressive hoop residual stress of about 205 MPa at the inner wall. For applications involving a very high working pressure, such as in the case of pressure vessels, a 100% autofrettage may be required for acquiring the maximum residual stresses, while for some applications, a lesser autofrettage might be enough. Therefore, it is important to economize the process by optimizing the autofrettage process to suit the vessel geometry for its intended application.

This chapter will briefly discuss the merits and demerits associated with each type of autofrettage in Section 8.2. After the comparison, a brief review of various studies that have been carried out for the optimization of autofrettage process is presented in Section 8.3. Section 8.4 presents case studies for the procedure for the optimization of hydraulic autofrettage in cylindrical and spherical vessels. Section 8.5 concludes the chapter.

8.2 Comparison of Various Autofrettage Processes

This section presents the comparison of the various types of autofrettage process. The basis for comparing the types are as follows: the efficiency of the process, applicability in the variety of vessels, overall cost and ease of operation. The comparison is listed in Table 8.1.

TABLE 8.1

Merits and Demerits of the Different Types of Autofrettage Processes

Type of autofrettage	Merits	Demerits
Hydraulic autofrettage	• The process can achieve the maximum increase in the pressure-carrying capacity. • It is suitable for the autofrettage of spherical as well as cylindrical vessels with varying end conditions.	• It requires a high-pressure hydraulic system. • The process is costly.
Swage autofrettage	• It requires less energy compared to hydraulic autofrettage and hence is comparatively more energy efficient. • It is more economical than hydraulic autofrettage. • The process is simple for practical implementation.	• It is not suitable for autofrettage of spherical or closed-ended cylindrical vessels.
Explosive autofrettage	• The process can be employed for the autofrettage of spherical as well as cylindrical vessels with varying end conditions.	• It involves explosives and needs a lot of safety measures. • Difficult to control and dangerous for practical implementations.
Thermal autofrettage	• The process can be employed for the autofrettage of spherical as well as cylindrical vessels with varying end conditions. • Very simple for practical implementation. • Cost effective.	• It has a limit in the achievable maximum increase in the pressure-carrying capacity.
Rotational autofrettage	• Very simple for practical implementation. • Cost effective.	• It is not suitable for the autofrettage of spherical vessels or very large components. • Proper balancing and adequate safety measures are needed due to high rotation.

8.3 Optimization of Autofrettage

According to Rees (1987), the optimum autofrettage pressure is the pressure that just avoids the reverse yielding in the cylinder due to compressive residual stresses after the unloading step. Reverse yielding is not necessarily harmful to the cylinder, but it implies that autofrettage has been carried out with a load whose magnitude is redundantly greater than an optimum magnitude. The theoretically calculated value of the magnitude of this optimum pressure is affected by three factors—the dimensions of the vessel, the strain hardening behavior and the Bauschinger effect. As already discussed in Chapter 3, assuming the Tresca yield criterion without hardening, the optimum autofrettage pressure in cylinders with wall-thickness ratios greater than 2.22 and spheres with wall-thickness ratios greater than 1.7, is equal to twice the yield pressure. With the consideration of the Bauschinger effect, the maximum employable autofrettage pressure will be lesser because of reduced compressive yield strength that affects the maximum achievable autofrettage. As a rule of thumb, Parker (2000) suggested that the maximum compressive residual stress obtained considering Bauschinger effect is 30% lesser than that obtainable based on ideal material behavior. According to Parry (1965), the maximum allowable autofrettage pressure for a cylinder roughly provides an elastic interface radius equal to the geometric mean of the inner and outer radii of the cylinder.

Varga (1991) minimized the weight of a closed cylindrical vessel with hemi-spherical ends subjected to hydraulic autofrettage. The basis was the minimization of the equivalent stress at the elastic–plastic interface after the application of working pressure in the hydraulically autofrettaged cylinder. Zhu and Yang (1998) also proposed an analytical optimization procedure to minimize the equivalent stress at the elastic–plastic interface radius after the pressurization of a hydraulically autofrettaged cylinder. They considered the Tresca yield criterion as well as a plane–strain von Mises yield criterion. Hardening was neglected for both cases. The optimum elastic–plastic interface radius obtained from the model based on the Tresca yield criterion was lesser than that obtained using von Mises criterion. Darijani et al. (2009) used the analytical solution by Mendelson (1968) to optimize the hoop stress, equivalent stress and the wall thickness ratio in a hydraulically autofrettaged cylinder. The optimization procedures for the hoop stress and equivalent stress were based on a graphical method whereas that for the cylinder wall thickness was analytical.

Hojjati and Hassani (2007) applied the analytical solution developed by Gao (1992) for plane–stress and Gao (2003) for plane–strain condition to optimize the elastic–plastic interface radius and the autofrettage pressure. Both the base models (Gao, 1992, 2003) were used with Hollomon's hardening law. An analytical procedure was derived for the plane–strain condition whereas a numerical procedure was required for the plane–stress condition. The optimum autofrettage pressures obtained from the two end conditions were almost the same. The optimum autofrettage pressures increased with the increase in the wall-thickness ratio and the hardening exponent. Çandar and Filiz (2017) also carried out a similar study to find out the optimum autofrettage pressure for the cylinder of a waterjet intensifier pump using the elastic–plastic interface radius as the design variable. The stress analysis was based on an analytical model using the von Mises yield criterion with linear kinematic hardening. For a 160 mm long AISI 4340 alloy steel cylinder of inner radius 14 mm and an outer radius of 38 mm, the optimum autofrettage pressure was 936 MPa.

Maleki et al. (2010) used a numerical method to determine the optimum autofrettage at which the compressive residual hoop stress at the inner wall was the maximum while the tensile residual hoop stress at the outer wall was the minimum in a spherical vessel. A non-linear hardening was assumed for the material behavior. Two cases of spherical vessels made of materials A723 and HB7 steels with a common wall-thickness ratio of 2 were investigated. For the sphere made of A723 steel, 40% was the recommended percentage of autofrettage. However, for HB7 steel, the procedure could not be applied due to considerable strain hardening behavior of the material.

Perl and Perry (2010) carried out a study based on the FDM to optimize the weight of a thick-walled spherical vessel. The model was based on the real material behavior incorporating kinematic hardening. The minimum pressure at which the autofrettaged spherical vessel yielded during pressurization was referred to as the safe minimum pressure (SMP) and the optimization was aimed at maximizing the specific SMP, which was the ratio of the SMP and the weight of the vessel. For a typical vessel with wall-thickness ratio of 1.5 with a prescribed SMP, it was shown that 60% autofrettage resulted in 57% reduction of the total weight of the vessel compared to the case without autofrettage designed to withstand a pressure equal to the SMP.

8.4 Case Study of the Optimization of Autofrettage in Cylindrical and Spherical Vessels

This section discusses some optimization procedures for minimizing the equivalent stress at the elastic–plastic interface radius of a hydraulically autofrettaged cylindrical and spherical vessel. For cylindrical vessels, the procedure is based on the analytically based study carried out by Zhu and Yang (1998). Two models, one based on the Tresca yield criterion and another based on a plane–strain von Mises yield criterion, were obtained. Material hardening was neglected in the study. These models will be derived in this section. This analytically based procedure is applied in the case of a spherical vessel as well. The mathematical procedure is explained as follows:

8.4.1 Optimization for a Hydraulically Autofrettaged Cylinder Based on the Tresca Yield Criterion (Zhu and Yang, 1998)

As already in Chapter 3, the expressions for the distribution of the residual radial and hoop stresses after unloading in a hydraulic autofrettage process are:

$$\sigma_r^{\text{res}} = -\frac{\sigma_Y}{2}\left(\frac{c^2}{a^2} - \frac{p}{p_Y}\right)\left(\frac{a^2}{r^2} - \frac{a^2}{b^2}\right), \tag{8.2}$$

$$\sigma_\theta^{\text{res}} = \frac{\sigma_Y}{2}\left(\frac{c^2}{a^2} - \frac{p}{p_Y}\right)\left(\frac{a^2}{r^2} + \frac{a^2}{b^2}\right), \tag{8.3}$$

where a denotes the inner radius, b denotes the outer radius, c denotes the elastic–plastic interface radius, p is the autofrettage pressure, p_Y is the yield pressure, σ_r^{res} denotes the radial component, and $\sigma_\theta^{\text{res}}$ denotes the hoop component of the residual stress. The expressions for p_Y and p as a function of c as shown in Chapter 3 are as follows:

$$p_Y = \frac{\sigma_Y}{2}\left(1 - \frac{a^2}{b^2}\right), \tag{8.4}$$

$$p = \frac{\sigma_Y}{2}\left(1 - \frac{c^2}{b^2} + \ln\frac{c^2}{a^2}\right). \tag{8.5}$$

Substituting Equations (8.4) and (8.5) in Equations (8.2) and (8.3) provides:

$$\sigma_r^{\text{res}} = \frac{\sigma_Y}{2}\left(1 - \frac{b^2}{r^2}\right)\left\{\frac{c^2}{b^2} - \frac{1}{k^2-1}\left(1 - \frac{c^2}{b^2} + \ln\frac{c^2}{a^2}\right)\right\}, \tag{8.6}$$

$$\sigma_\theta^{\text{res}} = \frac{\sigma_Y}{2}\left(1 + \frac{b^2}{r^2}\right)\left\{\frac{c^2}{b^2} - \frac{1}{k^2-1}\left(1 - \frac{c^2}{b^2} + \ln\frac{c^2}{a^2}\right)\right\}, \tag{8.7}$$

where $k = b/a$. At the radial position $r = c$, Equations (8.6) and (8.7) provide the following expressions for the radial and hoop residual stress, respectively, at the elastic–plastic interface:

$$\sigma_{rc}^{\text{res}} = \frac{\sigma_Y}{2}\left(1 - \frac{b^2}{c^2}\right)\left\{\frac{c^2}{b^2} - \frac{1}{k^2-1}\left(1 - \frac{c^2}{b^2} + \ln\frac{c^2}{a^2}\right)\right\}, \tag{8.8}$$

$$\sigma_{\theta c}^{\text{res}} = \frac{\sigma_Y}{2}\left(1+\frac{b^2}{c^2}\right)\left\{\frac{c^2}{b^2}-\frac{1}{k^2-1}\left(1-\frac{c^2}{b^2}+\ln\frac{c^2}{a^2}\right)\right\}. \tag{8.9}$$

The suffix "c" in the subscript signifies the values of the components of stresses at the radial position $r=c$. When the hydraulically autofrettaged cylinder is pressurized with the working pressure w, the radial and hoop stresses at the elastic–plastic interface radius are obtained from Equations (3.3) and (3.4) by substituting $r=c$, respectively as follows:

$$\sigma_{rc}^{w} = -\frac{w}{k^2-1}\left(\frac{b^2}{c^2}-1\right), \tag{8.10}$$

$$\sigma_{\theta c}^{w} = \frac{w}{k^2-1}\left(\frac{b^2}{c^2}+1\right). \tag{8.11}$$

where σ_{rc}^{w} denotes the radial component and σ_{θ}^{w} denotes the hoop component of stress due to the applied working pressure w at the elastic–plastic interface radius. The net value of the radial and hoop components of stress at the elastic–plastic interface are obtained as follows:

$$\sigma_{rc}^{\text{net}} = \sigma_{rc}^{w} + \sigma_{rc}^{\text{res}}, \tag{8.12}$$

$$\sigma_{\theta c}^{\text{net}} = \sigma_{\theta c}^{w} + \sigma_{\theta c}^{\text{res}}. \tag{8.13}$$

where σ_{rc}^{net} denotes the net radial stress and $\sigma_{\theta c}^{\text{net}}$ is the net hoop stress at the elastic–plastic interface radius. The equivalent stress at the elastic–plastic interface radius assuming the Tresca yield criterion is:

$$\sigma_{\text{eq}} = \sigma_{\theta c}^{\text{net}} - \sigma_{rc}^{\text{net}}. \tag{8.14}$$

Using Equations (8.8–8.13) in Equation (8.14) provides:

$$\sigma_{\text{eq}} = \sigma_Y - \frac{\sigma_Y}{k^2-1}\left(\frac{b^2}{c^2}-1+\frac{b^2}{c^2}\ln\frac{c^2}{a^2}\right)+\frac{b^2}{c^2}\left(\frac{2w}{k^2-1}\right). \tag{8.15}$$

Partial differentiation of Equation (8.15) provides:

$$\frac{\partial\sigma_{\text{eq}}}{\partial c} = -\frac{\sigma_Y}{k^2-1}\left(-\frac{2b^2}{c^3}\ln\frac{c^2}{a^2}\right)-\frac{2b^2}{c^3}\left(\frac{2w}{k^2-1}\right). \tag{8.16}$$

Equating Equation (8.16) to 0 provides:

$$\frac{\sigma_Y}{k^2-1}\left(\frac{4b^2}{c^3}\ln\frac{c}{a}\right)=\frac{4b^2}{c^3}\left(\frac{w}{k^2-1}\right), \tag{8.17}$$

which subsequently reduces to the expression for the optimum elastic–plastic interface radius $c_{\text{opt}}^{\text{Tresca}}$ based on the Tresca criterion as follows:

$$c_{\text{opt}}^{\text{Tresca}} = a\exp\left(\frac{w}{\sigma_Y}\right) \tag{8.18}$$

Partially integrating Equation (8.16) provides:

$$\frac{\partial^2 \sigma_{\text{eq}}}{\partial c^2} = \frac{\sigma_Y}{k^2-1}\left(\frac{2b^2}{c^4}\right)\left(2+3\ln\frac{c^2}{a^2}\right)+\frac{6b^2}{c^4}\left(\frac{2w}{k^2-1}\right) \tag{8.19}$$

that is always greater than zero. Therefore, for all values of c in the range $[a, b]$,

$$\frac{\partial^2 \sigma_{\text{eq}}}{\partial c^2} > 0. \tag{8.20}$$

Hence, σ_{eq} at the elastic–plastic interface radius is the minimum. Substituting Equation (8.18) into Equation (8.5), provides:

$$p = \frac{\sigma_Y}{2}\left\{1-\frac{1}{k^2}\exp\left(\frac{2w}{\sigma_Y}\right)\right\}+w. \tag{8.21}$$

Expressing Equation (8.4) as a function of k provides:

$$\frac{1}{k^2} = 1-\frac{2}{\sigma_Y}p_Y, \tag{8.22}$$

which is substituted in Equation (8.21) to provide the following expression for the optimum autofrettage pressure $p_{\text{opt}}^{\text{Tresca}}$ based on the Tresca yield criterion:

$$p_{\text{opt}}^{\text{Tresca}} = \frac{\sigma_Y}{2}\left\{1-\left(1-\frac{2}{\sigma_Y}p_Y\right)\exp\left(\frac{2w}{\sigma_Y}\right)\right\}+w. \tag{8.23}$$

8.4.2 Optimization for a Hydraulically Autofrettaged Cylinder Based on the von Mises Yield Criterion (Zhu And Yang, 1998)

It has been shown in Equation (3.66) that for the plane–strain case, the expression for the von Mises yield criterion is:

$$(\sigma_\theta - \sigma_r) = \frac{2}{\sqrt{3}}\sigma_Y. \tag{8.24}$$

Hence, the procedure for the optimization for a material based on the von Mises yield criterion is the same as that followed for the Tresca yield

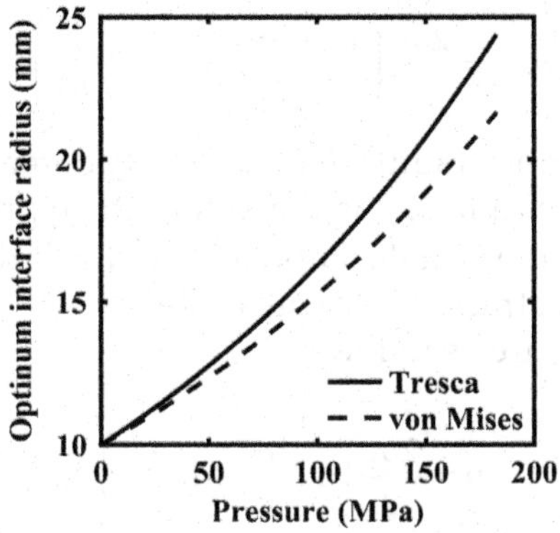

FIGURE 8.1
Variation of the elastic–plastic interface radius with the pressure in a cylindrical vessel.

criterion shown by Equations (8.2–8.23) except that the initial yield stress σ_Y is scaled by a factor of $2/\sqrt{3}$. Thus the optimum elastic–plastic interface radius $c_{\text{opt}}^{\text{von Mises}}$ for the assumption based on the von Mises yield criterion is:

$$c_{\text{opt}}^{\text{von Mises}} = a\exp\left(\frac{\sqrt{3}w}{2\sigma_Y}\right). \tag{8.25}$$

Similarly the expression for the optimum autofrettage pressure $p_{\text{opt}}^{\text{von Mises}}$ based on the von Mises yield criterion is obtained from Equation (8.23) as follows:

$$p_{\text{opt}}^{\text{von Mises}} = \frac{\sigma_Y}{\sqrt{3}}\left\{1-\left(1-\frac{\sqrt{3}}{\sigma_Y}p_Y\right)\exp\left(\frac{\sqrt{3}w}{\sigma_Y}\right)\right\}+w. \tag{8.26}$$

Figure 8.1 shows the variation of the optimum elastic–plastic interface radius with the applied pressure for the procedures based on the Tresca and von Mises yield criterion. At low pressures, the two models give almost identical values of the optimum elastic–plastic interface radius. At high pressures, the difference on the basis of the two yield criteria becomes more significant.

8.4.3 Optimization for a Hydraulically Autofrettaged Sphere

As already shown in Chapter 3, the expressions for the distribution of the residual radial and hoop stresses in a hydraulically autofrettaged sphere are:

$$\sigma_r^{\text{res}} = -\frac{2\sigma_Y}{3}\left(\frac{c^3}{a^3}-\frac{p}{p_Y}\right)\left(\frac{a^3}{r^3}-\frac{a^3}{b^3}\right), \tag{8.27}$$

$$\sigma_\theta^{\text{res}} = \frac{2\sigma_Y}{3}\left(\frac{c^3}{a^3} - \frac{p}{p_Y}\right)\left(\frac{a^3}{2r^3} + \frac{a^3}{b^3}\right), \tag{8.28}$$

where a denotes the inner radius, b denotes the outer radius, c denotes the elastic–plastic interface radius, p is the autofrettage pressure, p_Y is the yield pressure, σ_r^{res} denotes the radial component, and $\sigma_\theta^{\text{res}}$ denotes the hoop component of the residual stress. The expressions for p_Y and p as a function of c as shown in Chapter 3 are as follows:

$$p = \frac{2\sigma_Y}{3}\left(1 - \frac{c^3}{b^3} + \ln\frac{c^3}{a^3}\right), \tag{8.29}$$

$$p_Y = \frac{2\sigma_Y}{3}\left(1 - \frac{a^3}{b^3}\right). \tag{8.30}$$

Substituting Equations (8.29) and (8.30) in Equations (8.27) and (8.28) provides:

$$\sigma_r^{\text{res}} = -\frac{2\sigma_Y}{3}\left(\frac{b^3}{r^3} - 1\right)\left\{\frac{c^3}{b^3} - \frac{1}{k^3 - 1}\left(1 - \frac{c^3}{b^3} + \ln\frac{c^3}{a^3}\right)\right\}, \tag{8.31}$$

$$\sigma_\theta^{\text{res}} = \frac{2\sigma_Y}{3}\left(\frac{b^3}{r^3} + 1\right)\left\{\frac{c^3}{b^3} - \frac{1}{k^3 - 1}\left(1 - \frac{c^3}{b^3} + \ln\frac{c^3}{a^3}\right)\right\}, \tag{8.32}$$

where $k = b/a$. At the radial position $r = c$, Equations (8.31) and (8.32) provide the following expressions for the radial and hoop residual stress, respectively, at the elastic–plastic interface:

$$\sigma_{rc}^{\text{res}} = -\frac{2\sigma_Y}{3}\left(\frac{b^3}{c^3} - 1\right)\left\{\frac{c^3}{b^3} - \frac{1}{k^3 - 1}\left(1 - \frac{c^3}{b^3} + \ln\frac{c^3}{a^3}\right)\right\}, \tag{8.33}$$

$$\sigma_{\theta c}^{\text{res}} = \frac{2\sigma_Y}{3}\left(\frac{b^3}{c^3} + 1\right)\left\{\frac{c^3}{b^3} - \frac{1}{k^3 - 1}\left(1 - \frac{c^3}{b^3} + \ln\frac{c^3}{a^3}\right)\right\}. \tag{8.34}$$

The suffix "c" in the subscript signifies the values of the components of stresses at the radial position $r = c$. When the hydraulically autofrettaged cylinder is pressurized with the working pressure w, the radial and hoop component of stresses at the elastic–plastic interface radius are obtained from Equations (3.79) and (3.80) by substituting $r = c$, respectively, as follows:

$$\sigma_r^{w} = -\frac{w}{k^3 - 1}\left(\frac{b^3}{c^3} - 1\right), \tag{8.35}$$

$$\sigma_\theta^w = \frac{w}{k^3-1}\left(\frac{b^3}{2w^3}+1\right). \tag{8.36}$$

where σ_{rc}^w denotes the radial component and σ_θ^w denotes the hoop component of stress due to the applied working pressure w at the elastic–plastic interface radius.

Both the Tresca yield criterion and the von Mises yield criterion are identical for a spherical vessel. Hence, applying Equations (8.12–8.14) and using Equations (8.33–8.36) provides:

$$\sigma_{\text{eq}} = \frac{4\sigma_Y}{3} - \frac{4\sigma_Y}{3}\left(\frac{1}{k^3-1}\right)\left(\frac{b^3}{c^3} - 1 + \frac{b^3}{c^3}\ln\frac{c^3}{a^3}\right) + \frac{3wb^3}{2r^3}\left(\frac{1}{k^3-1}\right). \tag{8.37}$$

Partially differentiating Equation (8.38) with c and equating to zero provides:

$$\frac{\partial\sigma_{\text{eq}}}{\partial c} = \frac{4\sigma_Y}{3}\left(\frac{1}{k^3-1}\right)\left(\frac{3b^3}{c^4}\ln\frac{c^3}{a^3}\right) - \frac{9wb^3}{2c^4}\left(\frac{1}{k^3-1}\right) = 0, \tag{8.38}$$

which upon solving provides the following expressions for the optimum elastic–plastic interface radius c_{opt} and the optimum autofrettage pressure p_{opt}:

$$c_{\text{opt}} = a\exp\left(\frac{3w}{8\sigma_Y}\right), \tag{8.39}$$

$$p_{\text{opt}} = \frac{2\sigma_Y}{3}\left\{1 - \left(1 - \frac{3p_Y}{2\sigma_Y}\right)\exp\left(\frac{9w}{8\sigma_Y}\right)\right\} + \frac{3w}{4}. \tag{8.40}$$

Figure 8.2 shows the variation of the optimum elastic–plastic interface radius with the applied pressure.

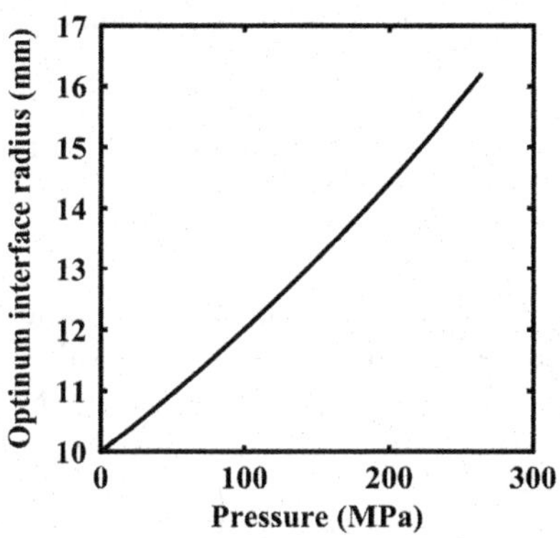

FIGURE 8.2
Variation of the elastic–plastic interface radius with the pressure in a spherical vessel.

8.5 Conclusion

The comparison of the different types of autofrettage processes were presented based on aspects such as the efficiency of the process, applicability in the variety of vessels, overall cost, and ease of operation. This comparison will help an industrialist to identify and select a particular type of autofrettage to suit her/his needs. This was followed by a discussion on the need to optimize an autofrettage process and a review on various studies that focused on optimizing the autofrettage process. After this, some case studies pertaining to hydraulically autofrettaged cylinders and spheres were discussed.

References

Bihamta, R., Movahhedy, M.R. and Mashreghi, A.R., (2007), A numerical study of swage autofrettage of thick-walled tubes, *Materials & Design*, **28**, 804–815. https://doi.org/10.1016/j.matdes.2005.11.012

Çandar, H. and Filiz, İ.H., (2017), Optimum autofrettage pressure for a high pressure cylinder of a waterjet intensifier pump, *Universal Journal of Engineering Science*, **5**, 44–55. https://doi.org/10.13189/ujes.2017.050302

Darijani, H., Kargarnovin, M.H. and Naghdabadi, R., (2009), Design of thick-walled cylindrical vessels under internal pressure based on elasto-plastic approach, *Materials & Design*, **30**, 3537–3544. https://doi.org/10.1016/j.matdes.2009.03.010

Faupel, J.H., (1955), Residual stresses in heavy-wall cylinders, *Journal of the Franklin Institute*, **259**, 405–419. https://doi.org/10.1016/0016-0032(55)90681-5

Gao, X., (1992), An exact elasto-plastic solution for an open-ended thick-walled cylinder of a strain-hardening material, *International Journal of Pressure Vessels and Piping*, **52**, 129–144. https://doi.org/10.1016/0308-0161(92)90064-M

Gao, X.-L., (2003), Elasto-plastic analysis of an internally pressurized thick-walled cylinder using a strain gradient plasticity theory, *International Journal of Solids and Structures*, **40**, 6445–6455. https://doi.org/10.1016/S0020-7683(03)00424-4

Hojjati, M.H. and Hassani, A., (2007), Theoretical and finite-element modeling of autofrettage process in strain-hardening thick-walled cylinders, *International Journal of Pressure Vessels and Piping*, **84**, 310–319. https://doi.org/10.1016/j.ijpvp.2006.10.004

Maleki, M., Farrahi, G.H., Haghpanah Jahromi, B. and Hosseinian, E., (2010), Residual stress analysis of autofrettaged thick-walled spherical pressure vessel, *International Journal of Pressure Vessels and Piping*, **87**, 396–401. https://doi.org/10.1016/j.ijpvp.2010.04.002

Mendelson, A., (1968), *Plasticity, Theory and Application*, Macmillan, New York.

Parker, A.P., (2000), Autofrettage of open-end tubes—pressures, stresses, strains, and code comparisons, *Journal of Pressure Vessel Technology*, **123**, 271–281. https://doi.org/10.1115/1.1359209

Parry, J.S.C., (1965), Fatigue of thick cylinders: further practical information, *Proceedings of the Institution of Mechanical Engineers*, **180**, 387–416. https://doi.org/10.1243/PIME_PROC_1965_180_030_02

Perl, M. and Perry, J., (2010), The beneficial contribution of realistic autofrettage to the load-carrying capacity of thick-walled spherical pressure vessels, *Journal of Pressure Vessel Technology*, **132**, 011204–011204-6. https://doi.org/10.1115/1.4000513

Rees, D.W.A., (1987), A theory of autofrettage with applications to creep and fatigue, *International Journal of Pressure Vessels and Piping*, **30**, 57–76. https://doi.org/10.1016/0308-0161(87)90093-7

Varga, L., (1991), Design of optimum high-pressure monobloc vessels, *International Journal of Pressure Vessels and Piping*, **48**, 93–110. https://doi.org/10.1016/0308-0161(91)90060-F

Zhu, R. and Yang, J., (1998), Autofrettage of thick cylinders, *International Journal of Pressure Vessels and Piping*, **75**, 443–446. https://doi.org/10.1016/S0308-0161(98)00030-1

9

Epilogue

Autofrettage is a process of inducing beneficial residual stresses in a pressure vessel by means of plastic deformation. Plastic deformation can be introduced by hydraulic pressure, thermal gradient, explosion, rotation, or by pushing a mandrel through the bore. Each method of autofrettage has its own pros and cons.

The hydraulic autofrettage process, where plastic deformation is initiated by the applying of hydraulic pressure, is an effective and widely practiced method of autofrettage used for the strengthening of high-pressure components in industry. In order to attain the desired level of plastic deformation in the wall of the pressure vessel, the method requires a very high amount of pressure. The pressure is generated by utilizing a high capacity hydraulic power pack, making the process highly expensive. Moreover, the handling of such a high pressure is a hazardous task, meaning that stringent safety norms need to be followed. Thus, the process requires skilled operators for the successful operation of the process. Usually the hydraulic autofrettage process is practiced for those vessels that require an extremely high pressure in service, e.g., a gun barrel. The shortcomings of the hydraulic autofrettage process may be mitigated to some extent by introducing the method of swage autofrettage, where the plastic deformation in the vessel is setup by pushing an extra-large mandrel through the inner side of the vessel.

The swage autofrettage requires relatively lesser pressure to push the mandrel. However, due to the localized nature of plastic deformation during its operation, the process cannot produce a high level of plastic deformation. This limits the pressure-carrying capacity of the swage autofrettaged vessel. The method of swage autofrettage is effectively practiced in industry for the strengthening of those vessels that require a moderate degree of high pressure in service, e.g., a chemical reactor.

The explosion method of autofrettage is not very popular due to the use of a hazardous explosive charge for achieving plastic deformation.

The methods of introducing plastic deformation for achieving autofrettage by means of thermal gradient and rotation are the more promising methods, known as thermal and rotational autofrettage, respectively. Both of these processes provide certain advantages over the practicing methods of hydraulic and swage autofrettage processes. The thermal and rotational autofrettage processes do not require ultra-high hydraulic pressure, making the operation of the processes safe when compared to the hydraulic and swage autofrettage methods.

The thermal autofrettage process requires a very simple setup with a heating and subsequent cooling arrangement. This method is less expensive and appears to be environmentally friendly, as it does not pour out anything harmful into the environment. The thermal autofrettage method may also be useful for enhancing the working temperature gradient of chemical reactors used in chemical industries.

The method of rotational autofrettage requires a high-power motor for the rotation of the vessel during its operation. This may make the process a little expensive. The rotational autofrettage could be a useful method for the strengthening of pressure vessels by enhancing the pressure-carrying capacity and working thermal gradient. It could also be useful in strengthening the rotating parts in many industrial components by enhancing the centrifugal load bearing capacity.

Both the thermal and rotational autofrettage processes have a strong potentiality to be used in industry for strengthening pressure vessels/rotating parts against different service loads.

Very often the pressure vessels are subjected to fluctuating loads in service. Thus, it is desirable to have a longer fatigue life of the vessels during service. The methods of autofrettage are also employed in order to increase the fatigue life of the vessels apart from increasing their load carrying capacities. Moreover, by employing the methods of autofrettage, the stress–corrosion cracking resistance of the vessels is also improved.

In this book, all the methods of autofrettage have been discussed in detail in chronological order. The autofrettage process was introduced in general in Chapter 1. The basis for any autofrettage process is nonhomogeneous plastic deformation. The foundation for understanding the mechanics of any autofrettage process is the plasticity theory. A brief review of the plasticity theory was presented in Chapter 2. In order to predict the beneficial residual stress level in the vessel, the accurate modeling of any autofrettage process is very important. In the subsequent chapters (Chapters 3 to 7), the modeling of different autofrettage methods was carried out, considering different material models and yield criteria. The detailed technology of each process has also been presented along with a schematic setup.

An attempt has been made to carry out an extensive study of all of the known autofrettage processes in this book. Most of the book's chapters were devoted to the theoretical aspects of the processes. Nevertheless, in Chapter 5, a detailed experimental procedure for the thermal autofrettage process was presented. In Chapter 6, a heat treatment process for mitigating the tensile residual stresses was described. This procedure may prove to be better than shot-peening or wire-winding. A procedure for shrink-fitting was also described. Chapter 7 described the rotational autofrettage process and its modeling. Chapter 8 discussed the selection and optimization issues.

Based on the works presented in this book, some key challenges and areas where future research can be directed to improve the process are identified as follows:

- It is well known that during cold working or thermal loading, the microstructure of the material may change. So far, no significant attention has been paid to this aspect. The microstructural aspects in the thermal autofrettage process were described in Chapter 6 of this book. These may be further explored for the other autofrettage processes, such as in the swage autofrettage process.
- So far, the different autofrettage methods have mostly been explored for cylindrical pressure vessels. A few studies have been carried out for strengthening spherical pressure vessels as well. Some description was provided in this book. In the future, the application of the autofrettage process can be extended for strengthening components other than cylindrical or spherical vessels.
- The experimental study of autofrettage of thick-walled spherical vessels is an important research area.
- A detailed study on the aspect of the stress–corrosion cracking of autofrettaged components is another interesting area.
- Research has been devoted to optimizing different process parameters of the hydraulic autofrettage process. The proper optimization of different process parameters can be explored for other autofrettage processes as well.

Index

M

N

O

P

V